# Enemies Within

# Enemies Within

*The Global Politics of Fifth Columns*

Edited by

HARRIS MYLONAS AND SCOTT RADNITZ

OXFORD
UNIVERSITY PRESS

# OXFORD
## UNIVERSITY PRESS

Oxford University Press is a department of the University of Oxford. It furthers
the University's objective of excellence in research, scholarship, and education
by publishing worldwide. Oxford is a registered trade mark of Oxford University
Press in the UK and certain other countries.

Published in the United States of America by Oxford University Press
198 Madison Avenue, New York, NY 10016, United States of America.

Library of Congress Cataloging-in-Publication Data
Names: Mylonas, Harris, 1978- author. | Radnitz, Scott, 1978- author.
Title: Enemies within : the global politics of fifth columns /
Harris Mylonas and Scott Radnitz.
Description: New York : Oxford University Press, [2022] |
Includes bibliographical references and index.
Identifiers: LCCN 2021058412 (print) | LCCN 2021058413 (ebook) |
ISBN 9780197627945 (Paperback) | ISBN 9780197627938 (Hardback) |
ISBN 9780197627969 (ePub)
Subjects: LCSH: Insurgency—Cross-cultural studies. |
Subversive activities.
Classification: LCC JC328.5 .M95 2022  (print) | LCC JC328.5 (ebook) |
DDC 355.02/18—dc23/eng/20220413
LC record available at https://lccn.loc.gov/2021058412
LC ebook record available at https://lccn.loc.gov/2021058413

DOI: 10.1093/oso/9780197627938.001.0001

1 3 5 7 9 8 6 4 2
Paperback printed by Lakeside Book Company, United States of America
Hardback printed by Bridgeport National Bindery, Inc., United States of America

# Contents

PART III: CHALLENGES TO FIFTH-COLUMN POLITICS

# List of Figures

# List of Tables

# Preface and Acknowledgments

This book grew of our distinct yet overlapping interests: Mylonas's previous work on externally backed non-core groups and nation-building, and Radnitz's work on post-Soviet conspiracy theories and authoritarian politics. It also arose out of the observation that talk of fifth columns—typically surrounding narratives of World War II and the Cold War—was increasingly in the news in the era of Trump, Brexit, and democratic backsliding across much of the world. Yet despite renewed interest, there was limited academic scholarship addressing fifth-column politics.

The melding of our minds took place in the pre-Covid days of unfettered travel, at a PONARS Eurasia workshop in Helsinki, and became a genuine collaboration in the least sinister sense of the word. The first fruits of this collaboration were presented at a PONARS Eurasia workshop in Kharkiv, Ukraine, in June 2017, where we received brilliant comments from Henry Hale, Volodymyr Ishchenko, Mark Kramer, Irina Soboleva, Josh Tucker, and Yuri Zhukov, among others.

In 2019, we convened a workshop on the topic at the University of Washington. The workshop was hosted by UW's Simpson Center for the Humanities with additional funding from George Washington University's Elliott School of International Affairs. We would like to thank the Ellison Center for Russian, East European, and Central Asian Studies; the Jackson School of International Studies; the Center for Global Studies; and the Near and Middle East Interdisciplinary Program for their contributions to the workshop. We are grateful to Lisa Wedeen, Laia Balcells, Keith Darden, Benjamin Tromly, Karam Dana, Reşat Kasaba, Daniel Bessner, Cabeiri Robinson, Daniel Chirot, James Caporaso, and James Long for participating in the workshop and enriching this project. Radnitz taught a Simpson Center-funded graduate seminar on fifth columns relating to the workshop and thanks Elena Bell, Roman Pomeshchikov, Bernard Loesi, and George Reynoldson for their contributions to the project.

Earlier versions of several chapters were presented during a Roundtable on *The Politics and Discourse of Fifth Columns in Eurasia* at the 2019 Annual Meeting of the American Political Science Association in Washington, DC. We presented our analytical framework at the GW Department of Political Science's Comparative Politics Workshop, where we received excellent comments from our discussant Elizabeth Grasmeder and the rest of the participants. We would also like to thank Glennys Young, Marko Žilović, Juris Pupcenoks, Bairavi Sundaram, and the two anonymous reviewers for their constructive feedback.

We would like to thank the editorial and production teams at Oxford University Press, in particular our editor David McBride and assistant editor Sarah Ebel, as well as Koperundevi Pugazhenthi, who ensured a smooth and constructive process. We are grateful to the talented George Kontos for creating the cover artwork for our book. We also thank two anonymous reviewers for their insightful comments. Maria Fusca, Shayon Moradi, and Annabel Hazrati were tremendously helpful in putting together our *Index*. Finally, it goes without saying that we could not have put together this volume without the fantastic contributions by Samer Anabtawi, Efe Murat Balıkçıoğlu, András Bozóki, H. Zeynep Bulutgil, Volha Charnysh, Kathryn Ciancia, Robert Crews, Sam Erkiletian, Kristin E. Fabbe, Lillian Frost, Erin K. Jenne, Kendrick Kuo, and Péter Visnovitz. We hope that *Enemies Within* becomes a building block for a new research program for studying fifth-column politics.

Harris Mylonas and Scott Radnitz<br>February 2022

# List of Contributors

## Editors

**Harris Mylonas** is Associate Professor of Political Science and International Affairs at the George Washington University and editor-in-chief of *Nationalities Papers*. He is coeditor of *The Microfoundations of Diaspora Politics* (Routledge, 2022) and the author of *The Politics of Nation-Building: Making Co-Nationals, Refugees, and Minorities* (Cambridge University Press, 2013), for which he won the 2014 European Studies Book Award by the Council for European Studies as well as the Peter Katzenstein Book Prize in 2013. Mylonas's work on nation- and state-building, diaspora policies, as well as political development has been published in the *Annual Review of Political Science, Perspectives on Politics, Comparative Political Studies, Security Studies, European Journal of Political Research, Journal of Ethnic and Migration Studies, Territory, Politics, Governance, Nations and Nationalism, Social Science Quarterly, Nationalities Papers, Ethnopolitics*, and various edited volumes. He is currently co-authoring a book with Maya Tudor tentatively entitled *Varieties of Nationalism* (Cambridge University Press, forthcoming).

**Scott Radnitz** is the Herbert J. Ellison Professor of Russian and Eurasian Studies in the Jackson School of International Studies at the University of Washington. His research focuses primarily on post-Soviet politics and topics including authoritarianism, identity, and informal politics. He is the author of *Revealing Schemes: The Politics of Conspiracy in Russia and the Post-Soviet Region* (Oxford University Press, 2021) and *Weapons of the Wealthy: Predatory Regimes and Elite-Led Protests in Central Asia* (Cornell University Press, 2010). His articles have appeared in journals including *International Studies Quarterly, Comparative Politics, Comparative Political Studies, British Journal of Political Science, Journal of Peace Research, Journal of Democracy, Political Geography, Political Communication*, and *Post-Soviet Affairs*. Public commentary has appeared in *Foreign Policy, The Guardian, Slate*, and the *Washington Post*. He is an associate editor of the journal *Communist and Post-Communist Studies*.

# Contributors

**Samer M. Anabtawi** is a lecturer in comparative politics at University College London (UCL). His primary research focuses on LGBTQ movements and attitudes toward marginalized groups in the Middle East and North Africa. Aside from his book manuscript on the rise of queer activism in Lebanon, Tunisia, and Palestine, he is currently working on other research projects on judicial autonomy under authoritarianism.

**Efe Murat Balıkçıoğlu** is a lecturer in Islamic History at Wellesley College's Department of Religion. He holds a BA degree in Philosophy from Princeton University, and AM and PhD degrees in History and Middle Eastern Studies from Harvard University. His research focuses on the intersections of philosophy and theology in the early modern Ottoman Empire, as well as the rise of political Islam in modern Turkey.

**András Bozóki** is Professor of Political Science at the Central European University and at the CEU Democracy Institute. He has received fellowships from the Institute for Advanced Study in Berlin, the European University Institute (EUI), and the Netherlands Institute for Advanced Study. His books include *The Roundtable Talks of 1989* (2002), *Political Pluralism in Hungary* (2003), *Anarchism in Hungary* (2006), and *Post-Communist Transition* (2016). He has published numerous book chapters and articles in journals including *Democratization*, *Comparative Sociology*, and *East European Politics*, among others. He has received the István Bibó Award.

**H. Zeynep Bulutgil** is Associate Professor in the Department of Political Science at University College London. Her research focuses on political violence, state formation, religion and politics, and inequality and ethnic mobilization. Her first book, *The Roots of Ethnic Cleansing in Europe* (Cambridge University Press, 2016), studies the causes of ethnic cleansing and received the 2017 Best Book Award in the European Politics and Society Section of APSA. Her second book, *The Origins of Secular Institutions: Ideas, Timing, and Organization* (Oxford University Press, 2022), combines ideational and organizational mechanisms to explain how institutional secularization occurs.

**Volha Charnysh** is Assistant Professor in the Department of Political Science at the Massachusetts Institute of Technology. She studies political attitudes and behavior in culturally diverse societies and is particularly interested in the political and economic legacies of mass displacement, genocide, and plunder in Central and Eastern Europe. Her work has appeared in the *American Political Science Review*, *British Journal of Political Science*, *Comparative Political Studies*, and *European Journal of International Relations*, and won awards from the American Political Science Association and the Council for European Studies. She received her PhD in Government from Harvard University in May 2017.

**Kathryn Ciancia** is Associate Professor of History at the University of Wisconsin–Madison. Her first book, *On Civilization's Edge: A Polish Borderland in the Interwar World* (Oxford University Press, 2020), traces how Poles attempted to simultaneously "civilize" and Polonize the multiethnic eastern province of Volhynia. Articles drawn from this book have also appeared in the *Journal of Modern History* and *Slavic Review*. She is working on a new book that explores how consulates regulated Polish citizenship across the world between the creation of the Polish state in 1918 and the early years of the Cold War.

**Robert D. Crews** is Professor of History at Stanford University. He is the author of *Afghan Modern: The History of a Global Nation* (Belknap Press of Harvard University Press, 2015) and *For Prophet and Tsar: Islam and Empire in Russia and Central Asia* (Harvard University Press, 2006) and coeditor of *Under the Drones: Modern Lives in the Afghanistan-Pakistan Borderlands* (Harvard University Press, 2012) and *The Taliban and the Crisis of Afghanistan* (Harvard University Press, 2008).

**Sam Erkiletian** is a PhD candidate at the Department of Political Science at University College London. His research focuses on how the military socialization processes of armed groups affect the behavior and postwar identity of former combatants. He holds an MSc in Security Studies from University College London and a BA in History and Ancient Studies from Saint Joseph's University in Philadelphia.

**Kristin E. Fabbe** is the Jakurski Family Associate Professor at Harvard Business School in the Business, Government, and the International Economy Unit. She is also faculty affiliate at the Middle East Initiative at the John F. Kennedy School of Government's Belfer Center, at the Harvard Center for Middle East Studies, and the Harvard Center for European Studies. Her first book, *Disciples of the State?: Religion and Statebuilding in the Former Ottoman World*, was published by Cambridge University Press in 2019. She received her PhD in Political Science from MIT.

**Lillian Frost** is Assistant Professor in the Department of Political Science at Virginia Tech. She specializes in citizenship, migration, and gender issues, particularly in the Arab world. She has held research fellow positions with the European University Institute's Max Weber Programme, United States Institute of Peace, Harvard Belfer Center for Science and International Affairs' Middle East Initiative, American Center of Research in Jordan, and Fulbright Program in Jordan. She received her PhD in Political Science from the George Washington University, and her dissertation won the 2021 Best Dissertation Award from the American Political Science Association's Middle East and North Africa Politics Section.

**Erin K. Jenne** is Professor at the International Relations Department at Central European University, specializing in nationalism, populism, and ethnic conflict management in Eastern Europe. She has received numerous grants and fellowships, including a MacArthur fellowship, Carnegie Corporation scholarship, a Senior Fernand Braudel fellowship at European University Institute (EUI), and a Minerva grant from the US Office of Naval Research. Her first book, *Ethnic Bargaining* (Cornell University Press, 2007), is the winner of Mershon Center's Edgar S. Furniss Book Award. Her work has appeared in *International Affairs, International Studies Quarterly, Security Studies, Regional and Federal Studies, Journal of Peace Research, Civil Wars, International Studies Review, Research and Politics*, and *Ethnopolitics*.

**Kendrick Kuo** is Assistant Professor at the US Naval War College. His research focuses on military effectiveness, innovation, and strategy, as well as nationalism and nation-building. He holds a PhD in Political Science from the George Washington University, an MA in International Affairs and International Economics from the Johns Hopkins School of Advanced International Studies, and a BA in International Affairs and Religion from the George Washington University.

**Péter Visnovitz** is a doctoral candidate at Central European University's PhD Program in Political Science. His research focuses on populist legitimacy in opposition and power, using content analysis and discourse analysis to track how populist actors construct group identities and security threats. As a research assistant he participated in a number of research projects at Yale, Princeton, and the Hungarian Center for Social Sciences. He also has substantial work experience with the media, having worked as a journalist for over eight years.

# Introduction

## Theorizing Fifth-Column Politics

*Scott Radnitz and Harris Mylonas\**

Secretary of State Hillary Clinton "set the tone for some actors in our country [Russia] and gave them a signal. They heard the signal and with the support of the US State Department [they] began active work." Vladimir Putin made this claim following Russia's parliamentary elections in December 2011. Putin had recently announced his intention to return as president after a term as prime minister and was expecting a smooth transition. Instead, he was met by Russia's largest protests in years. Drawing out the implications of the challenge as he saw it, Putin concluded, "We are obligated to protect our sovereignty and we will have to think about strengthening the law and holding more responsible those who carry out the tasks of a foreign government to influence internal political processes" ("Putin obvinyaet" 2011).

Putin's suggestion that his detractors, ostensibly taking to the streets to express their displeasure about fraudulent elections, were in fact acting in the interests of the United States, would have struck a familiar chord with his audience. Claims about *fifth columns*—when one's compatriots supposedly collaborate with a hostile foreign power to subvert the popular will—had a long pedigree in both the Soviet Union and the Russian Federation. In this instance, Putin apparently sought to discredit the demonstrations by implying that they did not reflect genuine grassroots sentiment but were instead being instigated by Russia's main international rival. The principles at stake in Putin's allegation, such as legitimacy, national identity, and sovereignty, are not particular to Russia, and similar claims have been witnessed recently around the world.

For example, in 2016 Turkish president Recep Tayyip Erdoğan asserted, "Unfortunately, these so-called academics claim that the [Turkish] state is carrying out a massacre. You, those so-called intellectuals! You are dark people. You are not intellectuals." He was reacting to a declaration of more than one thousand academics calling for the end of fighting between Turkey's security forces and the Kurdistan Workers' Party (PKK). Erdoğan linked this plea for peace to activities

---

\* The authors have contributed equally to the preparation of this volume.

Scott Radnitz and Harris Mylonas, *Introduction* In: *Enemies Within*. Edited by: Harris Mylonas and Scott Radnitz, Oxford University Press. © Oxford University Press 2022. DOI: 10.1093/oso/9780197627938.003.0001

that occurred during the collapse of the Ottoman Empire: "Turkey experienced betrayal [at the hands of] this mindset 100 years ago. Then there was a group of so-called intellectuals who preferred the protectorate of a great power with the belief that only foreigners could fix the [problems] in this country." He concluded, "Let the US Embassy in Ankara invite [Professor Noam] Chomsky to Turkey, let's host him in the [Kurdish] region. Let him see [the] facts [for] himself, not through those 'so-called academics' who act as a fifth column" (*Zaman* 2016).

Erdoğan's accusation was part of a broader effort to undercut his liberal critics within Turkey as part of an authoritarian power grab. He believed this rhetoric would be especially attractive to members of his conservative Justice and Development Party (AKP), who perceived themselves as being engaged in a struggle against the liberal and secular establishment. Erdoğan invoked the specter of fifth columns not only as a threat to his ongoing battles against Kurdish rebels, but also as a recurrence of the traumatic dismantling of the Ottoman Empire by European powers.

Turkey and Russia are both authoritarian systems, in which rulers are notorious for cynically playing on people's anxieties in order to rally support for their regimes. Yet the language of fifth columns also resonates in established democracies. In Greece, such rhetoric was frequently deployed after the 2009 financial crisis to attack internal party opposition or impugn political and ideological opponents. As then-Prime Minister Alexis Tsipras put it while addressing the Central Committee of his party in May of 2015: "We should never forget that the enemy is not just in Berlin, Brussels or Washington. The enemy, maybe the harshest one, is also within our borders" (Prime Minister 2015). He was referring to political forces in Greece that had voted for two bailout agreements and other austerity policies before the Coalition of the Radical Left (SYRIZA) came to power (Mylonas 2014). By the end of his term as prime minister, these types of fifth-column claims had diminished. Instead, the main opposition party, New Democracy (ND), flipped the script by accusing Tsipras and his government of colluding with external forces—the EU and the United States—over the Macedonia name dispute (Kambas and Maltezou 2018).

In the United States, there is a long and sordid history of leaders identifying, accusing, and punishing suspected fifth columns. There are no clearer examples than the forced internment of Japanese Americans during the Second World War and the House Un-American Activities Committee (HUAC), which hunted for Communists in the federal government. More recently, new dynamics reminiscent of the Cold War have been apparent in US-China relations. Thus, the FBI under the Trump administration began targeting Chinese researchers and Americans with reputed links to China for investigation on suspicion of espionage or theft of intellectual property. One representative in the California assembly expressed concern over what appeared to represent the reemergence of a troubling historical pattern, saying, "Today we face a dangerous narrative . . . that anyone of Chinese ancestry out there could be a national security threat and should be viewed with more scrutiny and suspicion than others" (Dilanian 2020).

As these disparate examples suggest, fifth-column claims have appeared in different parts of the globe and in the politics of both democratic *and* authoritarian states. The invocation of fifth columns in the political arena—whether contrived or based on real fears—has historically tended to recur periodically and appears to have intensified amid the "populist explosion" (Judis 2016). Accusations about fifth columns can have baleful effects on governance and trust in multiethnic states, as they call into question the loyalty and belonging of the targeted populations. History shows that sustained campaigns to target and persecute suspected fifth columns have been associated with human rights abuses, discriminatory practices, political repression, ethnic cleansing, and even genocide (Fein 1993; Straus 2010; Mylonas 2012; Bulutgil 2016; Suny 2017). Yet despite the enduring relevance of fifth-column politics, scholars have not studied the phenomenon systematically.

*Enemies Within* investigates the roots and implications of this important phenomenon. The chapters included in this volume address several questions: When are actors likely to employ fifth-column claims and against whom? What accounts for changes in fifth-column framing over time? How do the fifth-column discourses of governments differ from those of societal groups? How do accusations against ethnically or ideologically defined groups differ? Finally, how do actors labeled as fifth columns respond?

In the next sections, we discuss the history of fifth-column politics and situate our inquiry among previous studies on similar topics. Then we provide a theoretical framework to investigate the politics of fifth columns, introduce several conceptual interventions, and differentiate fifth columns from similar phenomena. Finally, we preview the other chapters and describe their contributions to this volume.

## A Brief History of Fifth-Column Politics

We understand "fifth columns" as domestic actors who work to undermine the national interest, in cooperation with external rivals of the state. Although there are historically documented cases of *actual* fifth columns, this volume instead focuses on how elite actors and societal groups make claims about, and engage with, *alleged* fifth columns. In other words, we are agnostic about the veracity of the claims.

The first documented use of the term "fifth column" is uncertain, but most sources suggest that it was originally coined in the context of the Spanish Civil War by William P. Carney in a 1936 article published in the *New York Times*.

He used the term in English as the translation of the Spanish *quinta columna*, used to describe the Nationalist supporters that were sent ahead to Madrid to assist, from within, the four columns of troops led by Nationalist rebel general Emilio Mola Vidal against the loyalist Republicans (Carney 1936).[1] Although the original usage refers to actors contending for power within a single country, in common and scholarly usage today the term typically involves an *internal-external linkage*. The dynamics of this form of politics is distinct from purely domestic accusations because it involves questions of sovereignty and geopolitics, and therefore merits special treatment, as we will elaborate.

Although the term fifth column is of recent vintage, the notion of *enemies within* goes back much further. In the post-Napoleonic era, episodes involving potentially disloyal internal groups grew out of state formation and the collapse of empires.[2] The rise of nationalism as a political principle, coupled with incomplete assimilation, gave rise to the concept of minorities, whose differences from the national majority within a polity made them inherently suspect. The insecurity of *ancien régime* rulers and mobilization from below led to cross-class alliances as elites embraced nationalism to form a linguistic or cultural majority (Wimmer 2018; Mylonas and Tudor 2021). Outliers from the majority nation were often relegated to a lesser status of membership in the political community. In a competitive international environment, co-ethnic or mercenary proxies sometimes served the interests of external patrons by working to undermine the state from within (Darden and Mylonas 2016). The belief that disloyal members of society, backed by outside powers, posed a persistent threat to the territorial integrity of a state has often unified (the rest of) the imagined community.

It is well documented that conflict can generate dynamics in which the alienation of insecure minorities accused of subversive activity can have a self-fulfilling effect by pushing the targeted population into the arms of an external adversary for protection (Mylonas 2012). Examples range from ethnic organizations of Sudeten Germans in interwar Czechoslovakia that aided the Third Reich just before the annexation of Sudetenland, to members of the Greek Orthodox millet in the Ottoman Empire who collaborated with the Greek authorities (Lippmann 1942; Bulutgil 2016). Massive post–World War II population transfers were also provoked, at least in part, by fears that non-core groups (especially Germans) could reemerge as fifth columns at some point in the future (Naimark 2002; Schechtman 1953).

Claims about fifth columns have also appeared outside of the context of active conflict. In the United States, the HUAC was created in 1938 to investigate subversive elements—Nazi and Soviet sympathizers—in American life. Opponents of the New Deal linked the introduction of social democracy in the United States to the purported threat of Soviet communism. Republican committee member

J. Parnell Thomas of New Jersey called the Roosevelt administration "the official or unofficial sponsor of the very Communist groups which the Committee is trying to investigate" (Ceplair 2011: 55).

The Cold War saw citizens targeted by elites and accused of disloyalty on both sides of the Atlantic, as the logic of insecurity coupled with political opportunism fed cycles of fear bordering on paranoia. The influential historian Arthur M. Schlesinger Jr. wrote that "the special Soviet advantage—the warhead—lies in the fifth column; and the fifth column is based on the local Communist parties" (1950: 92–93). Fears of fifth columns, whether as spies, moles, ideological agents, or mere sympathizers took root and were harnessed for political purposes in the United States. From Senator Joe McCarthy's claims of Communists within the government to the "Lavender Scare" targeting gay civil servants, citizens were harassed or persecuted on account of real or suspected Communist sympathies. The irony of a "free" society was that it was open to foreign penetration, necessitating, in the eyes of some, an intrusive and heavy-handed state. As a measure of this imperative, from 1938 to 1960, the FBI's budget grew from $6 to $116 million, and by 1954, it maintained 430,000 files on suspected subversives (Theoharis and Rosenfeld 1999: 182; Ceplair 2011: 85). FBI director J. Edgar Hoover opened cases and spied on thousands of Communist Party members, liberal activists, and other suspected subversives—most often on flimsy pretexts—over his fifty-five-year career (Weiner 2012).

In the Soviet Union, the search for fifth columns was a central operating principle of the regime. From the earliest days of the Bolshevik Revolution, fears about internal subversion gripped the new state-builders. The security apparatus of the fledgling Bolshevik regime, the Cheka, grew out of informal organizations of revolutionaries seeking to expose tsarist forces in their midst. The threat posed by a promised worldwide Communist revolution instilled fear in European capitals, and the united front of a hostile West gave the new regime in Moscow cause to fear enemies within, sponsored from without (Harris 2013). After all, Lenin would not have made it back to the Russian Empire in 1917 without active German assistance.

Under Stalin, the threat of fifth columns was inflated to new levels. Many officials who lost their lives in the Great Terror were accused of secret connivance with foreign intelligence services (Khlevniuk 2015). Ambitious apparatchiks signaled their ideological fidelity to the system by unmasking and persecuting capitalist "wreckers" (Rittersporn 1992). The populations of entire ethnic groups—Crimean Tatars, Chechens, Meskhetian Turks, Volga Germans—were forcibly and brutally transferred to barren Soviet hinterlands on suspicion of complicity with enemies during World War II (Burds 2007; Naimark 2002). Jewish Communists were purged on accusation of "Zionist" loyalties. Even after Stalin's death, while state violence subsided, the legacy of fifth-column fears persisted in the persecution of pro-West dissidents and the ferreting out of suspected spies.

Similar dynamics were evident during the Cold War in Asia. In 1960s Indonesia, one author suggests that the Chinese minority "could credibly be represented [by the Indonesian government] as communist menace, fifth column and economic saboteur" (Coppel 1983: 63). In the late 1970s in Cambodia, during the war with Vietnam, "Cambodians who had Vietnamese features, or who simply were too pale, became as much the victims of [the Pol Pot regime's] racist hatred as of the political will to eliminate a fifth column" (Ponchaud 1989: 58).

Following the Cold War, outbreaks of ethnic conflict gave rise to a new focus on nationalism and ethnicity as the main sources of exclusion (Huntington 1993; Posen 1993). In Rwanda, the loyalty of ethnic Tutsis came into question when a Tutsi army formed in exile in Uganda (the RPF) advanced into Rwanda prior to the genocide in 1994. As Prunier writes: "[F]or the Habyarimana regime, and especially for the extremist elements in its ranks, this meant that the front [RPF] was now a direct agent in Rwandan politics and that the whole Tutsi population inside the country could be viewed as potential 'fifth columnists'" (1998: 132).

Many scholars saw newly democratizing societies in the 1990s and 2000s as vulnerable to populist ethnic appeals that singled out historically suspect minority populations as a threat to the majority (Snyder 2000; Gagnon 2004; Kaufman 2001). Due to the complex ethnic patchwork of Eastern Europe and the post-Soviet region, "beached" minorities were connected through family, business ties, or simply ethnic affinity, with majorities in other states (Waterbury 2010 and 2020; Pogonyi, Kovács, and Körtvélyesi 2010; Csergő and Goldgeier 2013). When insecure minority populations felt threatened, there was a risk that an external patron state could intercede on their behalf, leading to spiraling dynamics of ethnic conflict (Mylonas 2012; Jenne 2007).

Despite these fears, and aside from the bloody wars in Yugoslavia, most governments in Eastern Europe and the post-Soviet region carefully avoided aggravating the fraught politics of state-minority relations, and potentially revanchist states and restive minorities did not come to blows. Several post-Communist Eastern European states resolved border disputes peacefully with the assistance and leverage of the European Union and implemented policies to accommodate minorities (Snyder 1998). The Baltic states joined the EU and NATO and worked out a modus vivendi with their Russian-speaking population (Mylonas 2012: 180–185). Even Russia, where ultranationalists made irredentist claims on Crimea in the 1990s, did not take any actions to revise post–Cold War borders until 2014.

In the third decade of the twenty-first century, we see two converging forces that are conducive to a renewal of the politics of fifth columns globally.

First is the decline in trust of government and the collapse of support for centrist political parties across the democratic world (Diamond 2020). Political entrepreneurs on the extremes use appeals that starkly counterpose "the people" against an elite that represents the interests of some "other" group (Mudde 2011). Right-wing populism in particular emphasizes ethnic and cultural differences, using the prospect of the disloyalty of non-core groups as fodder for fifth-column claims. For instance, far-right parties in Europe have depicted immigrants and refugees from Muslim-majority countries as threats to Christian civilization and European values (Caldwell 2009; Mudde 2019), while similar rhetoric was voiced by the Trump administration in the United States in relation to migrants from Latin America (Norris and Inglehart 2019). These examples may represent just the latest phase in the questioning and redefining of belonging that have occurred over the last century.

The second important development involves the decline of the hegemonic role the United States played in the 1990s and early 2000s, coupled with the rise of multipolarity. With global hegemony in question, nationalist or ethnic groups aspiring for self-determination or various forms of autonomy have become emboldened; there are more than fifty active secessionist movements around the world (Griffiths 2016). Within this context, rising powers seeking to improve their regional influence have often pursued extraterritorial policies to assist purported co-ethnics or allied non-core groups abroad (Ho 2018; Suryadinata 2017; Pradhan and Mohapatra 2020). These domestic and international factors combine to produce a volatile mix that at times takes the form of interstate disputes but more often manifests itself in the arena of domestic politics.

## The Politics of Fifth Columns: A Theoretical Intervention

Scholars and analysts in the mid-twentieth century took for granted the presence and presumed effectiveness of fifth columns during wartime (Lippmann 1942; Speier 1940). Others have approached claims of fifth columns skeptically, showing how fears of the time were not borne out by the evidence (De Jong 2019; Loeffel 2015; MacDonnell 1995). These historical accounts have typically focused on single cases at particular moments. Social scientists have examined episodes involving fifth columns theoretically and often comparatively. For example, Brubaker (1995, 1996), building on Weiner (1971), warned of dangerous dynamics involving clashes of nationalizing states, national minorities, and national homelands. Although not explicitly referring to fifth columns, this work gave rise to numerous studies on nationalism and politics involving ethnic minorities (Saideman 2002; Jenne 2007; Mylonas 2012; Bulutgil 2016, 2017). Another strand of research examines how narratives about a group's collaboration with an external enemy during past wars are used to advance contemporary political projects (Grinchenko and Narvselius 2018; Petit and Rousso 2002; Deák 1995).

It is evident that while many scholars have focused on fifth columns, they have largely neglected *fifth-column politics*. Inquiry into "enemies within" has been hindered by a divide between studies of alleged ideological fifth columns on the one hand, and ethnic groups targeted by political demagogues or caught up in great power struggles on the other, even though both sets of circumstances may involve similar dynamics. Additionally, scholars typically draw on a single episode, which impedes a comparative investigation of the features of fifth-column politics under different geopolitical or domestic conditions, or across time.

In *Enemies Within*, we seek to identify underlying similarities in the origins and consequences of fifth-column politics across varying historical, political, and social contexts. To do so, we first distinguish between "fifth column" as a term articulated by political actors and "fifth-column politics" that we propose as a category of analysis (Brubaker 2013). In our framework, fifth-column politics refers to public statements, state policies, or collective action targeting domestic actors on the purported basis of their acting nefariously in cooperation with external rivals of the state against the national interest. This definition enables us to locate instances of fifth-column politics whether or not accusers use the term "fifth column" and irrespective of the surrounding circumstances or the veracity of the claim.

Second, we examine the politics of both alleged ethnic and ideological fifth columns within the same framework. Despite their distinct manifestations, accusations toward either type of actor exhibit structural commonalities, as they appeal to deep-seated drives to protect the unity of one's imagined community against potential threats. As such, claimants can select from a menu of options in how to craft an accusation, which can incorporate both resonant cultural frames and immediate political priorities. The criteria for determining difference and defining a threat may be based on ethnic characteristics,[3] in ways that trigger group solidarity and mobilize a defensive response by the majority. But people may also be branded as fifth columns on the basis of non-ethnic attributes, including their opportunistic association with a foreign government, ideological or political opposition, or objectionable identities and behaviors, such as perceived deviant religious beliefs or sexual practices.

The last century has provided abundant examples of both types of fifth-column claims. Beginning in the interwar years and continuing throughout the Cold War, ideological claims were prominent amid struggles between states promoting communism, fascism, and liberalism. Coinciding with the end of communism in Eastern Europe and the move from a bipolar to a unipolar international system, ethnic fifth columns were frequently invoked in the 1990s.

Ideological accusations appear to have reemerged in the twenty-first century during the wave of authoritarian resurgence in states including the Russian Federation, Turkey, Hungary, and Venezuela. Similarly, movements against unpopular neoliberal reforms imposed by distant supranational bodies have given rise to ideologically suspect "enemies within," from Latin America to the European Union.

Upon closer inspection, the ethnic-ideological divide is not so clear-cut. Elites can strategically exploit blurred ethnic and ideological boundaries to communicate information to different audiences. They can, and do, use coded language that does not explicitly refer to particular ethnic or religious groups but is intended to single out a target for discerning listeners (see Jenne et al. in this volume). Moreover, sometimes a fifth-column claim that initially targets ideological foes can transform into discourses or policies aimed at a particular ethnic group (see Charnysh in this volume).

Third, we introduce an analytical distinction in fifth-column politics based on whether the threat to the national interest purportedly comes from below or above. Conventional conceptions of fifth-column accusations involve *subversive* activity emanating from below, depicting marginalized yet hostile citizens as the agents of actors outside the state.[4] Subversive fifth-column claims typically accuse targets of plotting to overthrow the government, carry out acts of sabotage, or otherwise damage the national interest from within to advance the interests of their external backers. Typical examples include Nazi sympathizers in the United States working on behalf of Germany during World War II (MacDonald 1995), or nongovernmental activists in Hungary who are reputed to act on behalf of George Soros and other "globalist" patrons.

A second type of fifth-column claim involves alleged hostile activity from above. We label this type of fifth-column claim *collusive*,[5] as it involves influential actors seeking to undermine the national interest from within the government in collusion with a foreign power (Radnitz 2021, chapter 3). Claimants depict damage being inflicted through insidious official actions or deliberately ruinous inaction. The polity may be looted, destabilized, or otherwise perverted stealthily and legally, as opposed to the disruptive and illicit efforts of subversive fifth columns who undermine the system from below. This type of fifth-column claim has received less attention and has not previously been examined under the umbrella of fifth-column politics. Yet, there are prominent cases in which social and political movements have leveled collusive claims against elite actors whose secretive actions appear to serve malign foreign interests whether they be governmental, non-governmental, or multinational.

Theoretically, one might expect collusive fifth-column claims to appear mainly in postcolonial or client states in which an external patron exercises undue influence over the political system and economy through a local political elite. But not every instance of external influence is politicized as a collusive fifth column. For example, in the French postcolonial space, the formal colonial power actively meddles in the affairs of former West African colonies. Citizens of these countries view France as scheming and intent on advancing its geopolitical interests, and presume that their corrupt leadership is at least somewhat willing to do the bidding of the external power, yet this reality does not usually produce politically potent collusive accusations (Chafer and Keese 2015). More generally, when states cannot provide for their own security, external intervention and informal political control by a regional hegemon may be viewed as an unavoidable reality and therefore not politicized (Gleason 2001). In other cases, accusations of collusion arise out of intense political struggle and rancor. In Lebanon, both Sunni parties and the Shia Hezbollah movement are backed by external patrons seeking to further their respective sectarian and regional interests. Transparent foreign influence generates resentment toward the patrons supporting the other side, along with public indignation toward the corrupt and captured political class as a whole (Leenders 2012).

Opposition groups have made collusive fifth-column claims to garner support against ideologically objectionable leaders or policies, especially when there is a plausible case to be made that international actors benefit from those policies. This was the case with accusations by Communists in Russia in the 1990s that President Boris Yeltsin was deliberately doing the bidding of the United States and the IMF to harm Russia, along with claims by American Democrats that President Trump was secretly serving the interests of Russian President Vladimir Putin (see Radnitz in this volume). In Greece, during the financial crisis in the 2010s, leftist as well as right-wing critics not only lambasted the EU and its wealthiest member, Germany, but also insinuated that the governments that agreed to austerity policies were fifth columns. This rhetoric drew on Greece's historical memory of susceptibility to outside influence, from Ottoman domination to dependence on Western powers since the founding of modern Greece (Mylonas 2014: 440).

Finally, another relevant distinction is between public and private actions. Public fifth-column politics includes open rhetorical attacks by state officials or regime opponents, whereas private, or covert, actions aimed at purported fifth columns may be harder to observe but have important consequences. As the chapters in this volume illustrate in greater detail, certain elements of the politics of fifth columns, namely discourse and mobilization, can only take public forms, yet policies can be carried out either in private or in public. For example, a government may begin treating members of an alleged fifth column covertly by implementing secret directives against them, but may choose at a later point to make those actions public and launch a campaign accusing the group of being a fifth column (see Kuo and Mylonas in this volume). Such shifts may be the result of domestic political considerations or changes in geopolitical conditions.

The analytical distinctions we propose are ideal types and cannot possibly capture the details of each case. Our framework enables the contributors in this volume to identify how these dimensions shape fifth-column politics, while also problematizing categorical distinctions at times.

## Unpacking the Politics of Fifth Columns

The "politics" of fifth columns can refer to any of three elements: discourse, policy, and mobilization. *Discourse* includes public claims made by political actors that members of a domestic group are disloyal, have illegitimate ties with a foreign power, and harbor sinister intentions. Discourse may be employed by government officials or their supporters to articulate subversive claims as well as by anti-incumbent actors to make collusive claims. Discourse also includes rhetoric by neighboring states, global or regional powers, and non-state actors in defense of ethnic or ideological groups whose rights are perceived to be violated, as well as claims by members of such groups in their own defense.

*Policies* include the enactment of laws and regulations that limit the political, social, or economic rights of group members based on an allegation of their disloyalty. Governments have recourse to intelligence that ordinary citizens do not, and they claim the authority to dictate who constitutes a threat. Because of the asymmetry of information, official claims often cannot be directly refuted. This uncertainty coupled with the state's obligation to provide security enables governments to make authoritative statements concerning fifth-column activity, and enact corresponding policies. For example, governments may exclude purported fifth-column members from employment in particularly sensitive sectors or impose border controls that restrict interactions and weaken ties across ethnically contiguous populations. They can also impose restrictions on practices that facilitate fifth-column activity, such as holding public assemblies, communicating with foreign actors, or associating with nongovernmental organizations. Governments also have the wherewithal to decide when to deal with threats overtly or covertly. Oftentimes, if a state does not want to publicly acknowledge the presence of a fifth column in the country, actions can occur in the private realm rather than in public forums. Such a covert situation may emerge when a state perceives that an external patron seeks a pretext to intervene on behalf of the alleged fifth column. Discretion would be preferred if public measures or fifth-column discourse threatened to provoke diplomatic meddling or military intervention.

*Mobilization* includes contentious actions by real or alleged fifth-column groups seeking to shift public opinion in their favor or lobby public officials. Mobilization, however, may also involve vigilante attacks against perceived fifth columns by self-anointed defenders of the majority or purported national interests. Finally, mobilization could take the form of state-led protests or pogroms intended to intimidate or weaken accused fifth columns or build support for a majoritarian coalition premised on the defense of the state or nation.

In order to merit inclusion in our analysis, these discursive claims, policies, and episodes of mobilization cannot be one-off or ad hoc occurrences. Instead, they must involve a sustained campaign in which the idea of fifth columns plays a significant part.

## What Is Not a Fifth Column (Claim)?

Because fifth-column politics touches on various literatures and shares certain "family resemblances" to similar phenomena, it is important to draw some distinctions (Collier and Mahon 1993). First, subversive fifth-column accusations are not the same as "scapegoating" (Brass 1991; Gagnon 1994/1995, 2004; Harff 1987; Snyder 2000). Scapegoating is commonly defined as the practice of blaming an innocent group for misfortunes that befall society (Levine and Hogg 2010: 723). When the narrative of blame involves support from outside actors, then scapegoating takes the form of a subversive fifth-column accusations. But scapegoating can occur even when the "scapegoat" has no external links. Thus, although there is an overlap between the two concepts, a subversive fifth-column accusation is not always an instance of scapegoating (Mylonas 2012: 171–172).

Fifth-column accusations can also take the form of conspiracy theories, or "the belief that an organization made up of individuals or groups was or is acting covertly to achieve some malevolent end" (Barkun 2003: 3). Both types of claims involve assertions that focus blame on selected out-groups and may be used instrumentally to heighten divisions, stoke fear, distract from flawed governance, and mobilize supporters. Yet conspiracy theories differ in at least two ways. First, conspiracy theories are "*prima facie* unwarranted" (Levy 2007: 182) requiring inferential leaps beyond what established evidence can show, whereas fifth-column claims may rest on a solid evidentiary basis. Second, conspiracy theories are a broader category of claim that does not require a linkage between internal and external agents, and can involve a variety of machinations beyond subverting the institutions of a particular state (Radnitz 2021).

Fifth-column politics must also be distinguished from rhetoric or policies directed against ethnic mobilization for autonomy or ordinary oppositional politics in the absence of external linkages. Incumbents may make false or incendiary claims intended to discredit either type of challenger. However, unless the speaker links the group to an external sponsor, such claims are not fifth-column accusations. The extraterritorial component of the claim enlarges the scale of the threat, provides an additional pretext to take punitive measures against it, and evokes concerns about sovereignty and territorial integrity.

## Plan of the Volume

*Enemies Within* includes meticulous case studies analyzing episodes of fifth-column politics. Each chapter makes theoretical and empirical contributions that build on our conceptual framework. The volume is organized according to the analytical breakdown between subversive and collusive fifth-column politics and then turns to cases that challenge theoretical simplifications and dominant discourses. Table I.1 depicts the cases based on a breakdown by type of claim (subversive or collusive) and target (ethnic or non-ethnic).

Part I addresses subversive fifth-column politics and includes case studies from Central Europe to western China. In Chapter 1, Charnysh explains why some groups and not others become targeted by fifth-column claims in Poland. She argues that fifth-column discourse helps redefine in-group boundaries, affecting cohesion and coordination among members of the original and reframed in-groups. Political entrepreneurs thus invoke fifth columns to divide their

**Table I.1** Breakdown of Cases by Type of Claim and Target

|  |  | Claim | |
|---|---|---|---|
|  |  | Subversive | Collusive |
|  | Ethnic | Private & public | Public |
|  |  | German Americans and Japanese Americans in US; Jews in Poland, Palestinians in Jordan; Uyghurs in the PRC | Afghanistan's monarchy |
| Target | Non-ethnic (ideology, sexual orientation, class) | Private & public | Public |
|  |  | Communists in Poland; Gülenists in Turkey; LGBTQ in the USA and Palestine | Boris Yeltsin and reformers in Russia as US agents; Donald Trump as a Russian stooge |

opponents and reshuffle existing alliances. Charnysh uses these insights to interpret the rhetoric and policies adopted by the left- and right-wing parties in twentieth-century Poland. Tracing fifth-column discourse within a single polity over time, she details how domestic political rivalries motivated fifth-column accusations. She also finds that most fifth-column narratives focused on Jews, regardless of the variation in size or political power of the Jewish minority in Poland. This finding speaks to the importance of preexisting cultural schemas in the construction of fifth-column narratives.

In Chapter 2, Jenne, Bozóki, and Visnovitz explain why fifth-column claims resonate. Focusing on Hungary, they ask why anti-globalist and antisemitic tropes centered on Jewish financier George Soros gained traction in the 2010s, but not in the early 1990s when anti-Soros conspiracy theories first made their appearance in the political sphere. They argue that the changing rhetoric can be understood as part of an effort by the right-wing Fidesz government to consolidate and retain political power during periods of heightened political competition. To appreciate its mass appeal requires investigating the antisemitic and anti-globalist *cultural code* that Orbán and his media allies used to project a vision of foreign subjugation by rapacious banking and international elites. Its effectiveness lay in the repeated, subliminal activation of this latent code in the context of rising right-wing sentiments.

In Chapter 3, Bulutgil and Erkiletian explain why fifth-column policies may be carried out against one part of a group but not another. The chapter looks at German Americans in the United States during the First World War and Japanese Americans during the Second World War and argues that civil society organizations are more likely to target specific groups as fifth columns if these groups arrive after the emergence of a robust civil society. Japanese migrants arrived on the West Coast in the late 1800s, where anti-Asian organizations were already well entrenched and were able to campaign for anti-Japanese legislation and halt further Japanese immigration. Japanese migrants started arriving in Hawaii in the 1860s before the establishment of a vibrant civil society, which limited public pressure against Japanese Americans. German Americans in Chicago were able to resist significant attempts to label them as a fifth column during World War I because of their prior involvement in influential civil society organizations. Thus, the timing of settlement, before or after the emergence of a robust civil society, affected perceptions and opportunities for opponents to lobby for government policies to treat groups as fifth columns.

Finally, in Chapter 4, Kuo and Mylonas highlight the transnational dimension of fifth-column claims. They describe the joint production of fifth-column framings by governments, external "enemies," and the international system. They empirically probe the plausibility of their framework through careful process-tracing in Xinjiang, where the Chinese Communist Party changed its fifth-column framing of the Uyghurs over time: from "counterrevolutionaries" to "ethnoreligious national separatists" and, more recently, to "transnational Jihadists." Kuo and Mylonas argue that patterns of covert or overt support by external powers (state and/or non-state actors) correlate with changes in government rhetoric and policy toward the actual or perceived subversive group. Once this process unfolds, it is hard for any of the parties involved to undo the state's perception of the group until there is a change in the structure of the international system or in the constitutive story of the alleged external backer or the state experiencing the conflict.

Part II is devoted to the politics of collusive fifth columns. Examining drastically different contexts, Radnitz and Crews illustrate the dynamics of accusations from below in response to perceived collusion involving national leaders. In Chapter 5, Radnitz compares the United States and Russia, two unlikely cases for the rhetoric of collusive fifth columns to take hold. In the 1990s, Russian president Boris Yeltsin and liberal reformers found themselves accused of colluding with Western powers to weaken Russia when they introduced wrenching economic reforms. Two decades later, a very different historical figure, Donald J. Trump, was labeled a "puppet" after demonstrating uncanny interest—financial, rhetorical, and diplomatic—in Putin's Russia. In both countries, opposition politicians promoted the notion that their president was secretly serving malign foreign interests, as part of a galvanizing appeal to their otherwise deflated supporters. This chapter explores the odd parallels and historical echoes of these two episodes, showing how collusive fifth-column discourses can emerge even in powerful states that might appear impervious to external penetration.

In Chapter 6, Crews takes a long-term perspective of a single case, Afghanistan, to explain how competing forces have utilized the language of collusion with the aim of discrediting rival political projects. The distinctive history of the making of the modern state left Afghans with a formative legacy: at crucial moments in the nineteenth century and again in the late twentieth century, great power intervention determined the outcome of contests for power in the country. The imperial context of Afghan state-building made the specter of dissident collaboration with outsiders a fixture of elite campaigns aimed at discrediting potential challengers. At the same time, the language of "treason" and "nation-selling" became a formidable weapon wielded "from below" by various opposition groups who sought to delegitimize Afghan rulers by demonizing their reliance on support from abroad. Such accusations operated as an index of a cohesive language of nationalism in a country where most scholars have claimed that deep ethnic and tribal divisions have stymied the imagination of a national community. Thus, conspiracy narratives about fifth columns in the Afghan context have been both an interpretive practice and political strategy.

Part III showcases challenges. The first two chapters of this section challenge conventional assumptions about fifth-column politics. They involve cases in which the insider/outsider binary is blurred, and where the idea of belonging is contested. They complicate the conceptual categories we lay out in this introduction by questioning what happens when citizenship and state borders do not align.

In Chapter 7, Frost analyzes Palestinians in Jordan to problematize the *insider* dimension of fifth-column groups. In the standard understanding of a fifth column, the targeted population is assumed to be composed of *citizen* insiders. Frost points out that sometimes the targeted population consists of *noncitizen* insiders who, despite living in the country for generations, lack formal citizen status. Frost illustrates variations in Jordan's framing of Palestinians as fifth columns by examining two cases of nationality law reform in Jordan. One case, from 2014, demonstrates how political elites used fifth-column rhetoric to avoid changes to the nationality law to satisfy domestic interests. The second case, from 1954, highlights how political elites refrained from engaging in fifth-column rhetoric, despite geopolitical incentives to do so, to facilitate the adoption of a new nationality law. In both cases, political elites were driven by domestic political interests.

Whereas Frost argues that territorial insiders might be noncitizens, in Chapter 8, Ciancia argues that citizens might be territorially distant from "their" state but still be considered insiders. She explores how interwar Polish officials attempted to control the behavior, mobility, and citizenship status of Polish citizens who had migrated to France and were deemed to be acting in ways that could ultimately threaten the Polish state upon their return. Ciancia makes two main arguments. First, although fifth columns are defined as groups that are physically internal to the state with links to that state's external enemies, there are instances in which the term can be applied to citizens who live beyond the state's geographical borders, but whose return appears likely. This example argues for a more careful look at concepts such as "external" and "internal"—or "foreign" and "domestic"—in regard to fifth columns. Second, in showing that Polish state officials were particularly keen to revoke citizenship from Polish Jews, based not only on their allegedly subversive activities but also on their supposed lack of commitment to the Polish state, Ciancia explores the ambiguity involved when ideological and ethnic factors intersect and highlights the hazard of attributing fifth-column politics to a single principle of categorization.

The last two chapters in Part III focus on challenges to the dominant discourses within a polity. They give voice to the targets of fifth-column accusations, as opposed to the earlier emphasis on the actions of the accusers. The focus is on how targeted groups internalize, reject, or contest fifth-column accusations by seeking to evade scrutiny or strategically working to deflect the burden of suspicion onto others. Taken together, the contributions in Part III upset the conventional wisdom and decenter the regime- or state-centric analytical perspectives that tend to prevail in the study of politics.

In Chapter 9, Anabtawi explores the challenges faced by groups whose outsider status is not based on either ethnic or partisan differences: LGBTQ movements. He focuses on two historical and geographic settings—the United States between 1950 and 1970 and Palestine from 1987 to 2020. Anabtawi argues that in both situations, a nascent LGBTQ movement strategically relied on "corrective discourse" as a tool to dislodge the stigma of betrayal and subversion. Both of these communities were able to legitimate their presence on the national stage and wade through the waves of political homophobia by methodically generating nationalist counter-frames. The chapter highlights how a serious treatment of fifth-column narratives can advance the study of social movement outcomes and enrich an evolving research program on marginalization and identity politics in the age of populism.

In Chapter 10, Fabbe and Balıkçıoğlu investigate what happens to proximate groups once an alleged fifth column experiences a crackdown. They identify a pattern whereby both state officials and leaders of religious orders turn against one another in an effort to signal loyalty to the regime. Fabbe and Balıkçıoğlu's contribution documents how authoritarian regimes reconfigure *collusive* claims about fifth columns into *subversive* ones as they consolidate power. During the AKP's initial period of alliance with Gülenists, political opponents of the two groups used collusive claims to suggest that an elite cabal of political Islamists with backing from Western governments was engineering a takeover of Turkish politics. Later, after the alliance broke down, the AKP recast those charges into claims that the Gülenists were operating as a subversive fifth column. This chapter highlights the strategy of indirect fifth-columnization—asserting a group's association with an alleged fifth column—and exposes how governments can remain above the fray by raising suspicions through informal leaks by allied actors.

In the concluding chapter, we summarize and synthesize the major insights drawn from earlier chapters and turn toward the future. We (1) outline a new research agenda and (2) use the book's findings to consider how global developments, including the Covid-19 pandemic, may influence fifth-column politics in the coming years.

## Contributions to the Field

*Enemies Within* is intended to spark new research by embracing the concept of fifth-column politics as an analytical category and cutting across conventionally siloed theoretical subfields. Within social science, the study of regimes and institutions on one hand, and the study of ethnic politics and nationalism on the other, have emerged as important and burgeoning fields of inquiry. While some scholars have examined the interplay of the two fields, practitioners of each subfield have mostly pursued distinct research agendas. This volume puts them in conversation, asking how geopolitics, political institutions, citizenship, belonging, and identity are implicated in the production of fifth-column politics. In doing so, the cross-fertilization that occurs throughout this volume yields insights into political institutions and the problem of membership in political communities—and how they influence each other. By bringing disparate strands of scholarship together, we hope not only to highlight the politics of fifth columns as an object of analysis but also to lay the groundwork for a broader research agenda.

This volume is centered in the methodological approach of comparative area studies, which aims to "split the difference between a context-bound narrative and universalizing comparison" (Ahram, Kollner, and Sil 2018: 14). It comprises case studies from across a broad swath of the world, ranging from Eastern Europe and the Middle East, to the United States and China. This geographic breadth and the use of primary sources in eight languages enable the authors to seek out common processes and mechanisms across states and regions with different political, cultural, and social characteristics. The chapters implement various research designs to uncover causal processes: some use a longitudinal approach and study how fifth-column claims change over time within a single country (e.g., Chapters 1, 2, 4, 6, 7, and 10), whereas others develop comparative case studies that cover a similar phenomenon in different countries (Chapters 5 and 9), examine a single case (Chapter 8), or explain the differential treatment of groups within a single country at different critical junctures (Chapter 3).

The contributors to this volume also draw attention to the fuzzy boundary between quotidian politics and the politics of crisis. Critical events such as armed rebellion and internationalized civil wars, in which external countries become engaged in a conflict, constitute fertile ground for fifth-column politics. Yet this volume demonstrates that fifth-column politics also resonates in times of "normal politics," under conditions of institutional, regularized competition between parties, and without significant international turmoil. The case studies highlight how quotidian political practices in stable times may nonetheless be characterized by a high level of tension, in which polarization is high, national identities and historical narratives are fiercely contested, and antagonistic movements contend for power. When it comes to fifth-column politics, the ostensible distinction between settled and unsettled times does not hold firm (Swidler 1986).

By centering our analysis on the concept of fifth columns, this volume touches on broader themes and contemporary challenges, including the rise of populism and authoritarianism, the resurgence of exclusionary nationalism, the weakening of democratic norms, and the persecution of marginalized communities and political dissidents. The contributors encounter these issues in the course of their analyses, and we address their broader implications in the Conclusion. Examining these issues through the lens of fifth-column politics enables us to see them in a new light.

## Notes

1. Historian Hugh Thomas (2001: 456) claims that the term was first used during the Russo-Turkish War in 1790, but does not provide evidence.
2. The concept arguably applies to several instances in ancient times. However, we maintain that contemporary fifth-column politics has important distinctive features, and this periodization helps delimit the scope of our project and case selection.
3. According to Chandra (2006: 397), "ethnic identities are a subset of identity categories in which membership is determined by attributes associated with, or believed to be associated with, descent."
4. A standard dictionary definition of "subversive" is sufficient for our purposes: "tending to weaken or destroy an established political system, organization, or authority" (https://dictionary.cambridge.org/us/dictionary/english/subversive).
5. Again, a dictionary definition is useful: "secret agreement or cooperation especially for an illegal or deceitful purpose" (https://www.merriam-webster.com/dictionary/collusion).

## References

Ahram, Ariel, Patrick Köllner, and Rudra Sil (eds). 2018. *Comparative Area Studies: Methodological Rationales and Cross-regional Applications*. New York: Oxford University Press.

Brass, Paul. 1991. *Ethnicity and Nationalist Theory and Comparison*. New Delhi and Newbury Park, CA: Sage.

Brubaker, Rogers. 1995. "National Minorities, Nationalizing States, and External National Homelands in the New Europe." *Daedalus* 124 (2): 107–132.

Brubaker, Rogers. 1996. *Nationalism Reframed: Nationhood and the National Question in the New Europe*. Cambridge: Cambridge University Press.

Brubaker, Rogers. 2013. "Categories of Analysis and Categories of Practice: A Note on the Study of Muslims in European Countries of Immigration." *Ethnic and Racial Studies* 36 (1): 1–8.

Bulutgil, H. Zeynep. 2016. *The Roots of Ethnic Cleansing in Europe*. New York: Cambridge University Press.

Bulutgil, H. Zeynep. 2017. "Ethnic Cleansing and Its Alternatives in Wartime: A Comparison of the Austro-Hungarian, Ottoman, and Russian Empires." *International Security* 41 (4): 169–201.

Burds, Jeffrey. 2007. "The Soviet War against 'Fifth Columnists': The Case of Chechnya, 1942–4." *Journal of Contemporary History* 42 (2): 267–314.

Caldwell, Christopher. 2009. *Reflections on the Revolution in Europe: Immigration, Islam, and the West*. New York: Anchor.

Carney, William P. 1936. "Boys and Men of 50 Guard Madrid Road." *New York Times*, August 17.

Ceplair, Larry. 2011. *Anti-Communism in Twentieth-Century America: A Critical History*. Santa Barbara, CA: ABC-CLIO.

Chafer, Tony, and Alexander Keese. 2015. "Introduction." In *Francophone Africa at Fifty*, edited by Chafer and Keese. Manchester: Manchester University Press, 1–14.

Chandra, Kanchan. 2006. "What Is Ethnic Identity and Does It Matter?" *Annual Review of Political Science* 9: 397–424.

Collier, David, and James E. Mahon Jr. 1993. "Conceptual 'Stretching' Revisited: Adapting Categories in Comparative Analysis." *American Political Science Review* 87 (4): 845–855.

Coppel, Charles A. 1983. *Indonesian Chinese in Crisis*. New York: Oxford University Press.

Csergő, Zsuzsa, and James M. Goldgeier. 2013. "Kin-State Activism in Hungary, Romania, and Russia: The Politics of Ethnic Demography." In *Divided Nations and European Integration*, edited by Tristan James Mabry, John McGarry, Margaret Moore, and Brendan O'Leary, 89–126. Philadelphia: University of Pennsylvania Press.

Darden, Keith, and Harris Mylonas. 2016. "Threats to Territorial Integrity, National Mass Schooling, and Linguistic Commonality." *Comparative Political Studies* 49 (11): 1446–1479.

Deák, István. 1995. "A Fatal Compromise?: The Debate over Collaboration and Resistance in Hungary." *East European Politics and Societies* 9 (2): 209–233.

De Jong, Louis. 2019. *The German Fifth Column in the Second World War*. New York: Routledge.

Diamond, Larry. 2020. *Ill Winds: Saving Democracy from Russian Rage, Chinese Ambition, and American Complacency*. N.p.: Penguin Books.

Dilanian, Ken. 2020. "American Universities are a Soft Target for China's Spies, Say U.S. Intelligence Officials." *nbcnews.com*, February 2. https://www.nbcnews.com/news/china/american-universities-are-soft-target-china-s-spies-say-u-n1104291.

Fein, Helen. 1993. "Revolutionary and Antirevolutionary Genocides: A Comparison of State Murders in Democratic Kampuchea, 1975 to 1979, and in Indonesia, 1965 to 1966." *Comparative Studies in Society and History* 35 (4): 796–823.

Gagnon, V. P., Jr. 1994/1995. "Ethnic Nationalism and International Conflict: The Case of Serbia." *International Security* 19, no. 3 (Winter): 132–168.

Gagnon, V. P., Jr. 2004. *The Myth of Ethnic War: Serbia and Croatia in the 1990s*. Ithaca, NY: Cornell University Press.

Gleason, Gregory. 2001. "Why Russia Is in Tajikistan." *Comparative Strategy* 20 (1): 77–89.

Griffiths, Ryan D. 2016. *Age of Secession*. Cambridge: Cambridge University Press.

Grinchenko, Gelinda, and Eleonora Narvselius. 2018. *Traitors, Collaborators and Deserters in Contemporary European Politics of Memory. Formulas of Betrayal*. Palgrave Macmillan Memory Studies. Cham: Palgrave Macmillan.

Harff, Barbara. 1987. "The Etiology of Genocides." In *Genocide and the Modern Age: Etiology and Case Studies of Mass Death*, edited by Isidor Walliman and Michael N. Dobkowski, 47–59. New York: Greenwood Press.

Harris, James. 2013. "Intelligence and Threat Perception: Defending the Revolution, 1917–1937." In *The Anatomy of Terror: Political Violence Under Stalin*, edited by James R. Harris, 29–43. Oxford: Oxford University Press.

Ho, Elaine Lynn-Ee. 2018. *Citizens in Motion: Emigration, Immigration, and Re-migration across China's Borders*. Stanford, CA: Stanford University Press.

Huntington, Samuel P. 1993. "The Clash of Civilizations?" *Foreign Affairs* 72 (3): 22–49.

Jenne, Erin K. 2007. *Ethnic Bargaining: The Paradox of Minority Empowerment*. Ithaca, NY: Cornell University Press.

Judis, John. 2016. *Populist Explosion: How the Great Recession Transformed American and European Politics*. New York: Columbia Global Reports.

Kambas, Michele, and Renee Maltezou. 2018. "Greek Parliament Debates Tsipras No-Confidence Motion after Macedonia Deal." *Reuters*, June 14. https://www.reuters.com/article/us-greece-macedonia-tsipras-noconfidence/greek-parliament-debates-tsipras-no-confidence-motion-after-macedonia-deal-idUSKBN1JA1B2.

Kaufman, Stuart J. 2001. *Modern Hatreds: The Symbolic Politics of Ethnic War*. Ithaca, NY: Cornell University Press.

Khlevniuk, Oleg V. 2015. *Stalin: New Biography of a Dictator*. New Haven: Yale University Press.

Leenders, Reinoud. 2012. *Spoils of Truce: Corruption and State-building in Postwar Lebanon*. Ithaca, NY: Cornell University Press.

Levine, John M., and Michael A. Hogg. 2010. *Encyclopedia of Group Processes and Intergroup Relations*. Thousand Oaks, CA: Sage.

Levy, Neil. 2007. "Radically Socialized Knowledge and Conspiracy Theories." *Episteme: A Journal of Social Epistemology* 4 (2): 181–192.

Lippmann, Walter. 1942. "The Fifth Column on the Coast." *Washington Post*, February 12.

Loeffel, Robert. 2015. *The Fifth Column in World War II: Suspected Subversives in the Pacific War and Australia*. New York: Palgrave Macmillan.

MacDonnell, Francis. 1995. *Insidious Foes: The Axis Fifth Column and the American Home Front*. New York: Oxford University Press.

Mudde, Cas. 2011. "Radical Right Parties in Europe: What, Who, Why?" *Participation* 34 (3): 12–15.

Mudde, Cas. 2019. *The Far Right Today*. Cambridge, UK: Polity.

Müller, Jan-Werner. 2017. *What Is Populism?* London: Penguin.

Mylonas, Harris. 2012. *The Politics of Nation-building: Making Co-nationals, Refugees, and Minorities*. Cambridge: Cambridge University Press.

Mylonas, Harris. 2014. "Democratic Politics in Times of Austerity: The Limits of Forced Reform in Greece." *Perspectives on Politics* 12, no. 2 (June): 435–443.

Mylonas, Harris, and Maya Tudor. 2021. "Nationalism: What We Know and What We Still Need to Know." *Annual Review of Political Science* 24: 109–132.

Naimark, Norman M. 2002. *Fires of Hatred*. Cambridge, MA: Harvard University Press.

Norris, Pippa, and Ronald Inglehart. 2019. *Cultural Backlash: Trump, Brexit, and Authoritarian Populism*. Cambridge: Cambridge University Press.

Petit, Philippe, and Henry Rousso. 2002. *The Haunting Past: History, Memory, and Justice in Contemporary France*. Philadelphia: University of Pennsylvania Press.

Pogonyi, Szabolcs, Mária M. Kovács, and Zsolt Körtvélyesi. 2010. "The Politics of External Kin-State Citizenship in East Central Europe." *EUDO Citizenship Observatory* 8. Robert Schuman Centre for Advanced Studies.

Ponchaud, François. 1989. "Social Change in the Vortex of Revolution." In *Cambodia 1975–1978*, edited by Karl Jackson, 151–178. Princeton, NJ: Princeton University Press.

Posen, Barry. 1993. "The Security Dilemma and Ethnic Conflict." *Survival* 35 (1): 27–47.

Pradhan, Ramakrushna, and Atanu Mohapatra. 2020. "India's Diaspora Policy: Evidence of Soft Power Diplomacy under Modi." *South Asian Diaspora* 12 (2): 145–161.

Prime Minister. 2015. Prime Minister Tsipras' Speech at SYRIZA's Central Committee Meeting. https://primeminister.gr/2015/05/23/13692.

Prunier, Gérard. 1998. "The Rwandan Patriotic Front." In *African Guerrillas*, edited by Christopher Clapham, 119–133. Oxford: James Currey Publishers.

"Putin obvinyaet SShA v provotsirovanii protestov." 2011. *BBC News*, December 8. https://www.bbc.com/russian/russia/2011/12/111208_putin_opposition_protests.

Radnitz, Scott. 2021. *Revealing Schemes: The Politics of Conspiracy in Russia and the Post-Soviet Region*. New York: Oxford University Press.

Rittersporn, Gábor Tamás. 1992. "The Omnipresent Conspiracy: On Soviet Imagery of Politics and Social Relations in the 1930s." In Nick Lampert and Gábor T. Rittersporn, eds., *Stalinism: Its Nature and Aftermath*, 101–120. London: Palgrave Macmillan.

Saideman, Stephen M. 2002. "Discrimination in International Relations: Analyzing External Support for Ethnic Groups." *Journal of Peace Research* 39 (1): 27–50.

Schechtman, Joseph B. 1953. "Postwar Population Transfers in Europe: A Survey." *The Review of Politics* 15 (2): 151–178.

Schlesinger, Arthur Meier, Jr. 1950. *The Politics of Freedom*. Boston: Houghton Mifflin.

Snyder, Jack L. 2000. *From Voting to Violence: Democratization and Nationalist Conflict*. New York: W. W. Norton.

Snyder, Timothy. 1998. "The Polish-Lithuanian Commonwealth since 1989: National Narratives in Relations among Poland, Lithuania, Belarus and Ukraine." *Nationalism and Ethnic Politics* 4 (3): 1–32.

Speier, Hans. 1940. "Treachery in War." *Social Research* 7 (3): 258–279.

Straus, Scott. 2010. "Political Science and Genocide." In *The Oxford Handbook of Genocide Studies*, edited by Donald Bloxham and Dirk A. Moses. New York: Oxford University Press.

Suny, Ronald Grigor. 2017. *"They Can Live in the Desert but Nowhere Else": A History of the Armenian Genocide*. Princeton, NJ: Princeton University Press.

Suryadinata, Leo. 2017. "The Rise of China and the Chinese Overseas: A Study of Beijing's Changing Policy in Southeast Asia and Beyond." ISEAS-Yusof Ishak Institute. Cambridge: Cambridge University Press.

Swidler, Ann. 1986. "Culture in Action: Symbols and Strategies." *American Sociological Review* 51 (April): 279–286.

Theoharis, Athan G., and Susan Rosenfeld. 1999. *The FBI: A Comprehensive Reference Guide*. N.p.: Greenwood Publishing Group.

Thomas, Hugh. 2001. *The Spanish Civil War*. New York: Modern Library.

Waterbury, Myra A. 2010. *Between State and Nation: Diaspora Politics and Kin-State Nationalism in Hungary*. New York: Palgrave Macmillan.

Waterbury, Myra A. 2020. "Kin-State Politics: Causes and Consequences." *Nationalities Papers* 48 (5): 1–10.

Weiner, Myron. 1971. "The Macedonian Syndrome: An Historical Model of International Relations and Political Development." *World Politics* 23 (4): 665–683.

Weiner, Tim. 2012. *Enemies: A History of the FBI*. New York: Random House.

Wimmer, Andreas. 2018. *Nation Building: Why Some Countries Come Together While Others Fall Apart*. Princeton, NJ: Princeton University Press.

*Zaman*. 2016. "Erdoğan Accuses Academics of Being Fifth Column." January 11. http://cibal. eu/digital-library/9-politics/5009-erdo-an-accuses-academics-of-being-fifth-column.

# PART I
## SUBVERSIVE
## FIFTH-COLUMN POLITICS

# 1

## The Enemy Within

### Divisive Political Discourse in Modern Poland

*Volha Charnysh*

Poland's ethnic demography, borders, and regime type changed multiple times in the course of the twentieth century. Ethnic Poles made up just 65 percent of the population in the interwar period, but today the country is one of the most ethnically homogeneous in Europe. Poland's borders were redrawn in 1918, 1921, 1939, and 1945, creating conflicts with neighboring states along the way. Poles also lived through a brief bout of parliamentary democracy, followed by military dictatorship, Nazi and Soviet occupation, a period of Stalinist totalitarian rule, four decades of authoritarian socialism, and a more extensive democratic period that, some argue, is eroding (Rohac 2018). Through all of these transformations, fifth-column accusations remained a staple of political discourse.[1] In 1922, Poland's very first president Gabriel Narutowicz was murdered following a collusive fifth-column claim that he represented the interests of world Jewry and owed his career to Jewish financial circles (Brykczynski 2016: 23). In the 1940s, Polish Communists framed the opposition as a subversive fifth column seeking to sell their country to Nazi Germany. In 1967, the First Secretary of the Polish United Workers Party railed against a Zionist "fifth column," demanding that the supporters of Israel and Western imperialists leave the country. In the 1980s, the authorities claimed that the Solidarity trade union was led by Jews "whose interests and goals were incompatible with the Polish national interests" (Michlic 2006: 259). "Jewish connections" continue to be evoked to delegitimize political opponents in post-1989 Poland (Forecki 2009; Charnysh 2015, 2017).

How do political actors choose whom to target with a public fifth-column accusation, and with what images, discourse, and metaphors? What explains the resonance of a particular fifth-column appeal? Drawing on research in social psychology, I argue that fifth-column accusations work by redefining in-group boundaries. Their indirect targets are not only the alleged fifth column, but in-group members who can be pressured into switching sides to dissociate themselves from the fifth-columnists. Fifth-column appeals are as much about reshuffling existing political alliances as they are about demonizing opponents or external enemies. Such appeals work best when they activate preexisting cultural

Volha Charnysh, *The Enemy Within* In: *Enemies Within*. Edited by: Harris Mylonas and Scott Radnitz, Oxford University Press. © Oxford University Press 2022. DOI: 10.1093/oso/9780197627938.003.0002

schemas. When choosing their targets, political actors thus draw on deeply held biases in their societies and instrumentalize preexisting cleavages based on ethnicity, religion, status, or experiences.

I use this theoretical framework to interpret fifth-column politics in twentieth-century Poland. I highlight similarities between the left- and right-wing parties' attacks against ethnic and ideological fifth columns in the interwar period, in the aftermath of World War II, and in the late 1960s through early 1980s. In all three periods, domestic political competition rather than security threats motivated fifth-column accusations. In each case, political entrepreneurs sought to divide their opponents and redraw existing political alliances in their favor. In the interwar period, the nationalist right framed Jews and anyone working with them as a fifth column. It articulated a collusive claim in order to delegitimize Polish parties allied with the Piłsudski government, getting them to switch sides. After World War II, the Communists claimed that the opposition betrayed the Polish nation by colluding with Nazi Germany in order to attract some members of the anti-Communist underground to the Communist cause. In the 1960s, the United Polish Workers' Party (PZPR) adopted right-wing language from the interwar period by framing "Zionists" as a fifth column serving Germany, seeking to divide societal opposition to its rule and to convince Polish functionaries to fall in line with Gomułka. Across three cases, political entrepreneurs adopted comparable discursive techniques despite their ideological differences and changing geopolitical environment. They used antisemitic stereotypes that have become an integral part of the nationalist narrative, activating latent fault lines within the Polish society.

## Fifth-Column Discourse from the Social Identity Perspective

Fifth columns are understood as "domestic actors who work to undermine the national interest, in cooperation with external rivals of the state" (Radnitz and Mylonas, Introduction, 3). A standard fifth-column narrative thus evokes threats emanating from inside and outside a social group at the same time. I argue that this property makes fifth-column accusations particularly useful to political entrepreneurs who wish to divide their opponents and convince some members their rival group to switch sides.

The immediate consequence of fifth-column rhetoric is stigmatization of select in-group members who are exposed as deviant and duplicitous for collaborating with the enemy. Individuals framed as disloyal to their in-group for serving the competing out-group incur heavy penalties. Research shows that in-group deviance is judged more harshly than out-group deviance because it threatens the positive image of the in-group, as demonstrated by studies of the "black sheep effect" (Marques, Yzerbyt, and Leyens 1988). Relatedly, fifth-column activity means not only betraying the in-group but also strengthening the rival out-group at its expense, which makes it an especially consequential transgression in the eyes of the in-group members (Travaglino et al. 2014). For instance, disclosing sensitive information to a domestic audience is perceived as much less problematic than disclosing it to a foreign actor, an act of treason that historically entailed the death penalty. Fifth-column accusations thus provoke greater moral outrage and punishment than other types of accusations.

By provoking outrage against purported betrayal, fifth-column accusations have the potential to transform group boundaries and reshuffle existing alliances. As McDermott (2020: 7) argues, political actors evoke out-group threats because this tactic "offers a very elegant solution to the very real organizational challenge of establishing collective action." Researchers have shown that priming threats from outside the group has different effects than priming threats from inside (Greenaway and Cruwys 2019). External threats increase the perceived homogeneity of the in-group (Rothgerber 1997), strengthen in-group identification and attachment (Brewer 2001), and facilitate coordination and cohesion among in-group members (Benard 2012; W. G. Stephan and C.W. Stephan 2000). Internal threats, by contrast, decrease the perceived homogeneity of the in-group, weaken in-group identification and attachment, and thus undermine coordination within the in-group (Greenaway and Cruwys 2019; Jetten and Hornsey 2014).

The invocation of an internal threat side by side with an external threat can be used to undermine cohesion within the accused rival group, which is purported to contain a fifth column, while erasing perceived differences between those members of the accused group who are willing to dissociate themselves from it and members of the accusing group. For example, by singling out Jews as a fifth column in the interwar period, Polish nationalists sought to split the center-left coalition by getting the center-left Piast party to switch sides, as this chapter will discuss. Similarly, the German right attacked the Social Democratic Party as serving Jews and international interests in order to split the left and attract German workers to a more conservative agenda following Germany's democratic transition in 1918 (Crim 2011: 627). The countervailing effects of fifth-column accusations on group identification and cohesion create tension that can be resolved by redefining the in-group/out-group boundary to exclude the purported fifth column. Switching alliances to dissociate from the fifth column restores in-group cohesion and homogeneity and resolves this tension. Effective fifth-column discourse thus undermines unity and coordination within the group that contains the alleged fifth column and increases identification and cohesion within the new coalition, which now excludes the fifth column. Figure 1.1 depicts how these dynamics operate.

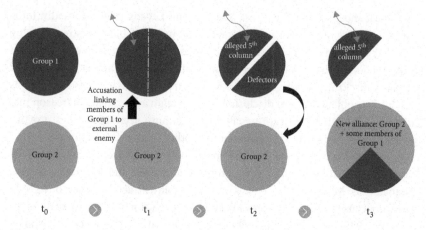

**Figure 1.1** Schematic representation of the argument.

Because of their potential for both inclusion and exclusion of in-group members, fifth-column accusations have frequently surfaced during the periods of nation- and state-building, regime change, or times of war. Branding ethnic minorities as serving external enemies has recast them as being outside of the national in-group and, at the same time, increased the perceived homogeneity and cohesion of the reconstituted in-group, now composed of ethnic majorities, by obscuring class and ideology distinctions among its members. For instance, the myth of Judeo-Bolshevism or Judeo-communism emerged in Central Europe in the aftermath of World War I, as a response to the dissolution of empires, military defeat, and the outbreak of revolutions of 1918–1919 (Gerwarth 2008). Fifth columns have also been invoked in times of internal instability because they place dissenters outside the national in-group, reducing opportunities for collective action against the regime, even as they rally popular support for the government among the rest of the population. This is the case in contemporary Russia, with anti-Putin protesters branded as Western agents. Fifth-column rhetoric can facilitate the reshuffling of existing political coalitions by excluding the alleged fifth column and by signaling greater opportunities for cooperation among actors who share internal and external enemies. In all these situations, fifth-column discourse shapes political outcomes by changing group boundaries and facilitating coordination and cohesion within the new alliance.

Fifth-column discourse is often unrelated to the actual level of internal or external threat; it does not necessarily respond to security risks presented by espionage, sabotage, or other subversive activity. In interwar Poland, German and Ukrainian minorities presented greater risks to the nascent Polish state than the Jewish minority, which became the main target of Polish nationalists. When the loyalty of some domestic groups appears suspect, political actors can act on their suspicions covertly—by increasing surveillance, restricting the right of assembly or freedom of movement, or excluding some groups from employment in sensitive occupations (Radnitz and Mylonas, Introduction). Airing their suspicions by openly confronting the alleged fifth-columnists can backfire by exacerbating intergroup tensions, antagonizing the purported fifth column's external allies, or putting traitorous groups on alert. When actors are instead interested in changing political alliances, they will pick a target that can activate preexisting cultural schemas and fragment their rivals. In Central Europe, Jews were a frequent victim of fifth-column accusations because antisemitism was widespread and because many center-left coalitions included Jews.

This discussion suggests the following expectations about the occurrence and content of fifth-column discourse. First, such discourse will be motivated by the desire to undermine or reinforce cohesion within social groups rather than by the genuine presence of internal or external threats. Political actors will publicly evoke fifth columns to reinterpret group boundaries and alter existing political alliances. Second, fifth-column discourse will target not only the alleged fifth column, but also the affiliated in-group members, whose allegiances such discourse will seek to influence. If fifth-column accusations are successful, these in-group members will switch sides. Third, fifth-column discourse will often draw on preexisting cultural schemas and prejudices, targeting peripheral in-group members.

## Fifth-Column Rhetoric in Poland

This section applies the insights from social identity theory to decode nationalist rhetoric in twentieth-century Poland. I draw parallels between (1) ethnically charged fifth-column discourse by the nationalist right in the interwar period (1921–1939); (2) fifth-column rhetoric focused on ideology by the Soviet-backed Polish Workers' Party immediately after World War II; and (3) the revival of ethnicity-based fifth-column accusations by the Communist Party in the post-Stalinist period (1956–1980).

## The Jewish Fifth Column in Interwar Poland

When Poland regained independence after 123 years of control by Russian, Prussian, and Austrian empires, ethnic Poles made up just 65 percent of its population. The country hosted sizable Ukrainian (14%), Jewish (10%), Belarusian (3%), and German (2%) minorities and bordered revolutionary

Russia and revisionist Germany. Forging a sense of national unity among diverse citizenry in a hostile international environment was a challenge. The Jewish minority pledged loyalty to the Polish state immediately after independence and lacked connections to external powers that could destabilize Poland. Other groups were much less governable, however. The Germans in the Prussian partition resented losing their dominant status and hoped for reunification with Germany (Chu 2012); the Ukrainians in Galicia sought to win more autonomy through terrorism (Horak 1961); and the Belarusians consumed the Soviet propaganda and voted Communist (Vakar 1956; Kopstein and Wittenberg 2003).

All three groups were potential fifth columns from the security standpoint; at one point or another all were under surveillance or infiltrated by government agents. Yet it was the less threatening Jewish minority and its Polish allies who bore the brunt of fifth-column accusations in the interwar period. This outcome is more consistent with the alliance-shifting functions of the fifth-column discourse. From a social identity perspective, ethnic minority groups that are willing to cooperate with the ethnic majority are better targets because they can be used for undermining existing interethnic coalitions and changing political alliances. Targeting the more cooperative minority delegitimizes some majority group members by association, motivating them to switch sides. I expand on this explanation later in this section.

The first decades of Poland's independence were marred by the stalemate between two political blocs: on the left, a broad coalition of socialists, liberals, and peasant radicals united by Marshal Józef Piłsudski, and on the right, the National Democrats (or Endecja) masterminded by Roman Dmowski.[2]

In the first democratic elections, in 1922, the Endecja movement, represented by the coalition of the National Populist Union, the Christian National Party, and the Christian Democratic Party, secured 39 percent of seats in the National Assembly. The left, incorporating the Polish Socialist Party (PPS), the radical peasant Emancipation party, and other smaller groups, secured 25 percent of seats in the Assembly. Together with center-left Piast, the pro-Piłsudski group held 40 percent of seats. Thus, neither the left nor the right could create a majority government (Brykczynski 2016: 84–85).

In this unstable situation, the National Minorities Bloc, comprising Jewish, Ukrainian, Belarusian, and German groups held together by the Zionist leader Yitzak Grünbaum, could serve as a kingmaker. The Bloc held 16 percent of the vote and 15 percent of seats in the National Assembly. While neither the left nor the right was happy about the situation, the left was more open to working with the Bloc than the openly antisemitic Endecja. An alliance with the Bloc would enable the left to form a majority government and secure the results of the upcoming presidential election (Brykczynski 2016: 87).[3] To prevent this outcome and split the pro-Piłsudski camp, the right sought to create an alliance with the Bloc untenable for the left by labeling Jews as traitors and those willing to work with Jews as betraying the Polish nation (Brykczynski 2016).

To that end, Endecja portrayed Gabriel Narutowicz, presidential candidate from the left and a Roman Catholic, as colluding with the world Jewry (Brykczynski 2016: 25–26). When Narutowicz was elected, the right sought to sabotage his inauguration by claiming that he owed his career to "Jewish financial circles" and would extend "Jewish-Masonic influence" over Poland (Brykczynski 2016: 25–26). The National Democrats unleashed their fury not only on Jews, but also on Piast, the party most likely to split away from the pro-Piłsudski camp. The party's leader, Wincenty Witos, was castigated for "marching under the command of the Jews" and "betraying Poland." Under pressure, Piast issued a public declaration that explained its support for Narutowicz as "not the result of some deal reached with any of the Polish left-wing parties or let alone with the national minorities" (Brykczynski 2016: 29–31). Narutowicz was assassinated just five days after assuming office, and the right succeeded in bringing Piast to its side. The center-right coalition between Piast and Endecja (Chjeno-Piast) ruled briefly and unstably in 1923 and again in 1926.

In May 1926, the Chjeno-Piast government was unseated by the coup organized by Piłsudski. Piłsudski soon created the Non-partisan Bloc for Cooperation with the Government (BBWR), which rested on electoral support from Poles, Jews, and other minorities. Endecja was now back in the opposition, facing a center-left alliance with ethnic minorities that it had tried to prevent years earlier. The National Democrats resorted to the old tactics to split their rival. They accused Piłsudski of protecting "Judeo-Polonia" and neglecting Polish national interests and claimed that support from ethnic minorities was evidence that Piłsudski "betrayed the nation" (Michlic 2006: 96). Jews themselves were charged with collaborating with the USSR and supporting communism. The Soviet Union was perceived as a key threat by Piłsudski's support base, which made Endecja's appeals more persuasive among this group. Endek activists referenced the recent Polish-Soviet war to reinforce their claims. For example, pro-Endecja Reverend Stanisław Trzeciak argued in a 1937 article: "The Jews betrayed the Polish Army. They did not participate in the defense of Lwów. They constituted 99% of those who acted against the Polish state during the Soviet-Polish War of 1920. Ninety-eight or 100 percent of Jews are communist revolutionaries" (cited in Michlich 2006: 90).

Communism was the enemy Dmowski's and Piłsudski's voters could agree on; linking Jews and communism facilitated coordination between some segments of the BBWR and Endecja. The Endeks questioned the loyalty of Piłsudski's Polish supporters for siding with the Jews and, by extension, with the Communists, in order to divide them and discredit the Piłsudski regime. Endecja's rhetoric was thus designed to rally ethnic Poles, particularly from the BBWR, against a common Jewish-Communist threat, to get them to join the right.

Endecja's approach succeeded after Piłsudski's death. In 1935, the BBWR disintegrated and its right-wing members founded the Camp of National Unity (Obóz Zjednoczenia Narodowego, OZN). The OZN adopted discrimination of Jews as its official policy, breaking away from the BBWR's tolerant legacies (Wynot 1971). In its 1937 *Theses on the Jewish Question*, the OZN warned that Jews belonged to "a universal, Jewish a-state group possessing separate national goals" from Poles and repeated Endecja's earlier rhetoric linking Jews to the Red Army and the Comintern (Wynot 1971: 1049). Antisemitism became a bridge between many former BBWR members and Endecja.

## Enemy Rhetoric in the Aftermath of World War II

In 1939, the Soviet Union and Germany invaded Poland. The ensuing five years of brutal occupation left a deep impression on Polish society. In contrast to other states of the Eastern Bloc that allied with Nazi Germany (Hungary, Romania, Bulgaria) or were occupied by Germany alone (Czechoslovakia), Poles felt neither guilt for collaborating with the Nazis nor gratitude for being "liberated" by the Soviet Union at the end of the war (Lewis 1982). During the conflict, ethnic Germans and Ukrainian nationalists collaborated with the Nazi regime and perpetrated violence against Poles and Jews.

The security perspective suggests that governments would be more concerned about ethnic minorities linked to the country's external enemy. From the social identity perspective, however, fifth-column discourse against the collaborating ethnic groups, already ostracized and soon to be expelled from Poland, would have limited political uses. In competition with diverse political groupings united to prevent the Communist takeover, the small but well-connected Communist forces decided to base their fifth-column rhetoric on ideological differences instead.

The anti-Communist opposition was composed of ethnic Poles from all sides of the political spectrum; some wound up on the German and others on the Soviet side of the Molotov-Ribbentrop border in 1939. To divide and conquer, the Communist Party framed some of the opposition members as fifth-columnists with ties to Nazi Germany, claiming that they fought the Red Army in order to deliver Poland into German hands. According to Gomułka, "These traitors dream[ed] about a fascist dictatorship in Poland; fearing . . . the growing strength of the Polish people, they want . . . to help Berlin by calling for the end of the fight against the Germans and by turning arms against their brothers fighting the occupiers, against the Polish Workers' Party" (quoted in Zaremba 2001: 122).

Both Home Army and the right-wing National Armed Forces (Narodowe Siły Zbrojne) were attacked as Nazi collaborators who murdered Jews and partisans (Steinlauf 1997: 49). The PPR also continuously emphasized the unity of Poland and the Soviet Union in the fight against Nazi Germany, a key enemy for the interwar Endecja movement, with a goal of attracting some of Endecja's former members. It offered members of the anti-Communist underground a choice to switch sides and emphasized that they shared some enemies with the PPR.[4]

The opposition did not reciprocate with similar accusations against the Communist Party. Instead it portrayed Communists as an external threat and called for national unity and putting away old political disagreements. It ridiculed the Communist Party's new name, Polish Workers' Party (Polska Partia Robotnicza, PPR), as standing for "Paid Lackeys of Russia" (Płatne Pachołki Rosji) and branded Communists "Stalin-Jews," sent by the Soviet Union in order to take over the Polish government (Behrends 2009: 452).

Although the PPR and its successor, the United Polish Workers Party (PZPR), failed to attract broad societal support, they succeeded in splitting the opposition and recruiting some of its members into their ranks. An early convert was Bolesław Piasecki, who headed the extreme right faction, ONR-Falanga, on the eve of World War II and joined the Home Army at the end of the conflict. Piasecki agreed to join the Communist side after the Soviet security forces arrested him in 1944. In 1947, Piasecki founded the PAX movement of pro-Communist Catholics, which eventually absorbed many members of the former Endecja (Behrends 2009). Endecja activists also flocked to the newly created Association for the Development of the Recovered Territories (Towarzystwo Rozwoju Ziem Zachodnich) and to the anti-German League of Fighters for Freedom and Democracy (ZBoWiD).

To be sure, the fifth-column rhetoric alone was insufficient to bring about the rapprochement of Communists and the nationalist right. The Red Army and Soviet NKVD were necessary to convince many opposition leaders that resistance was futile. Still, the fifth-column narratives increased fragmentation and infighting within the anti-Communist underground and attracted some of its members to the Communist side.

## Witch-Hunt against the Jewish Fifth Column in 1968

Perhaps the most notorious example of fifth-column discourse occurred in 1968, while Poland was a one-party state. Although elections were still held, there was no meaningful political competition in the country at that time. To understand political discourse, one needs to focus on the growing opposition to Communist rule and internal rivalries within the Communist Party. In this period, the party invoked the Jewish fifth column to sow disunity in response to mounting societal opposition to its rule, blaming Poland's economic problems on its Jewish members and at the same time framing the dissenters as encouraged by Israel. Portraying some party functionaries as a Zionist fifth column also resolved the long-standing rivalries within the PZPR by convincing many party functionaries to fall in line with its leader, Gomułka. Even the public at large embraced the campaign, as suggested by the secret reports to the Ministry of Interior Affairs (Zaremba 1998, 2001).

At first blush, the 1968 hunt for the Zionist fifth column can be attributed to changes in the international environment. In June 1967, Israel launched a surprise attack on Egypt and secured a decisive victory over the coalition of Arab states in six days, changing the balance of power in the Middle East. The USSR perceived the Israeli strike an act of aggression sponsored by Western imperialists and broke off diplomatic relations with Israel. Poland and the rest of the Eastern Bloc had to follow suit. Yet Polish society did not see the situation the same way. Poles sympathized with Israel's fight against the Arab coalition (Rozenbaum 1978). The Polish Ministry of Internal Affairs was particularly worried about Polish Jews, concluding that the majority "adopted pro-Israeli views, opposed to the Party's politics and the position of the Polish government and foreign to the Polish population." It registered numerous expressions of solidarity with the Jewish cause and eighty-one instances of Jews volunteering to join the Israeli army or transferring their savings to Israel (Stola 2000: 48).

To deal with the situation, Gomułka summoned the provincial party secretaries to coordinate an anti-Israeli campaign. The state media and the provincial and local party branches played a crucial role in framing the conflict. To make Israel more threatening to the Polish masses, propaganda linked it to a more traditional enemy, Germany. On June 13, daily *Zycie Warszawy* claimed that Israel had received military supplies and ideological direction from West Germany (Rozenbaum 1978).

On June 19, 1967, Gomułka for the first time invoked a fifth column. In his speech at the Congress of the Polish Trade Union, he exhorted: "We cannot remain indifferent toward people who in the face of a threat to world peace, that is, also to the security of Poland and the peaceful work of our nation, support the aggressor, wreckers of peace and imperialists. . . . We do not wish a 'fifth column' to be created in our country" (Michlic 2006: 247). This part of the speech was criticized by other Politburo members, who were taken by surprise by his decision to search for internal enemies (Stola 2000: 184). The "fifth column" sentence was censored in print, but the phrase became a "hit" following television and radio broadcasts of the speech (Michlic 2006: 247). Emboldened by praise from Moscow, Gomułka repeated the fifth-column accusations at the June 27 Politburo meeting. He framed those who did not stand firm against Israel as traitors ("with two souls and two homelands") and argued that they could bring Poland to the brink of a nuclear war if no action were taken (Stola 2000: Appendix 3). For the time being, however, the propaganda centered on external enemies and no policy changes occurred.

Only in March 1968, nearly a year after the Arab-Israeli war, did the fifth-column accusations become concrete and reach a broad audience. This development was prompted by student protests against the ban on Adam Mickiewicz's play *Dziady*. The play was banned as Russophobic in January 1968 and by March, the unrest spread from the Warsaw University to other educational institutions. Student rallies were brutally suppressed and soon blamed on a "Zionist conspiracy." The authorities mentioned prominent Jewish students as instigators. Among others, they named Antoni Zambrowski, the son of prominent PZPR Central Committee member Roman Zambrowski, and Ewa Zarzycka, the daughter of Chairman of the Warsaw National Council Janusz Zarzycki. Both were well known as Jewish and neither was in Poland at the time of the protests (Michlic 2006; Stola 2000). From then on, the anti-Israel campaign morphed into the campaign against the Zionist fifth column within Poland (Michlic 2006). The "Zionist" label was applied loosely, referring to Jewish descent, alleged disloyalty to Poland, or both (Michlic 2006: 244). Its use allowed the PZPR to deflect accusations of antisemitism while continuing to target Jews and Poles associated with them. Indeed, most Poles understood the term as the equivalent of "Jew," an interpretation recognized and encouraged by the authorities (Michlic 2006: 245–246).

For maximum effect, the party resorted to tropes from the interwar period. It framed Zionists as wealthy, powerful, and conspiratorial (Stola 2000: 154). The propaganda also contained elements from the myth of Judeo-communism, though Jews were now accused of a different crime: Stalinism (Glowinski 1991: 64; Michlic 2006: 257). The "Zionists" were framed as threatening because they served a broad range of foreign interests—Israeli, American, and West German. The link to Germany, Poland's traditional enemy, was especially prominent in the propaganda. For example, an article published in *Trybyna Mazowiecka* on March 25, 1968, claimed: "The Zionists . . . would like to impose upon the people of socialist Poland the policies of Israel, the German Federal Republic, and imperialism. . . . While they impute to us all kinds of barbarism and crimes, they smile at the 'German henchmen of their relatives' in West Germany" (cited in Michlic 2006: 249). Rumors circulated that one thousand former Nazis advised the Israeli army. Jews were blamed for launching an "anti-Polish offensive" in the West by slandering the Polish nation as antisemitic and blaming Poles for the Holocaust (Steinlauf 1997: 80). The propaganda claimed that Israel and the Zionists decided to absolve Germans of the crimes committed during World War II "in exchange for compensation in the amount of more than three billion marks" and to convince the world that these crimes were perpetrated by Poles instead (Stola 2000: 165). Reactions to these outrageous claims in the West were then broadcast in Poland to escalate the campaign.

Why did the party with many prominent Jewish members and a history of combating antisemitism now turn against Zionists? What explains the curious mixture of antisemitic tropes from the interwar period, distortions of the Holocaust, and claims about Jewish Stalinism?

These narratives start making sense when we consider fifth-column appeals as an attempt to redefine group boundaries by breaking some groups and strengthening others. By insinuating that the protests against censorship were a Zionist conspiracy, the authorities aimed to undermine the cohesion of the student movement and to prevent it from spreading. The purported Jewish connection implied that the protests did not represent public opinion and served Poland's external enemies. However, the propaganda did not simply blame individuals of Jewish origin for the unrest. In naming children of prominent Jewish Communists, the anti-Zionist campaign targeted segments of the party apparatus itself. The anti-Zionist campaign allowed some party elites, most notably the Partisan faction (Partyzanci), to weaken the cohesion of their rivals within PZPR and to redraw coalitions within the party in their favor. The fifth-column accusations exposed some party members and protected others, facilitating coordination by actors with diverse interests around a preexisting but dormant fault line that separated individuals of Jewish origin, their families, and their allies from the rest of the group. The PZPR's inclusion of Jews in governance and attempts to reduce popular antisemitism now allowed the party to accuse Jews of betraying their Polish in-group by colluding with the outside enemy. The campaign brought about the dismissals of many Jewish and Polish Communists, opening new career opportunities for their disgruntled colleagues and subordinates and thus creating shared interests among them (Stola 2000: 202). In doing so, the campaign increased cohesion within the PZPR.

Separating Jewish from Polish Communists also ensured that the party's problems could be recognized and punished without endangering the stability of the Communist regime as a whole (Stola 2000, 196). If the "Jewish Stalinists" had diverted the party's agenda in the past, then the party cleansed from the Zionist elements would truly serve the Polish nation. The ordinary Poles could denounce corruption, economic mismanagement, and police brutality, as long as they attributed these problems to the Zionist meddling (Stola 2000: 193–196; Zaremba 1998: 144–170).

To sum up, although many aspects of the anti-Zionist campaign were improvised, its main pieces came together in a politically effective way. The propaganda excluded Jews from the Polish in-group by presenting them as traitors. The use of Endecja's tropes underscored the continuities between interwar Poland and the Polish People's Republic and tapped into widespread antisemitism. Reliance on preexisting cultural schemas thus added credibility to the PZPR propaganda despite popular distrust in the party as a source of information. Similarly, the references to West Germany and the Holocaust served to rally Poles in defense of their in-group from both symbolic and military threats. West Germany was much more threatening than Israel, with which Poland had recently enjoyed good diplomatic relations. Germany did not recognize the redrawing of borders in 1945, and the Poles still remembered the brutality of Nazi occupation and the unease over the fate of Polish Jews. Connecting Zionists to West Germany and invoking the Holocaust thus anchored the Communist propaganda in powerful symbols and painful memories and signaled shared interests between the PZPR and Polish society. Whatever its failings in domestic politics, the party-state alone could defend Poland from external threats to its territory and reputation.

Marcin Zaremba's analysis (1998, 2001) of the secret reports to the Ministry of Interior Affairs suggests that large segments of Polish society embraced the campaign. Zaremba argues that during the anti-Zionist witch-hunt, the majority of Poles viewed the PZPR as supporting their national values (Zaremba 1998: 144–160). Additional evidence of the campaign's popularity comes from the local party gatherings. Stola (2000: 189–190) notes that party events were better attended than usual, lasted for many hours, and involved many more speeches by the rank and file during the anti-Zionist purge. Piotr Osęka (2008: 302) points out that in March 1968 party membership increased faster than in the previous months. Of course, the party branches also had an incentive to report higher participation to the center, lest they also be accused of supporting Israel.

The PZPR's own analysis of letters sent to the editorial office of *Polityka* in the period March–May 1968, reprinted in Stola (2000), suggests that the propaganda resonated with Polish society. The letters repeated stereotypes about Jews and discussed the special hatred of Jews for Poles together with their love for Germans. All references to Gomułka, on the other hand, were positive: the letter writers expected the First Secretary to address Poland's social and economic problems by removing Jews from power (Stola 2000: 358). Thus, even if some harbored doubts about the credibility of the PZPR, the fact that the propaganda referenced familiar stereotypes and evoked the threat from one of Poland's traditional enemies (Germany) ensured its popular resonance.

## Fifth-Column Appeals at the End of the Communist Period

The anti-Zionist campaign ended by 1969, but the PZPR continued to use the "Jewish fifth column" to divide the opposition in the 1970s and 1980s. The Communist press portrayed the Committee for the Defense of the Workers

(KOR), created in 1976, as alien to Polish society because of its "revisionist-Zionist" connections. A few years later, the authorities claimed that the Solidarity Trade Union was run by Jews and distributed fake leaflets warning that Solidarity's "Jewish leadership" planned to capture power in order to rule over Poles (Michlic 2006: 259). Posters distributed by the Polish Security Service depicted one of KOR's leaders, Bronisław Geremek, receiving instructions from an Israeli rabbi on the phone (Zawadzki 2010: 231). The growth of Solidarity was accompanied by the rise in antisemitic leaflets, brochures, and books sponsored by the state (Cała 2012: 513).

By presenting key opposition figures as fifth-columnists, the authorities hoped to split the movement and to convince the "true Poles" within it to negotiate with their government. The decision to repeat some of the claims from 1968 suggests that the party believed in their appeal among some segments of Polish society. In a 2008 interview, Adam Michnik, one of the Solidarity's advisors of Jewish origin, admitted his fears that "so-called true Poles" within the Solidarity move-ment would take over and use antisemitism to their advantage, isolating KOR and splitting the movement (Cała 2012: 513). His concerns were not unfounded. In 1980, some trade union members criticized KOR as dominated by Jews, and ominous graffiti, including "KOR and Jews away from Solidarity," appeared on buildings and fences near the 1981 Countrywide Meeting of the Solidarity Delegates in Gdańsk (Dobosz 1981: 8–9).

Ultimately, these attempts to undermine Solidarity by exposing real or imag-ined Jewish origins of its leaders failed. The trade union organizers resisted the fifth-column narrative created by the PZPR, perhaps due to their concerns about the union's international reputation. Likewise, the Polish society could no longer be persuaded to cooperate with the party-state. It seems that by the 1980s the PZPR's legitimacy had eroded to the point where its claims were not credible.

## Conclusion

Tracing the persistence of fifth-column claims in Poland suggests that they were driven by the need to undermine domestic political rivals rather than by gen-uine concerns about the disloyalty of alleged fifth-columnists. Social identity theory elucidates how fifth-column accusations can be useful for a political en-trepreneur. Such rhetoric stigmatizes select in-group members by associating them with the treacherous fifth column and implying that they are betraying the in-group. By priming threats from both inside and outside the in-group, fifth-column accusations also undermine the cohesion of the social groups they target and increase the likelihood that their members defect to other groups. In this way, fifth-column discourse serves to redefine and restructure existing coalitions.

The analysis also suggests that institutions, ideological differences, or ethnic demography alone cannot predict the intensity or the content of fifth-column discourse. Jews were a frequent target for political actors of all stripes, not only for the antisemitic Endecja. References to Jewish betrayal continued even when the size of the Jewish minority in Poland dwindled from over 3 million to under 20,000 people. The Communist Party defined the fifth column in ideological terms in the late 1940s but returned to antisemitic cues in the 1960s. The Jewish threat was reinforced by its purported connections to the USSR or to West Germany, regardless of the actual allegiances of Polish Jews. Political entrepreneurs in Poland evoked Jewish enemies in part because of widespread antisemitism. They leveraged the emotional power of preexisting cultural schemas, with Jews as a minority that could be conceptualized as both inside and outside the national in-group. Yet their ultimate aim was reshuffling existing alliances and dividing the opposition rather than demonizing the Jewish population or changing its behavior.

Poland was not alone in accusing the Jewish minority of disloyalty. The myth of Judeo-Bolshevism or Judeo-communism was prevalent in much of Central Europe between the two world wars. In the aftermath of World War I, Jews were charged with betraying their nations on the battlefield and conspiring with Bolsheviks to foment revolutions—this stab-in-the-back myth resembled claims of the Polish right that Jews supported the Soviet side in the Polish-Soviet war of 1919–1921. Jews were simultaneously portrayed as a revolutionary menace from "the East" that threatened the Christian nation and as agents of "Golden International" and Western democratization (Gerwarth 2008: 198). The assassination of Poland's first president in 1922 was just one of many antisemitic attacks on politicians and public figures, including German foreign minister Walter Rathenau, Rosa Luxemburg, Karl Liebknecht, and Matthias Erzberger.

The myth of the Jewish fifth column remains alive and well in contemporary Poland, despite the absence of both Jews and Communists. The arrival of democratic competition in the 1990s increased the incentives to frame political opponents as not sufficiently Polish in spirit. In the first presidential election after the transition, the anti-Communist opposition spread rumors that Lech Wałęsa's main opponent, Tadeusz Mazowiecki, was a hidden Jew (Gebert 1991). Since then, not a single electoral campaign was completely free from antisemitism (Forecki 2009: 163). Most recently, in the 2020 election, the ruling Law and Justice Party (PiS) insinuated that Rafał Trzaskowski of the liberal Civic Platform (PO) was beholden to Jewish and LGBT interests. In order to ensure the reelection of the incumbent president Andrzej Duda, PiS leader Jarosław Kaczyński accused Trzaskowski of colluding with billionaire George Soros and of supporting the restitution of prewar Jewish property. These allegations were repeated not only by the right-wing nationalist outlets, but also by the public television station, TVP, which is controlled by PiS. In the end, Trzaskowski lost to Duda by just 1.2 percent of the vote in the runoff.

The term "Jew" is increasingly used as a metaphor for fifth-column status. It no longer signifies Jewish heritage, but instead represents "anti-national" values (Michlic 2006: 10). As Forecki (2009: 160) argues, the image of the "Jew in Poland stands for the foreignness of power and its appointed representative." Correspondingly, the Jewish label is often used by unscrupulous politicians to mobilize voters, discredit political opponents, or influence public opinion on complex policy issues. The Jewish threat is constantly linked to political issues that have little to do with Israel or Judaism as such (Charnysh 2015). Understanding the persistence of such associations is an important task for future research.

## Notes

1. Discourse, policy, and mobilization are three possible forms of fifth-column politics that may or may not occur concurrently (Radnitz and Mylonas, Introduction). This chapter focuses primarily on *fifth-column discourse.*
2. Piłsudski and Dmowski stood for the two opposing conceptions of Polish nationalism. Piłsudski glorified Poland's past as the multiethnic Polish-Lithuanian Commonwealth (1569–1795) and embraced ethnic minorities as a part of the Polish nation. He led Poland through the destructive but eventually victorious Polish-Soviet war (1919–1921), which determined Poland's eastern borders and solidified the negative view of the Red Army and the USSR among the large segments of the population in the borderlands. Dmowski, by contrast, perceived Poland's diversity as its Achilles heel. He condemned the religious tolerance that attracted a large Jewish population to the Commonwealth and argued that the future belonged to the ethnically homogeneous Polish nation. In his view Germany, not Russia, was Poland's key external enemy. He glorified Poland of the Piasts (tenth to fourteenth centuries), which allied with Moscow to fight the Teutonic Order and controlled lands that were subsequently conquered by Prussia (Dabrowski 2011).
3. The Polish president in 1922 was elected indirectly, by members of the National Assembly.
4. In addition to discrediting their opponents, the PPR adapted its universalist ideology to incorporate many elements of Polish nationalism. In 1944, when their reputation was especially poor, Communist activists went as far as to hold a Catholic mass to celebrate the anniversary of the Polish victory against the Soviet Union in the battle of Warsaw (Zaremba 2001: 140). To justify their support for the Soviet annexation of the eastern territories, an extremely unpopular policy, the PPR invoked Dmowski's vision of "Piast Poland." The lands acquired by Poland from the West were presented as the "Recovered Territories," returned to their motherland after some nine hundred years of German exploitation (Kulczycki 2002). This narrative linked Poland's

new borders to its heroic past under the medieval Piast dynasty, reframing the loss of eastern borderlands as compatible with Poland's national interest (Zaremba 2001: 133–134).

# References

Behrends, Jan C. 2009. "Nation and Empire: Dilemmas of Legitimacy during Stalinism in Poland (1941–1956)." *Nationalities Papers* 37 (4): 443–466.

Benard, Stephen. 2012. "Cohesion from Conflict: Does Intergroup Conflict Motivate Intragroup Norm Enforcement and Support for Centralized Leadership?" *Social Psychology Quarterly* 75: 107–130.

Brewer, Marilynn B. 2001. "Ingroup Identification and Intergroup Conflict: When Does Ingroup Love Become Outgroup Hate?" In *Social Identity, Intergroup Conflict, and Conflict Reduction*, edited by Richard D. Ashmore, Lee Jussim, and David Wilder, 17–41. Oxford: Oxford University Press.

Brykczynski, Paul. 2016. *Primed for Violence: Murder, Antisemitism, and Democratic Politics in Interwar Poland*. Madison: University of Wisconsin Press.

Cała, Alina. 2012. *Żyd—wróg odwieczny. Antysemityzm w Polsce i jego żródla*. Warsaw: Wydawnictwo Nisza.

Charnysh, Volha. 2015. "Historical Legacies of Interethnic Competition: Anti-Semitism and the EU Referendum in Poland." *Comparative Political Studies* 48 (13): 1711–1745.

Charnysh, Volha. 2017. "The Rise of Poland's Far Right: How Extremism Is Going Mainstream." *Foreign Affairs*, December 18. www.foreignaffairs.com/articles/poland/2017-12-18/rise-polands-far-right.

Chu, Winson. 2012. *The German Minority in Interwar Poland*. Cambridge: Cambridge University Press.

Crim, Brian E. 2011. "'Our Most Serious Enemy': The Specter of Judeo-Bolshevism in the German Military Community, 1914–1923." *Central European History* 44 (4): 624–644.

Dabrowski, Patrice M. 2011. "Uses and Abuses of the Polish Past by Jozef Piłsudski and Roman Dmowski." *The Polish Review* 56 (1/2): 73–109.

Dobosz, Henryk. 1981. "Notatki na marginesie nadziei." *Samorzadność* 1 (177): 8–9.

Forecki, Piotr. 2009. "Stolzman w Belwederze?: Instrumentalizacja antysemityzmu w kampaniach prezydenckich w Polsce po roku 1989." *Srodkowoeuropejskie Studia Polityczne* 1 (2): 157–184.

Gebert, Konstanty. 1991. "Anti-Semitism in the 1990 Polish Presidential Election." *Social Research* 58 (4): 723–755.

Gerwarth, Robert. 2008. "The Central European Counter-Revolution: Paramilitary Violence in Germany, Austria, and Hungary after the Great War." *Past & Present* 1: 175–209.

Glowinski, Michał. 1991. *Nowomowa po polsku*. Warsaw: Wydawnictwo PEN.

Greenaway, Katharine H., and Tegan Cruwys. 2019. "The Source Model of Group Threat: Responding to Internal and External Threats." *American Psychologist* 74 (2): 218–231.

Horak, Stephan M. 1961. *Poland and Her National Minorities, 1919–39*. New York: Vantage Press.

Jetten, Jolanda, and Matthew J. Hornsey. 2014. "Deviance and Dissent in Groups." *Annual Review of Psychology* 65: 461–485.

Kopstein, Jeffrey S., and Jason Wittenberg. 2003. "Who Voted Communist?: Reconsidering the Social Bases of Radicalism in Interwar Poland." *Slavic Review* 62 (1): 87–109.

Kulczycki, John J. 2002. "The Soviet Union, Polish Communists, and the Creation of a Polish Nation-State." *Russian History/Histoire Russe* 29 (2–4): 251–276.

Lewis, Paul G. 1982. "Obstacles to the Establishment of Political Legitimacy in Communist Poland." *British Journal of Political Science* 12 (2): 125–147.

Marques, José M., Vincent Y. Yzerbyt, and Jacques-Philippe Leyens. 1988. "The 'Black Sheep Effect': Extremity of Judgments towards Ingroup Members as a Function of Group Identification." *European Journal of Social Psychology* 18: 1–16.

McDermott, Rose. 2020. "Leadership and the Strategic Emotional Manipulation of Political Identity: An Evolutionary Perspective." *The Leadership Quarterly* 31 (2): 1–11.

Michlic, Joanna Beata. 2006. *Poland's Threatening Other: The Image of the Jew from 1880 to the Present*. Lincoln: University of Nebraska Press.

Oseka, Piotr. 2008. *Marzec '68*. Kraków: Wydawnictwo Znak.

Rohac, Dalibor. 2018. "Hungary and Poland Aren't Democratic. They're Authoritarian." *Foreign Policy*, February 5. foreignpolicy.com/2018/02/05/hungary-and-poland-arent-democratic-theyre-authoritarian.

Rothgerber, Hank. 1997. "External Intergroup Threat as an Antecedent to Perceptions of In-group and Out-group Homogeneity." *Journal of Personality and Social Psychology* 73 (6): 1206–1212.

Rozenbaum, Włodzimierz. 1978. "The Anti-Zionist Campaign in Poland, June–December 1967." *Canadian Slavonic Papers* 20: 218–236.

Steinlauf, Michael S. 1997. *Bondage to the Dead: Poland and the Memory of the Holocaust*. Syracuse, NY: Syracuse University Press.

Stephan, W. G., and C. W. Stephan. 2000. "An Integrated Theory of Prejudice." In *Reducing Prejudice and Discrimination: The Claremont Symposium on Applied Social Psychology*, edited by S. Oskam, 23–45. Mahwah, NJ: Erlbaum.

Stola, Dariusz. 2000. *Kampania antysyjonistyczna w Polsce, 1967–1968*. Warsaw: Instytut Studiów Politycznych PAN.

Travaglino, Giovanni A., et al. 2014. "How Groups React to Disloyalty in the Context of Intergroup Competition: Evaluations of Group Deserters and Defectors." *Journal of Experimental Social Psychology* 54: 178–187.

Vakar, Nicholas P. 1956. *Belorussia: The Making of a Nation*. Cambridge, MA: Harvard University Press.

Wynot, Edward D., Jr. 1971. "A Necessary Cruelty: The Emergence of Official Anti-Semitism in Poland, 1936–39." *American Historical Review* 76 (4): 1035–1058.

Zaremba, Marcin. 1998. "Biedni Polacy '68. Społeczeństwo polskie wobec wydarzeń marcowych w świetle raportów KW i MSW dla kierownictwa PZPR." *Więź* 3: 161–172.

Zaremba, Marcin. 2001. *Komunizm, legitymizacja, nacjonalizm: Nacjonalistyczna Legitymizacja władzy komunistycznej w Polsce*. Warsaw: TRIO.

Zawadzki, Piotr. 2010. "Polska." In *Historia antysemityzmu 1945–1993*, edited by Léon Poliakov, 215–248. Kraków: Towarzystwo Autorów i Wydawców Prac Naukowych Universitas.

# 2

# Antisemitic Tropes, Fifth-Columnism, and "Soros-Bashing"

## The Curious Case of Central European University

*Erin K. Jenne, András Bozóki, and Péter Visnovitz*

At the end of the Cold War, regime change swept across the countries of the former Warsaw Pact. In Hungary, too, participants of the 1989 Roundtable Talks and József Antall's government (1990–1993) laid the ideological foundation for a market-oriented liberal democratic system. In the heady early days of triple transition (Offe 1991), parties across the political spectrum welcomed Western organizations, governments, and consultants to assist in the process of transforming their system from János Kádár's decrepit Communist regime to a liberal regime oriented toward NATO and the EU.

In the midst of liberalization and a blossoming of civil society activism, the Hungarian-American billionaire investor and philanthropist George Soros returned to the country of his birth—where he had nearly lost his life as a child in the 1944 Budapest Holocaust—to establish and finance the Open Society Institute (OSI) and Central European University (CEU) in the Hungarian capital of Budapest.[1] The aim was to teach democratic theory, market economics, and the principles of open society to students from former Communist countries, who would return to their home countries to cultivate a new generation of politicians, lawyers, professors, economists, and civil society activists, opening the societies of the former Communist world to the West. What was happening in Hungary was part and parcel of the post-1989 push for liberalization and democratization across East Central Europe.

However, even in the early 1990s, there were Hungarian figures who rejected the turn to the West. They include noted Hungarian playwright and populist politician István Csurka. He was originally a vice president of the governing party, the Hungarian Democratic Forum (MDF), but upon his expulsion from the party in 1993, he founded the far-right Hungarian Truth and Life Party (MIÉP). In his writings and political statements, Csurka depicted Soros as an anti-patriotic globalist threat—a Jewish financier capable of manipulating entire governments (Csurka 1993). Oddly enough, Viktor Orbán, then a young

Erin K. Jenne, András Bozóki, and Péter Visnovitz, *Antisemitic Tropes, Fifth-Columnism, and "Soros-Bashing"* In: *Enemies Within*. Edited by: Harris Mylonas and Scott Radnitz, Oxford University Press. © Oxford University Press 2022.
DOI: 10.1093/oso/9780197627938.003.0003

student dissident leader and the future Hungarian prime minister, was among those who denounced Csurka's rhetoric as antisemitic and beyond the pale of civil society. A former recipient of a Soros fellowship that funded his studies at Oxford University,[2] Orbán offered a full-throated defense of CEU at the time—citing its importance to the Hungarian people and to the region. In response to antisemitic attacks on Soros in 1992, Orbán's Fidesz party issued a statement that read in part: "We watch with astonishment the recent attacks on both the Soros Foundation and the person of George Soros," who "actively contributed to the development of a freer and more open intellectual atmosphere in Hungary by supporting the young generation and the vocational college movement."[3]

Twenty-five years later, Orbán turned on his former benefactor. In his third term as prime minister, he and his media allies accused Soros of hatching a secret "Soros Plan" to settle Syrian refugees in the country to deracinate the Hungarian nation. At the same time, the Fidesz-controlled parliament passed a law that established nearly impossible conditions for CEU to meet in order to continue operating its US-accredited programs in Hungary. Although no mention of the university was made in the actual legislation—which was short and framed in technical language—there was little doubt that the target of the new law was the CEU itself, leading critics to refer to the law as "Lex CEU" or "the CEU law." In Hungarian radio interviews, Orbán strengthened this impression by arguing that the university enjoyed unfair advantages vis-à-vis other Hungarian universities.

As part of Orbán's campaign, anti-Soros stories—once confined to the far right—began to appear in government-backed media, many of them reprising arguments that had originally been made by Csurka twenty-five years earlier. A large exposé was published on individual professors at CEU, together with their pictures, suggesting that they served Soros's anti-national, anti-family liberal globalist agenda. In the run-up to the 2018 elections, Orbán broadened his accusations against Soros, blaming him for raising an army of "Soros soldiers" who worked against the interests of ordinary Hungarians. What led to Orbán's remarkable reversal on Soros from the early post-Communist years to the present? Why did Orbán's government engage in a politically divisive and diplomatically costly public relations campaign against Soros, leading ultimately to the shuttering of CEU and the Open Society Foundation (OSF) in Hungary? More puzzling for observers of Hungarian politics, why did fifth-columnist, anti-globalist, and antisemitic tropes centered on Soros resonate with his supporters in the 2010s, but not in the early 1990s when anti-Soros conspiracy theories first made their appearance on the political scene?

There are several possible answers to these questions. The first and most general explanation is that these are the symptoms of an increasingly autocratic regime seeking to consolidate power. In this view, jettisoning the CEU and OSF was part of Orbán's long game to achieve ideological and even direct political dominance over a Fidesz-controlled society. Strictly regulating an independent foreign institution like CEU was of a piece with the ongoing wholesale reorganization of Hungary's educational system. A second possible explanation is that CEU was the casualty of ethnic outbidding. In this view, Orbán was merely using the specter of Soros in an electoral scheme to scare voters into supporting the Fidesz Party in the 2018 elections—with the predictable consequence of pushing CEU and OSF out of the country. After all, the government *had* won the 2018 elections decisively on the strength of an anti-migration "Stop Soros" campaign. A third possibility is that by ejecting these institutions, Orbán was seeking to curry favor with his Russian ally, Vladimir Putin, who had long vilified Soros as a destroyer of nations. A fourth possibility is that Orbán had simply taken the opportunity that came with President Trump's election to exact long-awaited revenge on his former liberal peers; indeed, Lex CEU was passed just two months after Trump was inaugurated in January 2017.[4]

We believe that there is truth in all of these accounts, but caution that elite interests do not explain why or how *Sorosozás* ("Soros-bashing")[5] was so effective in rallying popular support for Orbán's regime, which used it to win a third veto-proof supermajority in parliament in 2018. *Sorosozás* is a recent Hungarian neologism that is used to describe the government's practice of repetitively and reflexively associating all manner of negative local and international developments with Soros—using a tone of rejection, condemnation, and even ridicule. To appreciate the appeal of *Sorosozás* requires investigating the antisemitic *cultural code* that underpins it. Its effectiveness lies in the repeated, subliminal activation of this latent code in the context of rising right-wing sentiment. By unpacking the fifth-columnist logic in the government's rhetoric, we show that the government used antisemitic tropes to configure Soros and his "networks" as an ontological threat to Hungarian sovereignty, one that required radical defensive action. The curious case of CEU can be understood as a discursive performance by the Fidesz government to consolidate and retain political power during a period of heightened political competition.

This chapter proceeds as follows. First, we examine fifth-columnism as a rhetorical tool for consolidating political power. Second, we argue that the government's fifth-column claims against Soros and CEU were supported by antisemitic tropes, which rely for their meaning on a deeper cultural code. Third, we demonstrate our argument by decoding antisemitic tropes in the writings and statements by earlier Hungarian far right figures and web-scraped government-backed online news portals from 2015 to 2020, showing the systematic use of these tropes in conjunction with mentions of Soros in the Hungarian press. We conclude that *Sorosozás* in the government-controlled media played a key role in naturalizing and justifying the government's incremental expulsion of the university from the country, but that the efficacy of the campaign was dependent on both preexisting cultural understandings and a political context of converging crises.[6] This case study reveals that, while Hungarian governments of the 1990s rejected this type of rhetoric as extreme, in just twenty years, the changing political context, rising ethnonationalism, and the government's hegemonic control over the political discourse made it possible to move this rhetoric to the mainstream, making it an effective means for the Orbán regime to consolidate ideological and political power.

## Fifth-Columnism and Antisemitism

As Radnitz and Mylonas point out in this volume, "fifth columns" are understood as "domestic actors who work to undermine the national interest, in cooperation with external rivals of the state." (Introduction, 3) Volha Charnysh (this volume) unpacks different types of fifth-column claims in twentieth-century Poland, from the interwar period onwards. Likewise, Bulutgil and Erkiletian (this volume) show how this term was used to similar effect by US officials who feared that Japanese Americans could serve as a fifth column for the Japanese Imperial forces during World War II, interning hundreds of thousands of American citizens in camps on the West Coast. In Eastern Europe, meanwhile, the postwar Czechoslovak government of Edward Beneš justified the ethnic cleansing of millions of German Czechs on the charges of serving as a Nazi fifth column during the war (Naimark 1999).Today, the conceptual terrain of fifth columns extends far beyond internal military threats to include internal cultural and economic threats as well. As suggested earlier, political leaders have long used this potent war metaphor to rally their base in support of endangered regimes. By reframing the sovereign as a community perilously penetrated from without, this in-out dialectic serves to reinforce leader-follower bonds in an indefinite battle for survival against hostile foreign interests; it is a socio-psychological closed circuit.

Radnitz and Mylonas define fifth column *politics* (hereafter, *fifth-columnism*) as "public statements, state policies, or collective action targeting domestic actors on the purported basis of their acting nefariously in cooperation with external rivals of the state against the national interest" (Radnitz and Mylonas, Introduction, 8). At the discursive level, fifth-columnism consists of accusing individuals or groups in society of knowingly (and sometimes duplicitously) conspiring with hostile outside actors to subvert the interests of the national ingroup. There are at least four elements of fifth-column discourse. The first is that there is a powerful, hostile foreign actor with hostile intentions toward the nation. The second is that that actor has forged proxy relationships with domestic agents who execute their agenda of dividing and weakening the nation. The third is that, due to their nefarious intent, the details of the relationship and the agenda it serves are hidden from public view. Finally, the fourth is that countering the threat of the foreign power requires unmasking, indicting, and constraining—and possibly expelling—these domestic agents from the body politic.

But what makes these claims so impactful? If articulated by the wrong person or in the wrong setting, such claims will fail to resonate. Given the right societal conditions, however, the act of identifying individuals and organizations as "fifth columns" can be an explosive form of antagonistic political communication, akin to declaring civil war.[7] Fifth-column claims are usually made against groups believed to enjoy undeserved advantages in society; Jewish people have historically fit the bill in Christian and Islamic societies alike. Indeed, there are more than a few similarities between fifth-columnism and *antisemitism*, a term coined by the radical German writer and politician Wilhelm Marr in the late nineteenth century to describe the racist policies that he advocated (Langmuir 1990: 311; Bauer 1994: 11). Antisemitism is based on the pseudo-scientific theory that humans are fundamentally divided into discrete races and that individual behavior is biologically determined by one's racial makeup; the "Semitic" race, it was believed, is less evolved than the Aryan race and hence closer to animals (Langmuir 1990). With its seemingly scientific grounding, antisemitism was deemed more modern and acceptable than classical "Jew hatred," which denotes a general antipathy toward all members of the "Semitic" race.

The principal difference between the "old" (Jew-hatred) and the "new" antisemitism introduced by Marr is that earlier versions had a stronger connection to Christian dogma; antipathy toward Jews was more rooted in religious beliefs, such as the pre–Vatican II Catholic doctrine that Jews had killed Jesus Christ or the Blood Libel accusation that Jews kidnap and sacrifice Christian children as part of their religious rituals. As an interpretative framework, traditional antisemitism configured Jews as hostile religious out-groups, justifying periodic pogroms against *shtetls* and other Jewish communities.[8] The "new" antisemitism, by contrast, portrays Jews as *powerful* and hidden from view, allowing them to use their levers of social influence insidiously and nefariously, like puppet masters, to destroy the majority society from within.[9] Through the state of Israel, Jews are now widely seen as a sovereign people who enjoy access to levers of economic, cultural, and political power. In the modern antisemitic worldview, certain powerful Jews, like the Rothschilds, are even configured as "global" or "foreign" threats. The Jewish threat superseded any single Jewish person, as exemplified by the Nazi slogan "The Eternal Jew."

At the practical level, however, there is often little light between traditional "Jew hatred" and modern antisemitism. Gavin Langmuir points out that antisemitic attacks, pogroms, and discriminatory policies have had very similar characteristics from the European Middle Ages to the present day. According to Bauer, this means that "the antisemitic language and images used to attack Jews are descended directly from earlier Christian Jew-hatred. . . . Its different expressions are but the particular manifestations of that enduring [Western cultural] code" (Bauer 1994: 7). This code is transferred intergenerationally in the family and community through folk tales, religious traditions, and by the state in national literature and histories (for the latter, see Darden and Grzymala-Busse 2006). In this, it is similar to James Scott's "hidden transcripts" in society (Scott 1990), Anthony Smith's "ethno-symbolism" (Smith 2009), and Stuart Kaufmann's "myth-symbol complex" (Kaufman 2001).

How do cultural codes function in political speech? According to Nigel Rapport, cultural codes are a "means by which individuals attempt to make more concrete, graspable and therefore resolvable what is otherwise barely comprehensible in their experience and relations with the world." To activate a code, speakers draw on "metaphor[s] and other linguistic tropes (metonym, simile, synecdoche)," which are assembled into narratives that can be "recounted among others in attempts to compare their experiences with others" (Rapport 1996: 67). Because they work on a subconscious, emotional, even atavistic level, tropes activate codes that are embedded deep in the receiver's culture, lending a seemingly superficial communication a deeper, familiar interpretive meaning. Their power lies in the fact that the receiver never critically processes their content. Likewise, in the symbolic politics strategy described by Stuart Kaufmann, "Politicians manipulate symbols . . . in order to induce people to make choices based on the values they are promoting (which are evoked by the chosen symbols), or to associate themselves with those values," stimulating action (Kaufman 2001: 29).

Antisemitic codes contain powerful tropes that work on a subconscious level to render a specific fifth-column narrative more credible. While the specific cast of characters featured in these narratives vary across societies and historical epochs, the conspiratorial tales they tell are instantly recognizable (see Charnysh, this volume). They include some or all of the following claims: first, certain powerful Jews/financiers embedded in the global power structure have a secret plan to dilute or divide the nation in order to facilitate their plundering.[10] Second, they use their outsized cultural and economic resources to cultivate credulous or traitorous members of the national in-group to spread their ideology and change the laws. Third, they use members of national out-groups to weaken the nation demographically. Fourth, and finally, foiling this plot requires putting both the domestic agents and foreign interests "on the political map" to neutralize their combined threat.

Over the past decade, Orbán and his allies have constructed credible fifth-column claims in the Hungarian press by tapping into this latent antisemitic code, inviting a certain interpretation of events through surreptitious use of metaphors and similes. Meaning is thus coproduced by narratives they articulate and the code upon which these narratives depend. In the process, CEU and OSF were configured as Soros-based fifth-column threats, placing them outside the bounds of legitimate civil society and hence justifying their removal. Before delving into the case itself, a short historical review of antisemitism in Hungarian society is in order.

## The Cultural Antecedents of "Lex CEU"

Hungary has a long history of peaceful coexistence among religious communities. However, antisemitism increased markedly under the Austro-Hungarian monarchy in reaction to waves of Jewish refugees escaping Russian pogroms in the late nineteenth century. Nonetheless, in bourgeois urban centers, German, Czech, and Jewish inhabitants continued to play a significant role in the development of capitalism, industrialism, and modernity.

The end of World War I opened the door to a new era: assimilationist nationalism was replaced by an ethnically based antisemitism (Ungváry 2012). The 1920 Trianon Peace Treaty reduced the urban population of Hungary roughly to Budapest, and a political cleavage soon emerged between cosmopolitan Budapest and the traditional countryside. The right-wing, revanchist, electoral authoritarian Horthy regime routinely blamed liberals and Communists together as enemies of the people.[11] In the pro-regime press, Budapest was derided as a "sinful city," and the capital city was blamed for liberalism, for the failed Communist experiment of 1919 and, finally, for the loss of the war and the two-thirds of pre-Trianon Hungarian territories. In the coming years, the semi-feudal social structure in the countryside effectively slowed down, if not prevented, the rise of capitalism there, while the market economy flourished in the cities. A study by Ferenc Erdei, a prominent sociologist of that period, described Hungary as a country with a double social structure—a "national-historical society" and a "modern-bourgeois society" (Erdei 1944). Almost 90 percent of the Hungarian population belonged to the former, living in the traditions of feudalism, inequality, hierarchy, and clientelism. The Holocaust drastically sharpened social divisions when half a million Hungarian Jews in the countryside were deported to concentration camps with the assistance of state authorities (Molnár 2005). Their former neighbors occupied the now-vacant Jewish homes, motivated by resentment against the Jews and a thirst for property, effectively homogenizing the countryside.[12] The post-Holocaust widening of the urban-rural cleavage contributed to the survival of everyday antisemitism. Inhabitants of Budapest in particular were often portrayed as cosmopolitans, liberals, and radicals—characteristics that indicated their inherent "Jewishness" in contrast to the "real" Hungarians who lived in rural areas.

The postwar Communist regime appeared to turn the page on the virulent antisemitism of the interwar Horthy era and the short-lived, openly fascist Szálasi era. The new Communist regime sought to recruit from the lower classes, catapulting younger people with working class backgrounds into top political and economic positions and offering a path to political power for former out-groups. The top three Communist leaders of the totalitarian Rákosi regime (1948–1956) were popularly known as "Jewish Communists."[13] This reinforced the popular conception that communism was a "Jewish thing," based as it was on internationalism and subordination to the Soviet system—both of which were alien to average Hungarians. Following the Soviet crackdown of the 1956 Revolution, social resistance to communism effectively ceased to exist. People learned to accept a softer Hungarian version, popularly known as "goulash communism" under the leadership of János Kádár. Two main dissident groups emerged under the Kádár regime. The first consisted of an urban group of philosophers and sociologists considered to be "Jewish"—some of them coming from the Budapest School of the late Georg Lukács, which formed the core of dissident intellectuals (Dorahy 2019). Many of these activists had started out as Renaissance Marxists, others as empirical sociologists, but all of them were the children of cultural liberation associated with 1968. They formed friendships with Polish and Czech dissidents of the late 1970s and later learned the strategy of radical reformism from the Polish Solidarity movement (Mitrovits 2020). This group became increasingly open to pro-market reforms as well as Western liberal democracy—their position moved from left radicalism to liberalism (Kis 2014), and in late 1988 they formed the Alliance of Free Democrats (SZDSZ), a liberal party.

The second dissident group was made up of populist nationalist writers from the literary circles of Budapest as well as minor educational institutions and libraries of smaller towns. They identified themselves with the tradition of "*népi*" (populist or *völkisch*) writers who were concerned with the survival of the Hungarian nation and the fate of Hungarians living in the post-Trianon territories outside Hungary (Csurka 1988). Although anti-Communist in principle, they were ready to make compromises with the Communist authorities. This group was equally opposed to Soviet-type communism and Western-style capitalist democracy. Their utopian "small-is-beautiful" approach to economy and society held that communism and capitalism were both alien structures imposed by foreign powers; they promoted a "third way" between the two systems (Csurka 1991a). In 1987, they formed the Hungarian Democratic Forum (MDF) as a movement that later became a catch-all party of the moderate and radical right.

After winning the 1990 democratic elections, the MDF was catapulted into government. It represented three different voices and political directions for Hungary: József Antall, Péter Boross, and István Csurka. While these figures worked together for three years in the party, their ideas differed significantly.

Antall, the Hungarian prime minister between 1990 and 1993, was from a middle-class Budapest family. He lived in Budapest, taught at a high school, but later was removed from his position and forced to eke out a living as a historian and later director of the Museum of Medical History. Antall's brand of liberal conservatism was quite unusual for the Hungarian right; he valued constitutional democracy, party competition, and the rule of law (Osskó 2013). Boross—Antall's successor who ran the MDF-led coalition government for six months in 1993–1994—envisioned a strong, disciplined, and hierarchical society and extolled interwar military leaders of the Horthy era.

Csurka represented the populist nationalist flank of MDF. He was a well-known writer and playwright who ran in Budapest's cultural circles. Under the Kádár regime, he spent his days with his writer friends in various cafés where he talked politics, sometimes telling antisemitic jokes. However, as he was known as a writer and bohemian intellectual, whose political views were not widely known, the regime did not take him seriously. The situation changed in 1988 when Csurka became editor-in-chief of *Magyar Fórum*, the party newspaper. As one of the MDF vice presidents from 1989, Csurka was a radical critic of the liberal SZDSZ and later Antall himself. In early 1990, Csurka published an infamous short essay that he read on his weekend radio show. He lamented that a tiny minority with Marxist-Lukácsian (Jewish) roots had forced their truth on the Hungarian majority and hence Hungarians could not feel at home in their Fatherland, exhorting, "Wake up Hungarians! You are misled again! . . . The era of Béla Kun's[14] has already begun, even if the new Lenin boys reject Lenin himself. What comes next? Terror. Soldiers. Blood and final collapse" (Csurka 1991b: 37). Csurka's essay stirred up the democratic public even before the 1990 elections. In Csurka's worldview, communism, capitalism, and liberalism were all part of a globalist, Jewish conspiracy to destroy Hungarians on behalf of foreign interests. He saw the SZDSZ as a *Jewish* party that did not represent true Hungarian interests. When Antall formed a post-election alliance with SZDSZ, then the largest opposition party, Csurka sharply criticized the deal, claiming that secret agreements had undermined the trust of Hungarians in the government (Csurka 1992, 1996). A fellow writer friend, István Csoóri, also proclaimed that the Jews had used the elections to create a powerful liberal party, ensuring their dominance in the new regime (Csoóri 1990).

George Soros was on the radar of MDF radicals already in the early 1990s. In 1992, MDF MP Gyula Zacsek compared the "Soros empire" to "termites" that "were chewing our nation" (Zacsek 1992). Soros was a subject of considerable interest for Csurka as well, who described him as a Freemason, a provocateur, a stockbroker, a calamity, a cosmopolitan, and a broker-philosopher, among other things. In his worldview, "Soros agents" merely replaced the Communist nomenclature with a liberal one; each operated within the networks of the "Soros

Galaxy." For Csurka, Soros was a man of hidden pacts—an alien who used his media power to indoctrinate the people and ensure the survival of the former Communist bosses. He believed that during the transition Soros had misled Hungarians while promoting the interests of the global liberal hinterland as well as invisible foreign powers (Csurka 1993). This was the only occasion on which Soros responded directly to personal attacks from the Hungarian far right. In 1992 he wrote an open letter to Prime Minister Antall asking him four questions: (1) whether he agreed that Jews were alien to Hungary; (2) whether he accepted Csurka's statement that the global and Hungarian Jewry were in-volved in an anti-Hungarian conspiracy; (3) whether he believed that the Soros Foundation aimed to increase the power of Communists and Jews; and, finally, (4) whether he found it acceptable that an MDF MP had asked Soros to leave Hungary voluntarily (Soros 1992). Antall responded to the controversy by dis-tancing himself and his government from Csurka and Zacsek, stating that the Ministry of Culture and Education had already made an agreement with the Soros Foundation to ensure its future operations, so this was much ado about nothing (Antall 1992).

As Antall attempted to tamp down antisemitism within his party, Csurka mobilized its resentful base, urging the government to get out of the "sandbox of compromise" (Csurka 1992). Tensions between the two men escalated until Antall finally removed Csurka from his positions and pushed him out of the party. The party's electoral loss in 1994 led to four years of a socialist-liberal co-alition government, followed by the return of conservatism in 1998 with Viktor Orbán's Fidesz party. As prime minister, Orbán proclaimed himself the inher-itor of Antall's liberal nationalist legacy, championing a "civic Hungary" ("*polgári Magyarország*"), a country of *citoyens*. In the meantime, Csurka founded his own party (MIÉP), a radical right-wing offshoot that won over 5 percent of the vote and fourteen seats in Parliament in the 1998 elections. Csurka supported the Fidesz government from the outside—voting with the government in re-turn for concessions over cultural policy. Csurka was more concerned with implementing his radical nationalist ideas than with gaining direct political power. This strategy exacted a political price, however, as MIÉP failed to pass the parliamentary threshold in the 2002 elections and thereafter drifted into obscurity.

In the 2000s, Csurka continued to be Hungary's preeminent antisemitic voice.[15] In his party newspaper, he identified the nation's "enemies" as the global financial interests and the "Jewish ruling class" (Csurka 2005). Liberal politicians were "the local wisemen of the Jewish world liberalism" (Csurka 2010). and Jews were "the bankers of the present day" who operated from a "position of world power" (Csurka 2007). Csurka argued that historically there were good reasons for antisemitism because Jews—who considered themselves superior to other nations—despised non-Jewish people and cruelly plundered them in their relentless quest for financial profit (Csurka 2007). Csurka's writings included antisemitic tropes such as "world power" (*"világhatalom"*), "background power" (*"háttérhatalom"*), "colonizers" (*"gyarmatosítók"*), "liberal elite" (*"liberális elit"*), "bankers," and "global financial elites and institutions." Although MIÉP and its colorful leader were a spent political force by the mid-2000s, their ideas lived on, albeit in a subtler form. In the pages that follow, we will show that all of Csurka's tropes have enjoyed new life in the rhetoric of the new Jobbik party and the post-2010 Fidesz government. Readers will notice that this newest incarnation is more refined, making only coded references to the Jewish "threat."

## Exiling Central European University
## (A Long Play in Three Acts)

### Act One: The Rise of Jobbik (2006–2010)

In parallel with Csurka's declining fortunes, the Fidesz party unexpectedly lost to the Socialist Party in the 2002 elections, in part because its focus on Hungarians abroad was seen as out of step with Hungary's recent accession to NATO and imminent accession to the EU. By the mid-2000s, the liberal nationalism of the early Antall period had all but disappeared from conservative circles, which now moved in an increasingly populist, reactionary direction. Viktor Orbán himself was a vice president of the Liberal International (LI) as late as 2000, at which point the Fidesz party left the LI to join the European People's Party (EPP)—the most powerful conservative party family in the European parliament. Still, Péter Boross's nostalgic Horthyism also seemed ill-suited to a country that had recently joined Western liberal clubs. Csurka's fifth-column fantasies were viewed as particularly outdated, as MIÉP had failed to win a single electoral mandate in 2002. The Hungarian far right appeared to be a spent political force.

In the late 2000s, however, converging political and social crises led to a revival of the far right in the form of a dynamic and broad-based Jobbik party (*Jobbik Magyarországért Mozgalom*, or "Movement for a Better Hungary"). In the context of harsh austerity measures implemented by the ruling Socialist Party, the movement was to a great extent animated by widespread anti-Roma sentiment, specifically the belief that "Gypsy Crime" (*"cigánybűnözés"*) was sweeping the country. Anti-Roma rhetoric and the focus on "Gypsy Crime" in the media proved to be an important galvanizing issue on the far right. The ultra-conservative Jobbik leader, Gábor Vona, even founded the paramilitary Hungarian Guard (*Magyar Gárda*) in 2007 with the goal of strengthening national defense and maintaining public order against the threat posed by Roma "criminals" (Róna 2016; Pirro and Róna 2019). The group's members, which were overwhelmingly Jobbik supporters, wore boots and coats adorned with the red-and-white-striped Árpád flag that had been used by the fascist Arrow Cross Party in the 1940s. Although Hungarian courts had ordered that Magyar Gárda be disbanded in 2008, the group used legal loopholes to reorganize it into three separate associated groups (Bozóki and Cueva 2020). Jobbik officially denied that it condoned violence and racism, asserting that its actions were not against anyone but only "for Hungary" (Jobbik 2008). Members of paramilitary groups wore similar fascist-era uniforms and employed intimidation tactics that sometimes sparked outright violence. In March 2011, paramilitaries went to Gyöngyöspata, a village northeast of Budapest, to carry out "military exercises" and set up "security patrols" in a town heavily populated by Roma, prompting the evacuation of women and children. Like MIÉP voters before them, Jobbik supporters were drawn to the party more because of its anti-Roma positions than its antisemitic symbolism (Kovács 2012). Surveys from 2006 and 2011 showed that antisemitic attitudes were *more the consequence than cause* of Jobbik party support. What this means is that the rise of antisemitism in Hungary between 2006 and 2011 was the *outgrowth* of the meteoric rise of the Jobbik party support during the crisis period between 2006 and 2010 (Kovács 2012). In fact, antisemitism resonates only with a minority of the Hungarian population; the proportion of Hungarians with no antisemitic attitudes remained the same over these two surveys. Then as now, the best predictors of an individual's antisemitic beliefs are high levels of nationalism, xenophobia, authoritarianism, and anomie—all of which spike during crisis periods when the social fabric weakens (Kovács 2010: 55–60).

What finally sparked anti-regime unrest in 2006 was the leak of the scandal-plagued Socialist premier Ferenc Gyurcsány's internal party address, in which he used profane language to castigate fellow Socialist leaders for their incompetence and duplicity, saying that the party had lied to the Hungarian people "morning, evening, and night."[16] This set off mass protests and riots and eventually led to a collapse of the Socialist government and the institution of a caretaker government. With the Great Recession and a sovereign debt crisis subsequently driving the country nearly to default, Viktor Orbán's Fidesz party won the 2010 elections with a veto-proof two-thirds majority coalition in a unicameral parliament. Orbán won with a compelling message: the IMF, World Bank, Brussels, and foreign speculators had colluded with the traitorous Socialist government to subjugate the Hungarian nation. In building this narrative, the modern antisemitic code provided ready tropes for projecting the image of an embattled Hungarian nation—penetrated from within by traitorous liberal elites who did the bidding of foreign financial speculators (a term drawn from Csurka's writings in which Soros was a familiar target).

## Act Two: Viktor Orbán and "*Sorosozás*" (post-2010)

Fidesz took note of the resonance of Jobbik's exclusionary ethnonationalism and quickly appropriated it to preempt ethnic outbidding on the right (Krekó and Mayer 2015; Mudde 2015). War rhetoric against enemy "others" was a constant feature of government communications from the beginning. From 2010 to 2014, the government's main themes were the Great Recession, the international economic and financial crises, the budget deficit, and the state debt. In this fight, Orbán's antagonistic "others" began to morph into Csurka's internationalist, capitalist, financial elite. For the first time, Orbán began to demonize "foreign investors" who wanted to buy Hungarian land, "financial speculators" who "smelled blood," and big sponsors who launch "coordinated action" against Hungary (Orbán 2011). He claimed that foreign-owned banks, financial speculators, and foreign companies use their monopolistic positions to "extract unjust, unfair profits" from Hungary using "money pumps" (Orbán 2013). In 2012, mass rallies in support of the Orbán government put forward the slogan "We will not become a colony," a theme that was echoed in Orbán's speeches (Orbán 2013). Orbán had successfully incorporated Csurka's antisemitic tropes into his populist political discourse without paying the political price for doing so.

Indeed, until 2014, Jobbik served as a "reliable buffer" for Orbán against charges of antisemitism. Compared to the openly antisemitic statements of certain Jobbik leaders, Fidesz could always lay claim to being the moderate right-wing party (Spike 2020). As popular support for Fidesz began to wane, Orbán reinvented its enemy as a different and (initially) more visible threat—migrants and refugees from the Middle East who had begun to appear on the southern border in mid-2015. Orbán adroitly shifted the theme of his political communication from job protection to guardian of Europe and defender of European borders against Middle East migrants. Coming just a few months after the killings of journalists at the satirical weekly *Charlie Hebdo* in Paris in January 2015, this new narrative resonated well with the cultural code of "the shield of Europe" and "shield of Christendom," which is embedded in Hungarian historical memory from the period of Ottoman occupation. In his speeches, Orbán linked the threat of migrants to international elites, building the narrative of a two-front war against both immigrants and the "myopic European politics" that willfully ignored the threat (Orbán 2015). The government cultivated a media narrative that international, liberal, European bureaucrats, disconnected from the masses, had a plan to permanently settle migrants in Europe. A "Stop Brussels" campaign culminated in a referendum at the end of 2016, in which Hungarians were asked to vote on the idea of a centrally enforced EU-level migration quota system.

Brussels did not have a face, however, and proved a less useful discursively constructed enemy than expected, based on the lackluster response to the anti-Brussels referendum. In late 2017, roughly half a year before the next national elections, the government therefore reframed the threat by attaching a face to it. Instead of pro-migration proposals by nameless Brussels bureaucrats, the refugee crisis was reconfigured as a conspiracy involving the speculating, globalist moneyman, George Soros, who compelled nameless "Brusselites" like puppets on strings to carry out the "Soros Plan." With Soros highlighted as the main enemy, the Orbán government proved adept at making the same antisemitic fifth-column claims promulgated by Csurka, but doing so indirectly—through tropes that activated a latent antisemitic code suggesting that the nation was ontologically threatened by a powerful, globalist (Jewish) cabal. In the run-up to the 2018 elections, the government rolled out its "Stop Soros" campaign—blanketing highways, subways, as well as bus and tram stations with massive grayscale portraits of George Soros's wrinkled, grinning visage, accompanied by phrases such as "Don't let Soros have the last laugh." Soros was also portrayed as a puppet master who manipulates opposition leaders on strings. Still other billboards featured Soros standing with opposition politicians with wire cutters before a broken fence. Meanwhile, television spots explained the dangers of the Soros Plan: George Soros was conspiring against the Hungarian nation by manipulating the European Commission and domestic political actors to settle one million immigrants from Africa and the Middle East into Hungary annually.

The "Stop Soros" campaign was a significant innovation because it presents in full what Radnitz and Mylonas describe as fifth-column discourse, while illustrating why Soros is such a productive subject of such narratives. First, as a Hungarian-born American billionaire and financial speculator, Soros is very recognizable as a *powerful foreign actor*. Second, Soros is believed to exercise excessive political influence in Hungary through *proxy relationships with domestic actors* who have received funding from his Open Society Foundation; his extensive "Soros network" of "Soros agents" carry out his secretive "Soros Plan." Third, these alleged relationships are presumed to be secretive and *hidden from public view*. Fourth, it features the idea that Fidesz is the last defense against George Soros and his proxies with their "Stop Soros" package. Hence, the Orbán regime's anti-Soros political discourse in Hungary is a textbook example of antisemitic fifth-columnism.

This narrative was systematically reinforced in the pro-government media, which undertook a campaign to portray Soros and his organizations as part of a shadowy globalist power structure, drawing on modern antisemitic tropes of the "powerful Jew." To illustrate the full force of this campaign, we analyze domestic news articles in pro-government news outlets in the period between September 2015 and May 2020.[17] We used web-scraping to build a corpus composed of all articles in the domestic news sections of two publications, *Magyar Idők/Magyar Nemzet*, and *Origo.hu*. *Magyar Idők* was the main pro-government

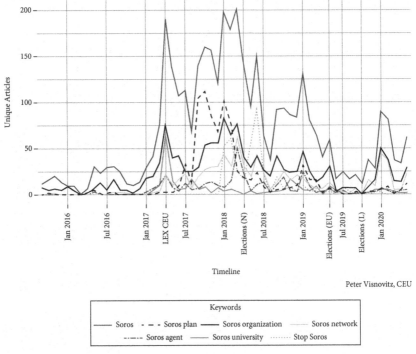

**Figure 2.1** Number of domestic news articles per month referencing "Soros" in *Magyar Idők/Magyar Nemzet*

printed political daily from September 2015; its articles were available on its website, magyaridok.hu; it changed its name to *Magyar Nemzet* in February 2019. *Origo.hu* was the leading Hungarian online news portal in the early 2010s, but after dramatic changes to its editorial team and ownership, it became a government mouthpiece by the end of 2015. In the domestic news sections of these two publications alone, 121,596 articles were published over a period of nearly five years, which we then searched for articles that mentioned George Soros.[18]

The results yield a revealing histogram of Soros-related articles (see Figures 2.1 and 2.2). The two publications, *Magyar Idők/Magyar Nemzet* and *Origo.hu*, show a remarkable similarity in how and when they begin to publish articles about Soros and associated fifth-column concepts. References to George Soros were only sporadically present through the end of 2015 and 2016; the mean number of articles per month mentioning "Soros" was 3.9 for *Origo.hu* and 16.7 for *Magyar Idők*. However, in early 2017, around the time when the government introduced "Lex CEU," both publications showed a significant spike in Soros-related articles: in April 2017 *Origo.hu* published 94, while *Magyar Idők* published

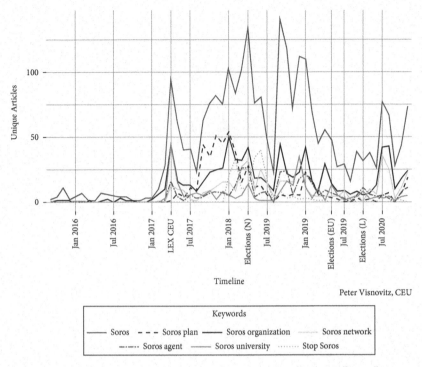

Peter Visnovitz, CEU

**Figure 2.2** Number of domestic news articles per month referencing "Soros" in Origo.hu

191 articles in their domestic news sections with a reference to Soros. About half of the *Origo.hu* articles (44) and more than one-third of *Magyar Idők* articles (65) referred to CEU as the "Soros university."

The data show that, while the preoccupation with CEU as a "Soros university" (*Soros-egyetem*) had almost disappeared by the summer of 2017 following the passage of Lex CEU, Soros-centered fifth-column discourse remained highly salient in government-backed media. Instead of "Soros university," however, the Soros fifth-column narrative was extended to the broader "Soros Plan," "Soros organizations," and "Soros agents." The number of Soros-related articles peaked again before and after the April 2018 national elections, when the pro-government outlets churned out literally hundreds of Soros-related articles: *Magyar Idők* reached 201 Soros-related articles in March, while *Origo* hit its high point with 134 Soros-related articles in April. This means that on average, including weekends, they ran 6.5 and 4.5 articles per day discussing George Soros, his dangerous plan, networks, agents, and organizations. In 2018, the focus shifted to the government response to the fifth-column threat, with an increased number of articles referencing the "Stop Soros" law package that was in the limelight for over six months, with a mean of 23.6 articles per month for *Origo* and 46.1 articles per month for *Magyar Idők* (with a June peak of 40 and 92 articles, respectively).[19]

The data tell the story of constantly shifting narrative construction by the pro-government media around the figure of George Soros. The initial focus on CEU—framed as the "Soros-university"—was followed by a discussion of the machinations of powerful foreign actors in the "Soros Plan," then their domestic proxies or "Soros agents," and finally the government's solution to these threats: the "Stop Soros" law. We can also see how government communication served to keep the fifth-column narrative active by endlessly interpellating new "enemies" into the "Soros network." For example, when MEP Judith Sargentini issued a report critical of the Hungarian government, she herself was declared to be an agent of Soros, and her report was labeled the "Soros-Sargentini report" in government media.[20] This shows the versatility of this fifth-column narrative as a tool for deflecting any kind of domestic or foreign criticism against the government.

## Act Three: Fighting the "Fifth Column" (2017–2019)

We now turn to the practical consequences of the government's fifth-columnism on one of their key targets—Central European University. For many years, *Sorosozás* in media outlets remained at an abstract level; writers focused on Soros's outsized influence in general, but did not explicitly name his "networks" in Hungary. All of this would change in 2017, when the Fidesz government set its sights on Central European University. With very little warning and no open debate, Fidesz MPs passed an amendment to the Higher Education Law in March 2017, introducing a far more stringent set of requirements that foreign universities would have to meet in order to operate in Hungary. Two conditions in particular appeared to set CEU up for failure. First, foreign universities could only operate on the basis of a treaty between the university's home and host countries, and second, foreign universities would have to run educational activities in their home country. (Since CEU had been founded in Hungary in 1991, it had no campus in the United States, unlike other foreign universities in Hungary.) Failure to meet any of these conditions by the January 2018 deadline meant losing their license to run educational programs in the country. That the amendment was intended to target CEU was indicated by its moniker, Lex CEU, as most of the new requirements affected one institution only (Enyedi 2019). Further proof of the law's intent lies in the fact that, after the amendment was passed, government officials publicly accused CEU of being a "fake" or "virtual" university that served as a conduit for Soros's influence in society.

The CEU leadership responded to the attack with a two-pronged strategy. First, they reached out to foreign governments, the European People's Party, and universities and organizations around the world to mobilize an international campaign to put pressure on the Hungarian government to compromise, using the classic "boomerang" strategy of transnational advocacy networks described by Margaret Keck and Kathryn Sikkink (1998). The result was an outpouring of support from Western governments like the United States and Germany, hundreds of universities, Nobel laureates, and even the conservative EPP, which sought to persuade the government to back down (Jan-Werner Müller 2017). Hungarian universities also mobilized in support of CEU, and many thousands of ordinary citizens turned out to protest on successive weekends, reaching a peak of 80,000 on April 9; "I stand with CEU" signs appeared on Hungarian storefronts. All evidence suggests that Orbán's government had not anticipated such a robust international and domestic backlash against their fight against the "Soros university" (Thorpe 2017). Despite the EPP's threats of expulsion and losing their voting rights in the EU under Article 7, the government doubled down on its position, arguing that the pro-CEU campaign was a coordinated attempt by George Soros through the European Commission to attack Hungarian sovereignty. Fidesz vice president Szilard Nemeth said that "every means must be employed to hold back such organizations, and I believe they should be cleaned out of here" (Németh 2017).

Their institution now in jeopardy, CEU administrators scrambled to meet the conditions set out by the new law before the January 2018 deadline set by the government. Their efforts to fulfill the new requirements were encouraged by members of the government. In communications with Minister for Information and Technology László Palkovics, CEU administrators were privately assured that the amendment was merely an electoral gambit, and that the pressure would be off after the hard-fought 2018 elections.[21] CEU therefore moved fast to set up and accredit a CEU campus in New York and obtain an agreement between the Hungarian government and the State of New York to meet the two most onerous requirements under the new law. The 2018 elections came and went, but the agreement remained on Orbán's desk unsigned. Rather than certify that the CEU had met the conditions set out in the new legislation, the government played a cat-and-mouse game with the institution, offering the university a one-year extension on meeting the conditions of the law. When the new deadline passed without the government certifying that CEU was in compliance, the university concluded that it could no longer legally run its programs in Hungary and prepared to leave the country.[22] In October 2020, the European Court of Justice (ECJ) ruled that "Lex CEU" violated European law. However, the Fidesz government neither changed the regulation to accommodate CEU nor withdrew CEU's license, allowing the government to argue that the university chose to leave of its own accord.

Returning to our media corpus, the data leave little doubt that the government used CEU to fulfill the promise of their Soros fifth-column narrative, which they would continually repurpose in the coming years to fight other political battles and further consolidate power. Figure 2.3 shows that news articles consistently linked CEU to Soros. Out of 984 news articles discussing CEU over the entire five-year period, 74 percent (729 articles) also mentioned Soros in some form, and 33 percent (327 articles) referred to CEU simply as the "Soros university."[23] This fifth-columnist *naming* occurred in a total of 482 articles, meaning that 155 individual articles did not even use the proper name of the university (CEU or *Közép-európai Egyetem*). When Lex CEU was introduced in parliament in late March 2017, there was an explosion of news headlines in both pro-government publications, showing a remarkable consistency in the linkage of CEU to George Soros—calling it the "Soros university" (*Soros-egyetem*) rather than using its proper name. At its peak in April 2017, *Magyar Idők* and *Origo.hu* together published 217 articles that mentioned CEU, out of which 156 also discussed

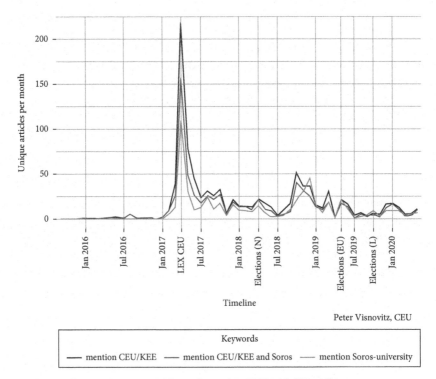

Peter Visnovitz, CEU

**Figure 2.3** Domestic news articles referencing CEU with "Soros"

Soros. In the subsequent three years, the CEU story remained highly visible, with an average of 18.9 articles per month mentioning CEU (75 percent of which also kept steadily referencing Soros), and an average of 10.4 articles per month using the term "Soros university" in the two news outlets.

Throughout the anti-CEU campaign, Soros never once publicly inserted himself in the fight, calculating that doing so would merely validate the government's accusation that Soros was a political actor with undue influence in Hungarian affairs. George Birnbaum, one of the US political consultants widely credited for having advised Orbán to follow the anti-Soros tack in 2010, said in an interview with a *Buzzfeed* reporter, "The perfect enemy is one that you can punch again and again and he won't punch back. . . . Soros was a perfect enemy. It was so obvious. It was the simplest of all products, you just had to pack it and market it" (Grassengger 2019). The antisemitic code that they used to build their fifth-column Soros narrative was already available on the Hungarian far right, as Csurka's writings plainly indicate.

In the end, the expulsion of CEU from Hungary was most likely overdeter-mined—part go-for-broke electoral maneuver, part authoritarian impulse to achieve ideological control over Hungarian civil society, and part overture to Russian president Vladimir Putin, who had ejected the Open Society Institute from Russia already in the 2000s. What is clear from the CEU Hungary *post mortem* is that, once the government recognized the potency of the fifth-column narrative for voters on the Hungarian right, and the ease with which it could be repackaged by the government to win successive political battles, Orbán could not be seen to back down from an opportunity to expel a key Soros institution from Hungarian territory.

## Epilogue

This chapter has investigated the ways in which fifth-columnism was used by the Orbán regime to maintain political power with devastating effects for one of his targets—the Central European University in Budapest. To show this communication strategy in action, we combined a discourse analysis of Csurka's earlier writing with a media content analysis of *Sorosozás* in Fidesz government-backed media outlets from 2015 to 2020. We showed that, from 2010, Orbán and his media allies interpellated numerous individuals and states as "financiers" and "global powers" into this discourse, configuring them as cogs in a unified global "Soros network." In this, he drew upon well-established fifth-column narratives that had been constructed and refined by Csurka and other ideologists from the Kádár era that relied on a latent antisemitic code. At one time vehemently rejecting such discourse, Orbán and his government allies now became its chief proponents (Haszán 2020).

Twenty-five years ago, the Hungarian public viewed George Soros as a pro-democracy, pro-market reform philanthropist and activist who had assisted the country in its historic transformation from a command economy to a market economy, from a Soviet Bloc autocracy to a liberal democracy. The 1990s were dominated by liberal democratic ideas in which external pro-market and pro-democracy assistance were widely respected and had the support of all major political parties. Hungarian media featured many competing views rather than a single government voice. In those heady pro-democracy days, Csurka's anti-Soros rhetoric remained marginalized; it was easily dismissed as backward and provincial, whereas Soros was viewed as a beacon of the future who symbolized Hungary's inevitable progression toward freedom and democracy.

By the 2010s, the political and economic situation had changed fundamentally—not just in Hungary, but the world over. The 2008 economic crisis made a global impact on the perceptions of liberal democracy, which was increasingly seen as an elitist project that ran counter to the interests of ordinary people. The rise of populism, ethnonationalism, and illiberalism offered ready-made legitimation strategies for autocratic-minded leaders, like Orbán, who exploited the resentments, frustrations, and anger of the so-called real people to consolidate political power. It was in this new setting that Csurka's long-running accusations against Soros began to resonate with the supporters of a growing far right movement, which needed "perfect enemies" to fuel their doomsday conspiracy theories. The Fidesz government also managed to "outbid" the far right Jobbik through its coded rhetoric. Having established control over the mainstream media in the 2010s, the government was able to use fifth-column narratives to counter all of its political opponents. Using antisemitic tropes, the Orbán government used state propaganda relentlessly to depict Soros as public enemy number one. By framing Soros as an agent of neoliberal capitalism, Orbán effectively granted Csurka a posthumous victory—both morally and politically.

Two conclusions emerge from this analysis. First, modern antisemitic tropes have served a vital tools for the Hungarian government to legitimize its campaign to "expose," "fight" and ultimately "clean" the Soros "fifth column"—CEU and the OSF—from the territory of Hungary in a period of under two years. Once configured as part of the Soros network, there was little political space for CEU to remain in the country. However, CEU was only the first of many casualties of *Sorosozás*. This rhetorical cocktail proved so potent that the government would reprise these claims whenever the government was in need of popular legitimation in the face of political challenges.

Second, our analysis sheds some light on why antisemitism may be a rising force on the right by showing why antisemitic fifth-column claims resonated more with conservative voters in Hungary today than they did in the 1990s. In the early post-transition period, the country was still in the throes of liberal democratic revolution and eager to join European and transatlantic clubs; the political *Zeitgeist* at the time was inhospitable to antisemitism. It took the converging political and economic crises of the late 2000s to galvanize a genuine *populist* right-wing movement in which antisemitic tropes became helpful for articulating perceived global threats. The political context created incentives for Orbán to appropriate these tropes as a means of preempting ethnic outbidding and consolidating political power. The fact that Orbán controls the media has given his fifth-column narratives almost unlimited potential as a means of deflecting challenges to the regime.

The implications of this case study extend far beyond the borders of Hungary and even Eastern Europe. Fifth-column narratives animate anti-liberal reactionary movements around the world. Lenka Bustikova argues that these narratives resonate with members of national in-groups who feel resentment at the perceived rise of ethnic minorities—a status reversal assisted by liberal, globalist ("Jewish") forces seeking to "break" nations (Bustikova 2020). In reactionary or identitarian movements, fifth-columnism is a tried-and-true political technique used to interpellate contemporary liberal parties, civil society organizations, foreign governments, international organizations, and ethnic outgroups into a unified globalist enemy that must be defended against at all costs. This obviously increases the level of threat facing out-groups in such societies.

While the scale of *Sorosozás* in Hungary is in some ways singular—drawing as it does on latent cultural codes and historical memories of Hungarian trauma—it bears a strong resemblance to recent "Stop Soros" campaigns in Slovakia, Poland, Russia, North Macedonia, the UK, and the United States (Plenta 2020). Volha Charnysh likewise shows how "antisemitic cues" by the populist right interacted with latent antisemitic prejudices to persuade a substantial minority of Polish voters to vote against EU accession in 2003 (Charnysh 2015). In each case, we see the outlines of a familiar narrative—the national in-group perilously interpenetrated by hostile foreign influences, wreaking havoc through disloyal societal elements. Fifth-columnism is a strategy used by political leaders to stitch together disparate threats to their regime by declaring them to be working to undermine their sovereignty in concert. Modern antisemitism provides a familiar local vocabulary for this project, providing resonant cultural tropes by which messages can be repackaged and transmitted to their constituents for ongoing political uses. The Hungarian government's "rejectionist" politicking—couched in barely concealed antisemitic tropes like the "laughing Jew"—clearly supports a form of ideological cleansing, with devastating knock-on effects for civil society.

The case of Central European University therefore merits very serious scrutiny by democracy theorists and human rights advocates alike.

## Notes

1. The Soros Foundation was already in Hungary in the 1980s, providing funds to schools, hospitals, libraries, civic initiatives, young people, artists, and anti-Communist dissidents. OSI was later renamed the Open Society Foundation (OSF).
2. Orbán received a nine-month fellowship from the Soros Foundation to study at Oxford University in 1988–1989 (Lendvai 2017: 23).
3. "Fidesz issued a statement in defense of György Soros," https://444.hu/2016/05/26/kozlemenyt-adott-ki-a-fidesz-soros-gyorgy-vedelmeben.
4. The possible reasons for Orbán's attack on the CEU are summarized by Bozóki (2017). For a broader discussion and insider account of the episode, see Enyedi (2018).
5. We thank Anna Szilágyi for this observation.
6. Although the record suggests that demagoguing Soros was a calculated choice by the Orbán government, our media analysis reveals the continuity of antisemitic and anti-globalist discourse on the Hungarian Right from the margins in the early 1990s to the political mainstream in which Jews are configured as "foreign" to the "Hungarian nation," in a "structured antisemitic worldview" (Kovács 2010: 21).
7. Fifth-columnism is central to super-agonistic *ethno-populist* claims that enemy elites are conspiring with "national others" to destroy the "people-nation" (Jenne 2018). For other treatments of populist nationalism, see Bonikowski et al. 2017; De Cleen 2017; Brubaker 2020; and De Cleen and Stavrakakis 2017.
8. For a critique of old anti-Judaism, see the work of the Hungarian populist writer Szabó 1938.
9. Jonathan Fox and Lev Topor argue that modern antisemitic tropes of powerful Jews manipulating states and societies through shadowy networks behind the scenes serve as potent fuel for populist conspiracy theories (Fox and Topor 2021: ch. 5).
10. In *The Culture of Critique*, one part of a trilogy read widely on the far right, evolu-tionary psychology professor Kevin McDonald set out to demonstrate that Jewish people promote multicultural values and pro-immigration policies as part of a group evolutionary strategy. He argued that because the naturally endogamous Jewish group thrives in multicultural societies, Jews are incentivized to promote liberalism to secure the future survival of their minority community, allowing them to "hide" in plain sight (MacDonald 1998).
11. During the White Terror against the short-lived Soviet Republic of Béla Kun in 1919–1920, contingents of Horthy's anti-Communist forces used violent repression against supporters of Kun. Many of the victims were of Jewish origin (Csunderlik 2019, 2020).
12. For similar stories of "intimate" depredations against Jewish communities in Poland, Lithuania, Romania, and Ukraine, see Kopstein and Wittenberg (2018).
13. These three Communist leaders were Mátyás Rákosi, Ernő Gerő, and József Révai.

14. Béla Kun (1886–1938) was a Marxist revolutionary and leader of the short-lived Hungarian Soviet Republic who came to power in a coup in 1919. He died in Moscow as a victim of Stalinist purges.

15. It should be noted that he himself rejected this label.

16. This was an oblique reference to a famous self-critical statement by Radio Kossuth during the 1956 Hungarian Revolution.

17. The starting date of the period coincides with the launch of one of the analyzed newspapers, *Magyar Idők* (Hungarian Times), which was renamed *Magyar Nemzet* (Hungarian Nation) in early 2019.

18. We compiled this database using the quanteda package in the statistical software R. We first ran searches for the keyword root "Soros," and then searched for expressions linked to fifth-column discourse. Keyword expressions included "Soros," "Soros-plan" (*Soros-terv*), "Soros-network" (*Soros-hálózat*), "Soros-organization" (*Soros-szervezet*), "Soros-agent" (*Soros ügynök/Soros embere*), and "Soros-university" (*Soros-egyetem*). In Hungarian, "Soros" has a literal meaning of "next in line" or "upcoming"; it can also mean "something with a certain number of lines." Hits with meanings not referring to the name of George Soros were filtered from the corpus by differentiating between the keyword upper-case "Soros" and lower-case "soros." Tests showed that the lower-case version of the keyword in the corpus mostly referred to the "rotating presidencies" of the EU (soros elnökség); only two types of lower-case occurrences, "*sorosozás*" and "*sorosista*" were relevant for our search. The latter two were retained in the corpus as relevant hits. The used keywords were found to be the most common expressions associated with the "Soros" root. To make sure that we captured in-text occurrences of these concepts even when they are not present in single hyphenated words (like "Soros-organization"), but in looser expressions ("an organization closely linked to George Soros"), we also ran the search in the corpus with quanteda's kwic ("keywords in context") function, checking the occurrences of these key elements (plan, network, organization, and so on) within a 10-word window before and after the keyword "Soros" in the article texts. Manual checking of the data showed false hits to be very low; hence, we believe this method yielded a reliable corpus of articles on Soros in the two publications.

19. Soros continued to feature continuously in government media, only dropping in the months around the 2019 EP elections and during the first wave of the coronavirus pandemic in March–April 2020. The 2019 drop may be explained by the conflict between Orbán's Fidesz party and the European People's Party, which escalated in February 2019 in reaction to anti-Soros billboard posters that featured George Soros together with EPP's European Commission president Jean-Claude Juncker. After the EPP started an investigation into the campaign and threatened to expel Fidesz from the party family, the Soros-Juncker posters disappeared from the streets. Our data show that pro-government outlets also scaled back their Soros-related articles in response. Still, even in this low period, *Origo* and *Magyar Nemzet* published an average of 35.3 and 32.4 articles about Soros per month, meaning an average of one article about Soros every day.

20. Judith Sargentini, Member of the European Parliament, was the rapporteur of the report on "the existence of a clear risk of a serious breach by Hungary of the values on which the Union is founded." The European Parliament resolution adopted the text on September 12, 2018.
21. One of the authors of this chapter received this information via personal communication.
22. Enyedi 2018.
23. To account for the typological variations, the calculation of the number mentions of "Soros-university" (*Soros-egyetem\**) also include hits for "the university of Soros" (*Soros egyetem\**) and "the university of George Soros" (*Soros György egyetem\**).

# References

Antall, József. 1992. "Válasz Soros Györgynek" [Reply to George Soros], *Új Magyarország*, October 1.

Bauer, Yehuda. 1994. "In Search of a Definition of Antisemitism." In *Approaches to Antisemitism: Context and Curriculum*, edited by Michael Brown, 10–21. New York: American Jewish Committee.

Bonikowski, Bart, et al. 2018. "Populism and Nationalism in a Comparative Perspective: A Scholarly Exchange." *Nations and Nationalism* 25 (1): 58–81.

Bozóki, András. 2017. "Hozzájárult a CEU, hogy Budapest cool hely lett" [CEU contributed to Budapest becoming a cool place]: An interview by Sándor Jászberényi. *24.hu.*, April 26. https://24.hu/belfold/2017/04/26/bozoki-andras-hozzajarult-a-ceu-hogy-budapest-cool-hely-lett/.

Bozóki, András, and Sarah Cueva. 2020. "Xenophobia and Power Politics: The Hungarian Far Right." In *Anti-Genderismus: Allianzen von Rechtspopulismus und religiösen Fundamentalismus*, edited by Sonja A. Strube et al., 109–120. Bielefeld: Transcript.

Brubaker, Rogers. 2020. "Populism and Nationalism." *Nations and Nationalism* 1: 44–66.

Bustikova, Lenka. 2020. *Extreme Reactions: Radical Right Mobilization in Eastern Europe.* Cambridge: Cambridge University Press.

Charnysh, Volha. 2015. "Historical Legacies of Interethnic Competition: Antisemitism and the EU Referendum in Poland." *Comparative Political Studies* 48 (13): 1711–1745.

Csoóri, Sándor. 1990. *Nappali hold*. [Daytime moon]. Budapest: Püski.

Csunderlik, Péter. 2019. A "vörös farsangtól" a "vörös tatárjárásig": a Tanácsköztársaság a korai Horthy-korszak pamflet- és visszaemlékezés-irodalmában [From red carnival to red occupation: The Council Republic in the pamphlets and memoirs of the Horthy era]. Budapest: Napvilág.

Csunderlik, Péter. 2020. "A judeobolsevizmus vörös tengere" [The Red Sea of Judeo-Bolshevism]. *Mozgó Világ* 6: 96–104.

Csurka, István. 1988. *Közép-Európa hó alatt* [Central Europe under snow]. New York: Püski–Smikk.

Csurka, István. 1991a. *Új magyar önépítés* [New Hungarian self-reconstruction]. Budapest: Püski–Magyar Fórum.

Csurka, István. 1991b. *Vasárnapi jegyzetek* [Sunday notes]. Budapest: Püski–Magyar Fórum.

Csurka, István. 1992. "*Néhány gondolat*" [Some thoughts]. Magyar Fórum, August 20.

Csurka, István. 1993. *Keserű Hátország* [Bitter Hinterland]. Budapest: Magyar Fórum Kiadó.

Csurka, István. 1996. "Some Thoughts." In *From Stalinism to Pluralism: A Documentary History of Eastern Europe since 1945*, edited by Gale Stokes, 265–269. New York and Oxford: Oxford University Press.

Csurka, István. 2005. "Egy nemzet- és népgyilkos kormány" [A government which is the murderer of the nation and the people]. http://www.eredetimiep.hu/fuggetlenseg/2015/januar/23/20.htm.

Csurka, István. 2007. "Dr. Fejér Lajos és a zsidóság" [Dr. Lajos Fejér and the Jewry]. http://www.eredetimiep.hu/fuggetlenseg/2013/junius/07/19.htm.

Csurka, István. 2010. "Ma mindenki demokrata" [Today everybody is a democrat]. http://www.eredetimiep.hu/fuggetlenseg/2015/aprilis/17/20.htm.

Darden, Keith, and Anna Grzymala-Busse. 2006. "The Great Divide: Literacy, Nationalism, and the Communist Collapse." *World Politics* 59 (1): 83–115.

De Cleen, Benjamin. 2017. "Populism and Nationalism." In The Oxford *Handbook of Populism*, edited by Cristóbal Rovira Kaltwasser et al., 342–362. Oxford: Oxford University Press.

De Cleen, Benjamin, and Yannis Stavrakakis. 2017. "Distinctions and Articulations: A Discourse Theoretical Framework for the Study of Populism and Nationalism." *Javnost— The Public* 24 (7): 301–319.

Dorahy, J. F. 2019. *The Budapest School: Beyond Marxism*. Leiden: Brill.

Enyedi, Zsolt. 2018. "Democratic Backsliding and Academic Freedom in Hungary." *Perspectives on Politics* 16 (4): 1067–1074.

Enyedi, Zsolt. 2019. "The Central European University in the Trenches." In *Brave New Hungary: Mapping the System of National Cooperation*, edited by János M. Kovács and Balázs Trencsényi, 243–266. Lanham, MD: Lexington Books.

Erdei, Ferenc. 1944. *A magyar társadalom*. [The Hungarian society]. Budapest: Magvető.

Fox, Jonathan, and Lev Topor. 2021. *Why Do People Discriminate against Jews?* New York: Oxford University Press.

Grassengger, Hannes. 2019. "The Unbelievable Story of the Plot against George Soros." *Buzzfeed*, January 20.

Haszán, Zoltán. 2020. "Orbán útja az összeesküvés-elméletektől a valódi összeesküvésekig" [Orbán's road from conspiracy theories to real conspiracies]. *444.hu*, May 15.

Jenne, Erin K. 2018. "Is Nationalism or Ethnopopulism on the Rise Today?" *Ethnopolitics* 5: 546–552.

Jobbik. 2008. "Manifesto." *jobbik.com*.

Kaufman, Stuart J. 2001. *Modern Hatreds: The Symbolic Politics of Ethnic War*. Ithaca, NY: Cornell University Press.

Keck, Margaret, and Kathryn Sikkink. 1998. *Activists beyond Borders*. Ithaca, NY: Cornell University Press.

Kis, János. 2014. "Hogyan lettünk liberálisok?" [How we became liberals?]. In *Mi a liberalizmus?* [What is liberalism?], 337–341. Bratislava: Kalligram.

Kopstein, Jeffrey S., and Jason Wittenberg. 2018. *Intimate Violence: Anti-Jewish Pogroms on the Eve of the Holocaust*. Ithaca, NY: Cornell University Press.

Kovács, András. 2012. "Antisemitic Prejudice and Political Antisemitism in Present-day Hungary." *Journal for the Study of Antisemitism* 4: 443–467.

Kovács, András. 2010. *The Stranger at Hand: Antisemitic Prejudices in Post-communist Hungary*. Leiden: Brill.

Krekó, Péter, and Gregor Mayer. 2015. "Transforming Hungary—Together?: An Analysis of the Fidesz-Jobbik Relationship." In *Transforming the Transformation?: The East European Radical Right in the Political Process*, edited by Michael Minkenberg. London: Routledge.

Langmuir, Gavin I. 1990. *Toward a Definition of Antisemitism*. Berkeley: University of California Press.

Lendvai, Paul. 2017. *Orbán: Europe's New Strongman*. London: Hurst & Company

MacDonald, Kevin. 1998. *The Culture of Critique: An Evolutionary Analysis of Jewish Involvement in Twentieth-Century Intellectual and Political Movements*. New York: Praeger.

Mitrovits, Miklós. 2020. *Tiltott kapcsolat: A magyar lengyel ellenzéki együttműködés, 1976–1989* [Forbidden relationship: The cooperation of the Hungarian and Polish opposition, 1976–1989]. Budapest: Jaffa.

Molnár, Judit. 2005. *A Holokauszt Magyarországon európai perspektívában* [The Holocaust in Hungary in European perspective]. Budapest: Balassi.

Mudde, Cas. 2015. "Is Hungary Run by the Radical Right?" *Washington Post*, August 10.

Müller, Jan-Werner. 2017. "Hungary: The War on Education." *New York Review of Books*, May 20.

Naimark, Norman. 1999. *The Fires of Hatred: Ethnic Cleansing in the Twentieth Century*. Cambridge, MA: Harvard University Press.

Németh, Szilárd. 2017. "Szilard Nemeth Said Exactly Who Should Be Cleaned Up: The HCLU, Helsinki, Transparency." https://444.hu/2017/01/11/nemeth-szilard-elmon dta-konkretan-kiket-kell-eltakaritani-tasz-helsinki-transparency?fbclid=IwAR24Yf SyYG0K1ag1GBTi_UlG2-go3kxH_Xy-CT3nXnGaL11KNcioahc4bTs.

Offe, Claus. 1991. "Capitalism by Democratic Design?: Democratic Theory Facing the Triple Transition in East Central Europe." *Social Research* 58, no. 4 (Winter): 865–892.

Orbán, Viktor. 2011. Speech at the annual meeting of Hungarian ambassadors.

Orbán, Viktor. 2013. Speech at the Baile Tusnad Summer University.

Orbán, Viktor. 2015. Opening speech at the beginning of the Fall Session of the Hungarian Parliament.

Orbán, Viktor. 2018. Speech at the commemoration of the 1848 revolution.

Osskó, Judit. 2013. *Antall József—kései memoár* [József Antall—belated memoir]. Budapest: Corvina.

Pirro, Andrea L. P. , and Dániel Róna. 2019. "Far Right Activism in Hungary: Youth Participation in Jobbik and Its Network." *European Societies* 21 (4): 603–626.

Plenta, Peter. 2020. "Conspiracy Theories as a Political Instrument: Utilization of Anti-Soros Narratives." *Contemporary Politics* 26 (5): 512–530.

Rapport, Nigel. 1996. *Social and Cultural Anthropology: The Key Concepts*. 3rd ed. London: Routledge.

Róna, Dániel. 2016. *A Jobbik-jelenség* [The Jobbik phenomenon]. Budapest: Könyv & Kávé.

Scott, James C. 1990. *Domination and the Arts of Resistance: Hidden Transcripts*. New Haven, CT: Yale University Press.

Smith, Anthony D. 2009. *Ethno-Symbolism and Nationalism: A Cultural Approach*. London and New York: Routledge.

Soros, György. 1992. "Nyílt levél Antall József miniszterelnökhöz" [An open letter to prime minister József Antall]. *Népszabadság.*, September 15.

Spike, Justin. 2020. "When the Prime Minister Said 'They Really Love Interest' Everyone in Hungary Knew Who He Meant." *Inside Hungary*, June 16. https://insighthungary.444. hu/2020/06/16/when-the-prime-minister-said-they-really-love-interest-everyone-in-hungary-knew-who-he-meant.

Szabó, Dezső. 1938. *Az antijudaizmus bírálata* [The critique of anti-Judaism]. Budapest: Ludas Mátyás.

Thorpe, Nick. 2017. "Hungary CEU: Protesters Rally to Save University." *BBC*, April 3. https://www.bbc.com/news/world-europe-39479398.

Ungváry, Krisztián. 2012. *A Horthy-rendszer mérlege: Diszkrimináció, szociálpolitika és antiszemitizmus Magyarországon* [The balance-sheet of the Horthy regime: Discrimination, social policy and anti-Semitism in Hungary]. Pécs and Budapest: Jelenkor.

Zacsek, Gyula. 1992. "Termeszek rágják a nemzetet, avagy gondolatok a Soros-kurzusról és a Soros birodalomról" [Termites are chewing our nation, or some thoughts about the Soros course and Soros empire]. *Magyar Fórum* 4 (36): 9–16.

# 3

# Civil Society, Fifth-Column Perceptions, and Wartime Deportations

## Japanese and German Americans

### *H. Zeynep Bulutgil and Sam Erkiletian*

Why do politicians perceive certain groups as fifth columns in some contexts but not others? What, if any, is the role of civil society organizations in this process? Groups identified as fifth columns are citizens or immigrants that parts of the political establishment suspect of cooperating with external rivals against national interests (Radnitz and Mylonas, Introduction). Based on these real or imagined strategic concerns, these groups often become targets of deportation, internment, economic expropriation, and physical violence. Understanding why some immigrant groups are depicted as fifth columns remains a pressing concern given that, in response to the threat of international terrorism, many liberal democracies have framed their immigration policies as issues of national security (Bigo 2002; Adamson 2006; Huysman 2006).

A prominent example of this phenomenon is President Trump's controversial Executive Order 13769 enacted in January 2017, titled "Protecting the Nation from Foreign Terrorist Entry into the Unites States," which banned immigration from seven Muslim-majority states (Exec. Order No. 12769, 2017). As a presidential candidate in 2015, Trump defended his proposals for a travel ban by citing President Roosevelt's notorious Executive Order 9066 issued in February 1942, which authorized the deportation and internment of over 120,000 Japanese Americans during World War II for "protection against espionage and sabotage to national defense" (Exec. Order No. 9066, 1942; Keneally December 8, 2015; Liptak 2018). At the time, candidate Trump stated, "what I'm doing is no different than FDR."[1] Trump's executive order did not go as far as targeting American citizens like Roosevelt's. Nevertheless, his allusion to the forced relocation of Japanese Americans is a reminder of how immigrant groups perceived as threatening can quickly become subject to harsh measures legitimized by national security concerns.

This chapter offers a two-step argument on why and how political leaders come to view certain groups as fifth columns that cannot be solely accounted

H. Zeynep Bulutgil and Sam Erkiletian, *Civil Society, Fifth-Column Perceptions, and Wartime Deportations* In: *Enemies Within.* Edited by: Harris Mylonas and Scott Radnitz, Oxford University Press. © Oxford University Press 2022. DOI: 10.1093/oso/9780197627938.003.0004

for by strategic arguments that emphasize state security during times of conflict (Downes 2008; Valentino et al. 2004). First, in democratic and semi-democratic contexts, one of the key determinants of this outcome is the extent to which there are influential civil society organizations that are devoted to the exclusion of the potential fifth-column group. Such organizations can galvanize public opinion by influencing the media and organize events that disseminate misinformation about the target. To the extent that they are successful in mobilizing public opinion, these organizations can also pressure politicians to enact exclusionary policies that restrict the immigration of the group and/or label them as a fifth column regardless of whether they pose a viable security threat.

Second, this chapter also argues that the emergence of these exclusionary organizations depends on the sequence between the development of civic associations and the timing of when the immigrant group in question has access to these associations. In short, civil society organizations are more likely to develop exclusionary ideologies that target specific groups if these groups have historically been outside the civic and political institutions of the country. This state of affairs appears if the immigrant group arrives on the territory of the country after the politically influential civil society organizations have already emerged.

We evaluate this argument through a controlled comparison of three cases: German Americans during World War I, Japanese Americans on the West Coast during World War II, and Japanese Americans in Hawaii during World War II. The selection of cases makes sense for two reasons. First, these cases provide variation on wartime targeting and deportations. On the West Coast, Japanese Americans were deported *en masse*, whereas in the case of Japanese Americans in Hawaii and German Americans, there were only very limited and selective deportations. Second, these cases allow the analysis to control for security-related considerations. In all cases, enemy states (Germany and Japan) could recruit members of these groups for intelligence or military purposes and, as a result, the groups were all at risk of being perceived and targeted as wartime security risks. Additionally, the enemy country and the ethnic group in question is the same for the Japanese on the West Coast and Hawaii, so these ethno-cultural factors do not explain the differences between these contexts.

The argument presented in this chapter provides an explanation for variation in fifth-column depictions and deportations during wartime. On the West Coast, Japanese migrants arrived after the emergence of robust civil society associations and were excluded from important economic and political organizations. This organizational disadvantage prevented them from effectively challenging exclusionary policies, leading to extensive wartime deportations. In the other cases, both Japanese and German migrants arrived before the formation of civil society associations, ensuring their inclusion during the development of these organizations, which later provided them with the influence necessary to resist fifth-columnization and wartime targeting.[2]

The findings of this chapter have implications for two strands of literature: one focusing on the securitization of immigration, when policymakers "make links between migration policy and national security," and the other on mass violence (Adamson 2006: 165).[3] The existing literature on immigration suggests that actors who seek to securitize immigration policies link certain migrant groups to hostile actors and create negative perceptions of the target group as threats to the state (Bigo 2002; Huysman 2006: 45–62). Under these circumstances, migrant groups with ethnic or national ties to rival states are particularly vulnerable to fifth-column depictions (Adamson 2006: 166). Additionally, scholars argue that when immigration is politicized as a matter of security, it can provide preexisting anti-immigration groups with a larger media platform and enable them to more effectively lobby for exclusionary policies (Lahav and Courtemanche 2012: 478; Sniderman et al. 2004; Hainmueller and Hopkins 2014; Hopkins 2010). Our study contributes to this literature by laying out and evaluating a theory of why securitization of immigration works against some groups but not others.

This chapter also contributes to the existing literature on mass violence. This literature can be divided into studies that highlight the importance of security concerns during wars and those that emphasize domestic sociopolitical factors (Bulutgil 2020). The former suggest that states become more likely to turn to ethnic cleansing or genocide when they are concerned about rebellion and sabotage behind the frontlines or when they anticipate and/or observe the members of an ethnic group joining the army of the rival state (Downes 2008; Valentino et al. 2004; Mylonas 2012). The studies in the latter category highlight that among contexts that fulfill these security-related conditions, there is still significant variation in how states treat minority ethnic groups during wars (Bulutgil 2017; Straus 2012; Maynard forthcoming). Following the rich literature on crosscutting social cleavages and political identities, they argue that this variation can be explained by the extent to which minority groups had organizational ties to the dominant group in their society.

This chapter advances our understanding of fifth-column politics in two ways. First, it provides a theoretical approach based on the timing of economic and political inclusion that explains why certain types of immigrant groups are unable to form these linkages with the dominant groups and accumulate organizational strength. Second, it tests the idea that domestic organizational arrangements influence the decision to use mass deportations during wars by using empirical evidence from a consolidated democracy, the United States. In doing so, the chapter also offers implications for the debate on whether democratization prevents or at least decreases the chances of mass violence against ethnic groups. Based on the findings of this chapter, the answer to this question hinges on the type of civil society in place, which in turn depends on the timing of when minority groups are included in it.

The rest of the chapter proceeds in four steps. First, we present the argument. Then, we discuss alternative explanations and case selection in more detail. Third, we compare the cases in terms of: the arrival of migrant communities, the involvement of these communities in civil society organizations, the evolution of exclusionary organizations, and wartime policies. Finally, we conclude by discussing the implications of our argument for current US and European immigration policies.

## Timing of Civil Society Organization and Fifth-Column Depictions

Civil society institutions such as trade unions, religious organizations, and business associations can influence the perception and treatment of ethnic groups in two ways. On the one hand, organizations that cross-cut ethnic, religious, or racial divisions in the society might decrease the likelihood of violence against specific ethnic groups, even if these groups have links to enemy actors (Bulutgil 2017; Bulutgil 2016; Straus 2012; Gubler and Selway 2012; Varshney 2003). Cross-cutting organizations might question the accuracy of reports that depict all members of the group as collaborators, highlight the costs of mass violence against the group, endorse selective alternatives to mass violence, and mobilize public opinion in support of the group in question (Bulutgil 2017). On the other hand, civil society organizations might contribute to the negative perception of minority ethnic groups if they reinforce ethnic identities and actively advocate for the exclusion of the group. If such organizations wield significant influence on public opinion, then they might be able to successfully portray specific groups as fifth columns and pressure elected leaders to target the group as a whole regardless of whether they pose a genuine security threat.

How do such exclusionist organizations emerge? Why do they emerge in some contexts and not others? Part of the answer to these questions lies in the timing of when influential civil society organizations emerge and when the members of the target group have the opportunity to participate in them (Figure 3.1 provides a summary of the main argument). In contexts in which civil society organizations emerge before the arrival of the immigrant group, they are likely to serve as instruments to sustain existing privileges by excluding the newcomers from achieving political, demographic, and economic ascendancy. The specific policies that these organizations advocate might include restricting political participation of the target group, putting up legal barriers against further immigration, and limiting access to public services and education. These restrictions not only

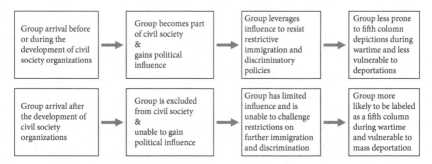

**Figure 3.1**  Timing of civil society development and wartime treatment of group

contribute to negative perceptions of the group but also diminish their organizational strength and in turn their ability to challenge exclusionary policies.

In contexts in which civil society organizations emerge at the same time as or after the arrival of the immigrant group, the organizational basis for excluding newcomers would be less likely. First, under these conditions, other groups in society would not have the grassroots organization to advocate for the exclusion and/or persistent mistreatment of the group from the beginning. As a result, the group in question could accumulate sufficient economic and political influence to prevent restrictive immigration policies and continue to grow demographically. Second, when groups in society generate civic associations such as trade unions or business associations at the same time rather than sequentially, they either become members of the same organizations, cooperating directly with each other, or simultaneously develop their own parallel organizations for influencing public policy. We argue that the direct involvement of a group in the early development of civil society and the subsequent political influence that stems from this development reduces the likelihood of the group later being depicted and targeted as a fifth column.

## Case Selection

The analysis compares German Americans during World War I, Japanese Americans in Hawaii during World War II, and Japanese Americans on the West Coast during the same period. For the German community, we pay particular attention to Chicago, the city with the largest concentration of German Americans at the time. The three cases have divergent outcomes for deportations and fifth-column depictions when controlling for strategic security concerns. In the case of the Japanese on the West Coast, President Roosevelt's signing of Executive Order 9066 expelled 120,000 Japanese (including 78,000 who were citizens) to internment camps (Hayashi 2004; Daniels 1988). In contrast, only some 2,000 (out of the population of about 160,000) Japanese Hawaiians and Japanese migrants suffered the same fate during World War II (United States Commission on Wartime Relocation and Internment of Civilians 1983: 19; Kashima 2003; Okihiro 1991). During World War I, fewer than 300 German Americans and German migrants were deported from Chicago, which had a German popula- tion of 450,000, and nationally, only 2,058 out of a population over 8 million Americans of German ancestry were interned (Tischauser 1981; Nagler 1993; Krammer 1997). This case selection provides variation on deportation outcomes, as German Americans in Chicago and Japanese Americans in Hawaii experi- enced selective deportation, while Japanese Americans on the West Coast were subjected to mass deportation.

In analyzing this variation, we control for wartime security/strategic concerns in a number of ways (see Table 3.1). Both German Americans and Japanese Americans had ethnic ties to prominent strategic rivals of the United States and therefore posed a potential security risk during wartime. While the German case is temporally distinct from the Japanese cases, the conditions were mark- edly similar. Provocations from the German Empire ended the period of mili- tary neutrality and forced America's entry into World War I, just as the attack on Pearl Harbor led directly to America's entry into World War II.[4] Germany's rising influence in the early 1900s challenged America's strategic interests in Europe and made it a significant international rival. Furthermore, during the period of neutrality, US officials uncovered plots sanctioned by the German government

Table 3.1  Summary of Cases: Similarities and Divergent Outcomes

|  |  | Japanese on the West Coast | Japanese in Hawaii | Germans in Chicago |
|---|---|---|---|---|
| Similarities | Strategic importance of territory | Medium | High | Medium |
|  | Ethnic links to rival state | Yes | Yes | Yes |
| Divergent outcomes | Early disruption of immigration and demographic growth | Yes | No | No |
|  | Limited political influence during war & mass internment | Yes | No | No |

to sabotage wartime infrastructure and subvert public opinion (Luebke 1974; Nagler 1993). The overtly pro-German behavior of some German American communities further fueled fears of a potential fifth column. For example, in Chicago, many German American communities openly supported Germany through fundraisers, parades, and even direct enlistments into the Imperial German Army during the period of American neutrality (Tischauser 1981). US officials were particularly worried about a German fifth column in Chicago, as the city was strategically vital because of its munitions factories and transport networks (Guistatis 2016). Despite these security threats and links to a rival power, German Americans were not deported in large numbers.

In the West Coast and Hawaii cases, both groups had ethnic links to Imperial Japan, a rival actor that directly challenged US interests in the Pacific. Many US officials and media outlets, particularly on the West Coast, raised the possibility that the Japanese populations on the mainland and in Hawaii could act as a fifth column (Daniels 1977: 27; Daniels 1988). Anti-Japanese attitudes increased after Japan defeated Russia in 1905 and further deteriorated after Japan left the League of Nations in 1933, invaded China in 1937, and signed the Tripartite Pact in 1940 (McClain 2002). While fears of a Japanese fifth column were present in both cases, US military officials were particularly concerned about the possibility of the large Japanese population in Hawaii sabotaging the critical naval installations across the islands (Kashima 2003). From a purely strategic perspective, it would have been easier to depict the Japanese in Hawaii as a potential fifth column, as they were geographically more vulnerable to Japanese invasion and lived close to a major naval base. Despite these security concerns, fifth-column depictions of Japanese Americans in Hawaii were relatively limited, and extensive deportations did not materialize even after the attack on Pearl Harbor (Okihiro 1991 United States Commission on Wartime Relocation and Internment of Civilians 1983). It is clear from both the German and Japanese cases that strategic considerations alone are not sufficient to explain the variation in fifth-column depictions and the extent of deportations.

An additional reason we focus on the comparison of these cases is that simple demographic arguments are not sufficient to explain the significant variation in the outcomes. In brief, the demographic argument would suggest that the Germans in Chicago and the Japanese in Hawaii were not deported because, given their numbers and share of the population, these policies would have been logistically too costly and disruptive. The problem with this argument is that it does not explain the extraordinarily small number of deportations that took place in the case of the Germans and the Japanese in Hawaii. US officials had the capability to deport and intern a far greater number of German Americans and had already allocated the necessary resources to do so (Nagler 1993; Krammer 1997).[5] Furthermore, during World War II, if the United States could deport over a hundred thousand Japanese Americans on the West Coast, they could have deported considerably more than two thousand Japanese Americans from Hawaii. The fact that in both cases the leaders did not expand the deportations suggests that population statistics alone do not explain the outcome.

Second, the population growth of these groups was at least partially endogenous to political decisions influenced by nativist civil society associations. Japanese migrants on the West Coast immediately encountered an established civil society comprising anti-Asian associations that were able to pressure West Coast politicians to pass anti-Japanese legislation (Daniels 1977; Daniels 1988; Fong and Markham 2002; Ichioka 1977). In contrast, no local counterparts to the mainland anti-Japanese organizations or immigration restrictions existed in Hawaii when the Japanese population first arrived, enabling the group to grow unimpeded. Similarly, anti-German nativist associations in Chicago emerged after German communities had already established themselves and were unable to restrict German immigration. In the empirical section, we discuss the relationship between the timing of immigration, the strength of nativist organizations, and group demography in more detail.

The chapter now proceeds with a comparative analysis of the cases in three steps. We first focus on the initial arrival and demographics of the groups and the development of civil society organizations. We then turn to nativist responses and exclusionary policies targeting the group during peacetime. Finally, we analyze the states' different treatment of the groups during wartime.

## The Arrival of New Ethnic Groups and Timing of Civil Society Development

### Japanese on the West Coast (1898–1920)

Japanese immigration to the Pacific Coast started after 1898. By 1908, the Japanese population had reached 40,000 and eventually expanded to 110,000 by 1920 (Daniels 1977; Lillquist 2007: 7). The driving factors for this trend were the decline in cheap Chinese labor, which was reduced after the Anti-Chinese Act of 1882, and the willingness of the Japanese in Hawaii to migrate to the West Coast after the latter became a US territory in 1898 (Daniels 1988). Japanese migrants gravitated primarily to California, specifically to the San Francisco Bay Area, in search of economic opportunities (Lillquist 2007). At the beginning of Japanese immigration, California already boasted a number of civil society organizations that eventually became the hub of anti-Japanese agitation.

Two types of institutions served as the precursor to anti-Japanese organization on the West Coast. The first included the anti-Chinese clubs that were already popular by the 1870s (Daniels 1977). The Chinese had started arriving in California in the 1840s, and their population peaked at around 60,000 (9% of the population in California in 1860). The anti-Chinese clubs emerged in the late 1840s among white mining communities and then became popular in large urban centers such as San Francisco (Fong and Markham 2002: 189). By 1870, they were strong enough to organize their first mass meeting in San Francisco, which advocated for excluding the Chinese from the political sphere and banning further immigration.

The second type of organization that fed into anti-Japanese sentiments was the labor movement (Daniels 1988; Hayashi 2004). The emerging labor movement had already been actively cooperating with the anti-Chinese movement starting from the 1870s. For example, labor organizations were among the prominent organizers of the 1870 anti-Chinese mass assembly and successfully appealed to the local Democratic Party and the Workingmen's Party to adopt anti-Chinese legislation. The anti-Chinese clubs and labor organizations successfully lobbied for the signing of the Chinese Exclusion Act of 1882, which barred Chinese immigration. Thus, when the Japanese started arriving to the West Coast in significant numbers, there was already a robust anti-Asian movement in place that could prevent them from gaining access to prominent civil society organizations (United States Commission on Wartime Relocation and Internment of Civilians 1983: 31; Ichioka 1977).

## Japanese in Hawaii (1860–1900)

Japanese immigrants arrived on the Hawaiian Islands from the 1860s onward as contract workers. Due to working conditions and disease, by 1850, the indigenous population had shrunk from about a million to 81,000 (Tamura 1994). To compensate for the loss of indigenous labor, the white landlords who owned the sugar plantations began to import an ethnically diverse contingent of workers. The first wave of foreign laborers consisted of Chinese, Filipino, and Portuguese migrants, and a second wave of Japanese workers arrived in the 1870s and 1880s. Between 1885 and 1924, over 200,000 Japanese migrants moved to Hawaii, making them one of the largest ethnic groups there (Okamura 2014).

Until 1898, Hawaii was a nominally independent country with a legal structure that exclusively served the interests of the white landowners, enabling them to curb labor organizations and develop an oppressive contract labor system that exploited workers and confined them to closely monitored residential areas (Okamura 2014; Guevarra 2012). One feature of this system was that the existing working class population did not have access to organizations akin to the labor movement on the West Coast, which could have allowed them to advocate against the growing numbers of Japanese migrants (Odo 2004). Thus, the Japanese population in independent Hawaii grew unimpeded by anti-Japanese legislation and by 1900 constituted 40 percent of the population (Tamura 1994).

Once Hawaii officially became a US territory in 1898, contract labor on sugar plantations was no longer allowed and labor organizations became legal (Okamura 2014). Japanese leaders utilized their extensive labor networks to establish some of the first labor organizations in Hawaii and challenged the oppressive plantation system, participating in over twenty strikes in 1900 alone (Okamura 2014: 32). Between 1900 and 1921, a number of unions organized by ethnic groups emerged in Hawaii. These ethnic unions coordinated with each other during strikes, such as that of 1920 orchestrated by Japanese and Filipino unions, which "worked to promote interethnic cooperation" and succeeded in improving wages and conditions (Guevarra 2012: 175; Jung 1999). The landowning elite at times pitted different ethnic communities against each other by using some as strikebreakers but were largely unable to weaken the growing influence of Japanese labor organizations (Jung 2003; Tamura 1994; Okihiro 1992: 70–71; Azuma 1998).

## German Americans in Chicago (1840–1890)

German-speaking migrants traveled to Chicago in two waves in the 1840s and 1880s. They consistently represented around 25 to 30 percent of the city's population during its rapid growth from 100,000 people in 1850 to 2.4 million in 1910 (Keil 1985: 191; Suhrbur 1983). The initial influx of German immigrants were known as the "Forty-Eighters" because they were predominantly political dissidents fleeing the unsuccessful 1848 revolutions across German lands (Levine 1992). The Forty-Eighters were a combination of intellectuals and skilled craftsmen experienced in forming political and labor organizations (Levine 1983: 167). When they arrived in Chicago, they re-established the Workers' Clubs and Workers' Halls they had left in Europe, which would become the foundation for cultural and political activity in German neighborhoods (Güney 2015).

The Forty-Eighters provided natural leadership in Chicago's nascent labor movement and their expertise quickly elevated them to key positions across various associations (Levine 1983; Keil 1985). German American involvement in the citywide carpenters' union demonstrates how they were able to collaborate with a coalition of different ethnic groups to achieve their political objectives. Local 21, the largest carpenters' union in Chicago with majority-German membership, "allowed the various nationalities to organize ethnic branches" that cooperated but maintained their ethnic identity, "foster[ing] a strong class-conscious camaraderie among its members" (Suhrbur 1983: 93–95). By building a cohesive, multiethnic coalition, the carpenters' union was able to effectively strike for higher wages and improved conditions. These ethnically diverse labor unions enabled traditionally oppositional groups, like Germans and Anglo-Saxons, who were often divided on prohibition politics, to interact and work alongside each other.

## Nativist Responses and Exclusionary Policies

### Japanese Americans on the West Coast (1900–1930)

When Japanese migrants arrived on the West Coast, they immediately faced anti-Asian organizations that excluded them from influential economic and political positions. The anti-Japanese movement was further strengthened by the labor movement in California (Daniels 1977: 19). In 1905, several trade unions, such as the prominent Sailors and Builders Unions, established the Asiatic Exclusion League (AEL).[6] In addition to the AEL, more local anti-Japanese organizations such as Anti-Jap Laundry League, which organized a boycott of Japanese laundries, started to emerge. By 1919, a number of other anti-Japanese organizations that adhered to the AEL's political goals had mushroomed across the West Coast (Daniels 1988). The most prominent were the Native Sons of the Golden West, which also published a monthly journal called *The Grizzly Bear*, the California State Federation of Labor, and the Japanese Exclusion League (later named the California Joint Immigration Committee) (Daniels 1977: 85, 105). The AEL and other organizations actively sought to exclude the Japanese (including second-generation Japanese born in the United States known as the "*Nisei*") from becoming citizens and to prohibit Japanese immigration. Once the Japanese population grew prosperous enough to acquire their own land, agricultural organizations such as the California State Grange also began to advocate for anti-Japanese legislation and pressured politicians to pass the Alien Land Law in 1913, which prevented the Japanese from purchasing land (Lillquist 2007; Daniels 1977: 85).

Anti-Japanese organizations were also able to mobilize their supporters and exert pressure on local and national politicians to pass more restrictive immigration policies (Yui 1992). In 1906, the AEL successfully advocated for the Board of Education to exclude Japanese students from public schools and confine them to the Oriental School, and in 1907, the governor of California affirmed the right of California to segregate "Orientals" (Daniels 1977: 34, 41; Lillquist 2007). After 1908, Democratic politicians running for office in California started to raise the Japanese as an issue to mobilize support and began to appeal to the AEL (Daniels 1977: 49, 50). Republicans, who initially did not consider the Japanese migrants a threat, also came under pressure to advocate for anti-Japanese legislation. By 1919, Republican politicians such as State Senator J. M. Inman and Democratic US Senator Hiram Johnson had become leaders in anti-Japanese organizations such as the California Oriental Exclusion League and the AEL (Daniels 1977: 84). Entrenched nativist organizations across the West Coast successfully pressured their local politicians to advocate for the passage of the informal Gentleman's Agreement in 1907 and the Johnson-Reed Act in 1924, which effectively ended immigration from Asia (Daniels 1988). Furthermore, these nativist organizations were also directly involved in the legislative process. For example, key nativist leaders from the West Coast such as Valentine McClatchy from the AEL and Attorney General Ulysses Webb of California testified before the Senate Committee on Immigration in the weeks preceding the passage of the Johnson-Reed Act in May 1924 (Daniels 1988). As a result, unlike Hawaii, where the Japanese population could grow unimpeded for sixty years after their arrival, the pressure from anti-Japanese organizations first restricted and then completely halted Japanese demographic growth on the West Coast within two decades.

The resistance to the anti-Japanese organizations and legislation was limited. In 1920s, some politicians, particularly at the federal level, had reservations about anti-Japanese legislation. Chief among them were President Theodore Roosevelt and some members of the Senate, who sought to prevent anti-Japanese legislation that would strain relations with Japan. Some business interests and missionary Protestant Churches that were active in Japan and among the Japanese Americans also opposed anti-Japanese legislation (Griffith 2018; Shaffer 1999). Various Japanese led associations across the West Coast, such as the Japanese American Citizens League, belatedly founded in San Francisco in 1929, also attempted to lobby against nativist policies (Daniels 1988; Ichioka 1977). Yet such organizations lacked the strength to reverse restrictive immigration legislation that already severely limited their demographic growth and in turn their long-term economic and political influence. By the 1930s, the politically isolated Japanese enclaves across the West Coast "occupied a nearly powerless position" to resist local nativist organizations (Kashima 2003: 67).

## Japanese Americans in Hawaii (1900–1930)

In Hawaii, the white landed elite attempted to limit the influence of the Japanese population but was unable to effectively restrict their rights or limit their numbers because of the advent of the Hawaiian labor movement and its multiethnic character. With the abolition of the contract labor system, the various ethnic groups in Hawaii had the opportunity to establish unions around the same time, preventing any one group from dominating Hawaii's labor movement

(Guevarra 2012). This failure was not due to lack of trying or the absence of racist attitudes. The Portuguese in Hawaii insisted that they alone were equal to white landholders and superior to the Japanese and other Asian migrants (Jung 2003: 380). However, unlike the white workers on the West Coast, they never successfully organized to bring about corresponding exclusionist policies. Building upon the previous successes of multiethnic strikes, by the 1930s and 1940s, Hawaii saw the emergence of powerful interracial trade unions such as the International Longshoremen's and Warehousemen's Unions, which included the Japanese, Portuguese, as well as other ethnic groups as members (Jung 2003; Odo 2004).

The main drivers of anti-Japanese feeling on the island came from the white landholding elite. This feeling was heightened by strikes in 1909 and 1920, during which Japanese labor leaders effectively mobilized multiethnic planta-tion workers, 60 percent of whom were of Japanese descent (Okihiro 1992: 56). Nevertheless, even in this period, the priority of the landed elite was to assim-ilate or "Americanize" the Japanese on the island while also keeping them as a docile workforce rather than excluding them. The white elite sought to achieve these outcomes by imposing "patriotic" programs at largely Japanese schools, restricting the teaching of Japanese language, and trying to exclude the Japanese from predominantly white public and private schools (Tamura 1994: 59). Starting in the 1920s, the local branches of mainland anti-Japanese organizations began to emerge in Hawaii. However, when these organizations clamored for stricter anti-Japanese legislation in Hawaii, they typically failed due to the objections of both Japanese and non-Japanese groups. For example, in 1921, the Republican territorial senator Lawrence Judd introduced an American Legion–sponsored bill that would have forced Japanese newspapers in Hawaii to publish English translations of their articles, with the ultimate goal of forcing these newspapers to close (Tamura 1994: 73–74). The opposition to this bill came from Japanese organizations on the island as well as the main church of the white elite, the Congregationalist Central Union Church in Honolulu (Tamura 1994: 59, 74). Ultimately, a watered-down version of the bill that required articles only about the government to be translated came into force.

## German Americans in Chicago (1850–1900)

In Chicago, the cultural and political institutions created by the Forty-Eighters resulted in organized German neighborhoods that provided German Americans an internal base of political power to resist early nativist groups. In response to the influx of immigrants arriving in the 1840s and 1850s, national nativist or-ganizations like the Know-Nothing Movement and Anti-Saloon League gained support across America and developed active branches in Chicago (Levine 2001). These nativist organizations feared that the politically radical ideas of European immigrants like the Forty-Eighters and their drinking cultures would ruin American society (Levine 1992: 237–249). While the Anti-Saloon League was ostensibly a prohibition pressure group, their membership was composed largely of the same nativists who supported the Know-Nothing Movement and their activities were often thinly veiled attacks on German and Irish immigrant communities. Increasing anti-German pressure culminated in the Lager Beer Riot of 1855, when the newly elected Know-Nothing mayor Levi Boone, who ran on an anti-immigration and temperance platform, directly attempted to enforce Sunday drinking laws (Güney 2015). German communities rallied in protest along with their Irish neighbors and marched throughout Chicago, effectively forcing the mayor to suspend temperance laws. The Lager Beer Riot and its aftermath "marked the demise of the Know-Nothing Party" in Chicago, which had recently dominated local elections (Spinney 2000: 73). After the riot, German political groups continued to mobilize voters and exerted pressure to rollback temperance policies and ensured that the Know-Nothing Party was removed from office in the following elections (Gems 2009: 1929).

Nativist organizations resurged in the 1880s and 1890s in response to another influx of immigrants. The American Protective Association was formed in 1887 and the depression of 1893 further exacerbated nativist sentiments and resulted in the creation of the Immigration Restriction League in 1894. While these nativist organizations tended to be anti-Catholic and prohibitionist like their predecessors, their primary objective was to restrict immigration and make naturalization more difficult (Ellis and Panayi 1994). They actively lobbied for federal legislation to curb the flow of European migrants, but also emphasized other ideals not largely seen in preceding nativist organizations, such as full Americanism, which called for American citizens to sever cultural ties with their countries of origin, and anti-radicalism (Higham 2002).

By 1890, Chicago was considered the "center of American organized labor" and socialism, as well as the primary destination for the second wave of German immigrants (Ensslen and Ickstadt 1983: 238). German Americans were increasingly associated with radicalism in Chicago because of publicized incidents like the Haymarket Riot in 1886, where German immigrant-anarchists were accused of throwing explosives against a group of policemen (Higham 2002). For the nativists in the city, the Haymarket Riot confirmed their suspicions that immigrants, particularly Germans, were radical and "un-American" (Spinney 2000: 113).

Nativist organizations exploited this fear of radicalism in an attempt to weaken German American influence. While they were successful in demonizing the minority of Germans who were anarchists, they were unable to break German-led unions or diminish their overall influence across the city (Güney 2015). German American involvement in ethnically cross-cutting labor organizations and the strength of their now established communities enabled them to sway the majority of mayoral elections from 1879 to 1905, and to once again weather the storm of nativism (Spinney 2000). Nativist pressure in the 1890s had failed to restrict immigration and had the unintended effect of galvanizing German communities nationally, leading to the creation of the National German American Alliance in 1901 (Ellis and Panayi 1994: 240).

## Wartime Experiences and Policies

### Japanese Americans on the West Coast (1941–1945)

After the attack on Pearl Harbor on December 7, 1941, military and civilian leaders on the West Coast and in Washington, DC, did not immediately converge to support mass deportations. During the first couple of weeks, California's governor Culbert Olson acted as an honorary chair of the Fair Play Committee, a group formed by scholars from the University of California, Berkeley, to endorse investigating individuals rather than resorting to mass measures (Eisenberg 2003; Kashima 2003). The attorney general of California, Earl Warren,[7] also advised the state personnel board not to exclude the Japanese Americans from public service (United States Commission on Wartime Relocation and Internment of Civilians 1983: 97, 98). Furthermore, a series of FBI-, Navy-, and White House–commissioned intelligence reports found no credible evidence of large-scale espionage or sabotage at the time of the attack and did not see any security grounds for the mass removal the Japanese in Hawaii or on the West Coast (Kashima 2003).

By early February, however, Governor Olson and Attorney General Warren had become avid supporters for the mass deportation of the Japanese to internment camps away from the West Coast (United States Commission on Wartime Relocation and Internment of Civilians 1983: 98; Hayashi 2004). Given the lack of evidence of a Japanese fifth column from multiple intelligence agencies, as well as the relatively calm reaction after the attack, why did the local politicians on the West Coast change their positions? Why did the pro-removal faction at the federal level gain the upper hand during the period between December 1941 and mid-February 1942? We argue that the reason lies in the active agitation by popular anti-Japanese organizations on the West Coast that turned politicians and the media against the Japanese (Yui 1992; Daniels 1988).

After the attack, there were differing opinions at the federal level. In particular, the Justice Department was in favor of only focusing on individual Japanese who were involved in suspicious activities. The Department of War, by contrast, favored collective measures against the Japanese, though even they were not yet clear on whether the policy would include citizens or all territories on the West Coast and Hawaii (Hayashi 2004). For example, earlier suggestions for mass removal focused on small strategic areas around airports or military bases (United States Commission on Wartime Relocation and Internment of Civilians 1983: 84). At one point, Governor Olson and the Commander of the West Coast General DeWitt also considered a voluntary (as opposed to compulsory) deportation of the Japanese to other areas (United States Commission on Wartime Relocation and Internment of Civilians 1983). One of the early proponents of mass deportation was the Secretary of the Navy William Knox, who accused the Japanese Americans in Hawaii without evidence of engaging in espionage and sabotage before and during the Pearl Harbor attack (Kashima 2003: 75). His remarks triggered vocal opposition from former and current military leaders in Hawaii and from United States attorney general Francis Biddle. Biddle repeatedly sent reports and personally met with President Roosevelt between January and mid-February 1942 in an effort to prevent the mass deportation of Japanese Americans.

Before the mass deportation, there were four sources of information that refuted the claim that Japanese Americans on Hawaii or the West Coast acted (or intended to act) as a fifth column. The first source came from FBI agents, who had already begun investigating small lists of individual Japanese Americans suspected of treason and found no evidence of large-scale espionage or sabotage (Kashima 2003). The head of the FBI at the time, J. Edgar Hoover, openly expressed the belief that local and federal level leaders that supported mass removal were acting in reaction to public pressure rather than a rational analysis of facts on the ground (United States Commission on Wartime Relocation and Internment of Civilians 1983: 55).

A second source of information came from a Navy intelligence officer (K. D. Ringle) who had worked in Japan and prepared reports on Japanese Americans in Hawaii and the West Coast. He also concluded that the Japanese community did not constitute a noteworthy threat to American wartime security and that the Japanese government considered them "cultural traitors" who could not be relied upon as a fifth column (Daniels 1988: 212). Third was a report prepared by Curtis Munson, a businessman President Roosevelt had commissioned to investigate the loyalties of Japanese Americans on the West Coast. The Munson Report was made available before the Pearl Harbor attack in October 1941 and concluded that the overwhelming majority of the Japanese were loyal American citizens, prophetically noting that the "Japanese here are in more danger from us than we are from them" (Daniels 1988: 212).

Finally, the Roberts Commission, which was convened after the attack under the leadership of Supreme Court Justice Owen Roberts, reached a more ambiguous conclusion on January 23. The commission identified espionage activity by Japanese consular agents and went on to argue that there had also been activity by individuals not affiliated with the Japanese consulate, without specifying who these individuals were or whether they were Japanese Americans (Hayashi 2004). The vague accusations in the Roberts Commission report became ammunition for those who wanted to remove the Japanese on the West Coast and Hawaii.

In this atmosphere of uncertainty, anti-Japanese organizations on the West Coast played a significant role in pressuring politicians to converge on mass deportation. On January 2, even before the Roberts Report, the Joint Immigration Committee sent a memo to California newspapers, labeling the Japanese as a dangerous and unassimilable fifth column (United States Commission on Wartime Relocation and Internment of Civilians 1983: 67, 68). The Immigration Committee had the support of other preexisting anti-Japanese civil society organizations including the Native Sons and Daughters of the Golden West and the California Department of the American Legion (Eisenberg 2003; United States Commission on Wartime Relocation and Internment of Civilians 1983: 68). *The Grizzly Bear* blamed the politicians and businesses who did not support anti-Japanese measures in the pre–World War II period for Pearl Harbor and called for the removal of all Japanese Americans (United States Commission on Wartime Relocation and Internment of Civilians 1983: 69, 70). These organizations also influenced local level measures. For example, some seventeen counties on the West Coast passed resolutions urging the federal government to remove the Japanese and dismissed all Japanese Americans from employment.

By early February, anti-Japanese organizations had the backing of prominent California politicians such as Attorney General Earl Warren, who started to work with the Joint Immigration Committee to convince the federal government to expel Japanese Americans to internment camps (Hayashi 2004; United States Commission on Wartime Relocation and Internment of Civilians 1983: 68). Several congressmen from the West Coast also wrote to the Secretaries of War and the Navy as well as to the FBI asking for the Japanese to be placed in concentration camps. House members from the West Coast urged the president to use mass deportations and attacked Attorney General Biddle, who opposed this measure. Moderate newspapers changed their positions and began publishing editorials that supported anti-Japanese propaganda (United States Commission on Wartime Relocation and Internment of Civilians 1983: 80–81). Eventually, Governor Olson and General DeWitt also switched from advocating voluntary deportations to mass removal (Kurashige 2016: 175; United States Commission on Wartime Relocation and Internment of Civilians 1983: 72–74).

Against this tide of anti-Japanese sentiment, there were isolated voices of opposition within civil society organizations, including from some liberal Protestant churches, major universities such as UC Berkeley, and the Socialist Party (Eisenberg 2003; Robinson 2009, 109; Griffith 2018). In February, Japanese Americans themselves also tried to form an organization to avoid mass removal (Grodzins 1949: 73). However, these efforts were unable to counter the pervasive influence of anti-Japanese organizations, which had convinced most of the public and local officials that the Japanese on the West Coast posed a viable security threat.

## Japanese Americans in Hawaii (1941–1945)

When martial law was declared in Hawaii after the attack on Pearl Harbor, it gave the United States military sweeping powers (more so than in West Coast states, including California, which never declared martial law) to detain and repress Japanese Americans across the islands. Yet only around two thousand Japanese Hawaiians were detained during the war (Hayashi 2004). Despite the relatively vulnerable security situation, there was no comparable anti-Japanese movement devoted to removing the Japanese population because the Hawaiian Japanese were better organized and more entrenched in their civil society organizations than their counterparts on the West Coast (Kashima 2003: 67).

The leaders in Hawaii, including General Delos Emmons, who was responsible for the island under martial law, maintained that the Japanese Americans in Hawaii did not constitute a military threat.[8] Thus instead of mass deportations, they preferred to carry out selective and limited removal based on thorough investigations (Kashima 2003). The few hundred Japanese Hawaiians detained at the prison camp on Sand Island or deported to the mainland were carefully vetted based on the intelligence reports compiled by the FBI and the Navy before the war (Okihiro 1991; Kashima 2003).

In the face of mounting pressure from the mainland to deport the Japanese *en masse*, including from President Roosevelt, General Emmons rejected this policy. When the pressure mounted further, Emmons resorted to making promises about deportation but not delivering on them (Okihiro 1991). For example, he made the deportation of the Japanese population an official order but ensured that it was assigned as a lower priority or delayed so that it would not be carried out (Kashima 2002: 77). General Emmons and his supporting staff resisted calls for large-scale deportations because they worked closely with non-Japanese community leaders who did not believe that the Japanese were a fifth column (Kashima 2003: 69). They also maintained that Japanese Americans were not a strategic threat, but "indispensable for the successful conduct of the war" (Okamura 2014: 41; Odo 2004).

Japanese Americans indeed occupied key positions in multiethnic labor organizations that were vital to Hawaii's economy and the war effort (Odo 2004). For example, Japanese Hawaiians were members and leaders of the International Longshoremen's and Warehousemen's Union (ILWU), one of the largest labor organizations in Hawaii, which was primarily composed of Japanese, Portuguese, and Filipino workers (Jung 2003). While martial law officially suspended labor unions, wartime industries were still organized along these prewar labor associations and these multiethnic relationships persisted. Wartime support for Japanese Hawaiians from other ethnic groups was evident, for example, when non-Japanese leaders of organizations like the ILWU expelled anti-Japanese members (Daniels 1988: 294).

Japanese leaders mobilized their multiethnic ties to form other organizations such as the Committee for Inter-Racial Unity in December 1940, which garnered support from other ethnic communities to counter anti-Japanese sentiment (Okihiro 1991: 202). Japanese Hawaiian leaders also established the Oahu Citizens' Committee for Home Defense in early 1941 to prepare their communities to assist the military and hosted a number of events to signal their American patriotism (Okihiro 1991: 202). Furthermore, at the start of the war, six members of the territorial house of representatives and one member of the territorial senate were Japanese Americans and could directly lobby on behalf of their constituents (Okamura 2014). Japanese Americans were also visibly involved in the defense of the islands. Following the attack on Pearl Harbor, Japanese ROTC members served in the Hawaiian Territorial Defense and Japanese civilians formed a number of ad hoc volunteer groups to assist the war effort (Odo 2004). These organizations enabled Japanese communities to forge further contacts with local politicians and leaders and allowed them to ease tensions and advocate for themselves in the immediate aftermath of Pearl Harbor.

While the Japanese communities in Hawaii suffered injustices and were subject to harsh restrictions under martial law, unlike the Japanese on the mainland, they avoided mass internment in the aftermath of the Executive Order 9066. Instead, the leaders in Hawaii opted for selective deportations that targeted only about 1 percent of the Japanese population, as opposed to the entirety of the Japanese Americans on the West Coast. This outcome is particularly significant as the Hawaiian leadership was under pressure from the mainland political establishment, including from President Roosevelt himself, to engage in large-scale deportations.

## German Americans in Chicago (1917–1918)

Why were German Americans not interned in larger numbers during World War I? Before America entered the war, thousands of German Americans enlisted in the German army, many German American communities openly supported the Kaiser's war effort, and German agents sabotaged munitions factories and actively subverted public opinion (Luebke 1974; Nagler 1993). Germany's foreign minister even threatened that "500,000 German American men" would rise up if the United States entered the war (Guistatis 2016: 261). Chicago's German community, the largest in America during World War I, posed a viable security threat to the strategically important munitions factories throughout the city (Higham 2002: 195; Tischauser 1981: 45–46). Given the widespread concerns of a German fifth column and preexisting anti-German organizations before the United States entered the war, it is puzzling that only 2,058 German Americans were deported to internment camps during World War I (Jensen 2013: 457; Nagler 1993).

During the period of American neutrality from 1914 to April 1917, the Preparedness Movement, comprised of pro-war nativist organizations across the United States, called for citizens to be "100% American" (Luebke 1974: 140). The largest Preparedness organization was the nativist and antilabor National Security League (NSL), founded in 1914 by wealthy members of the Immigration Restriction League (Ward 1960). When the United States entered the war, many Americans believed that their fellow German Americans were a "disloyal fifth column acting under the orders of a foreign despot" because of these prewar organizations (Higham 2002: 201).

Following the declaration of war, unprecedented anti-German sentiment spread across the United States and immediately put pressure on German American communities, particularly those in Chicago (Güney 2015). Anti-German sentiment was intensified by the policies and rhetoric of President Woodrow Wilson and his administration.[9] Wilson formed the Committee of Public Information, which demonized Germany and reinforced the notion of a domestic German spy network (Luebke 1974: 213).

Anti-German policies diffused from the national to the local level. Wilson encouraged states to form Councils of Defense and granted these councils "sweeping legal powers" to persecute German Americans and German-born immigrants labeled as "enemy aliens" (Luebke 1974: 214). He also enacted policies that restricted the civil liberties of German Americans. He urged the passing of the Espionage Act in June 1917, which permitted the imprisonment of any individuals accused of hindering the war effort, and later the Enemy Alien Act in November 1917, which ordered all "enemy aliens," or noncitizens from Germany to register with local officials.[10] Approximately 260,000 German "alien" men and 220,000 women were registered and interrogated nationwide

(Jensen 2013: 457). Enemy aliens faced the brunt of wartime persecution because they were unprotected by constitutional rights and could be interned or have their property seized without due process (Nagler 1993; Krammer 1997). Wilson's administration periodically revisited the policy of mass internment of enemy aliens, and there were debates in Congress on whether to denaturalize German citizens deemed disloyal or to declare all German Americans as enemy aliens subject to internment (Nagler 1993: 209). Despite this intense environment of anti-German sentiment and the legislative framework to persecute German Americans as a whole, only 6,300 German enemy aliens were arrested during the war (Krammer 1997).

Chicago was a wellspring for nativist organizations created during the war, yet Chicago and the state of Illinois had some of the lowest internment rates in the country (Nagler 1993: 211). Both the American Protective League (APL) and the Four Minute Men were founded in Chicago in 1917 by wealthy nativists with connections to the Immigration Restriction League (Gustaitis 2016). The Four Minute Men were a group of orators who traveled around cities to deliver anti-German and prowar speeches, while the APL was an organization of volunteer detectives sanctioned by the Department of Justice to surveil and report on German Americans suspected of disloyalty (Tischauser 1981: 31). APL "agents" often used extrajudicial actions to harass and investigate German Americans. By the end of the war, there were officially thirteen thousand APL agents operating in Chicago (Gustaitis 2016: 125). Prewar organizations like the NSL also took advantage of the wartime environment and directly pressured German schools and newspapers to stop using the German language. The NSL's anti-language campaign effectively ended bilingual public-school curricula and was one of the few successful nativist efforts in Chicago (Ellis and Panayi 1994).

German Americans in Chicago responded to this unprecedented onslaught of nativism by falling back on their well-established civil society organizations, many of which were multiethnic, and by mobilizing their political influence as they had done in the 1850s and 1890s. The strongest bulwark against anti-German attacks in Chicago was the local and citywide political influence of German Americans. German Americans sought political protection from their local ward leaders and relied on pro-German mayor William Hale Thompson. Thompson was able to resist some of the harsher polices pushed by Wilson and state officials (Tischauser 1981: 48). He openly demonstrated his antiwar stance when he was the only mayor in America to offer his city for a peace conference. While demonized nationally as "Kaiser Bill," in Chicago, he maintained his pro-German and pro-immigrant platform and easily won re-election after the war (Gustaitis 2016: 44). On the state level, the majority of Illinois' politicians also advocated for neutrality and continued to support their German constituents throughout the war.

At the neighborhood level, social and cultural German associations across Chicago eased local tensions and provided German Americans with a platform to signal their loyalty and intentions (Luebke 1974: 286). German involvement in multiethnic institutions also garnered them vital local support and sympathy from other ethnic groups. Irish Americans in particular defended German communities because of their largely similar social and political views and mutual disregard for prohibitionist policies (Güney 2015: 159). Economically, German leadership within labor unions provided job security against nativists, who threatened to fire German workers. These multiethnic relationships were also invaluable for Germans accused of disloyalty and subjected to invasive loyalty investigations carried out by APL agents, as they could call upon their non-German neighbors to vouch for their allegiances (Nagler 1993).

German-Chicagoan communities resisted anti-German policies and internment because of their involvement in influential civil society organizations before the emergence of anti-German, nativist groups. While anti-German policies during World War I weakened German culture, German Americans were able to maintain their communities and thwart attempts to have them deported in much larger numbers.

## Conclusion

We have argued that whether politicians target an immigrant ethnic group as a fifth column partially depends on the timing of the group's arrival and the resulting extent of their involvement in civil society organizations, at least in democracies. If the group arrives before or during the creation of civil society associations, its members can actively participate in their development and foster cross-cutting ties with other ethnic groups, which help them challenge exclusionary policies. If the group arrives after a robust civil society has already emerged, it becomes harder for its members to become an integral part of these organizations. As a result, they struggle to prevent nativist organizations from effectively lobbying for exclusionary policies.

We evaluated this argument through a comparative analysis of three cases— German Americans in Chicago, Japanese Americans on the West Coast, and Japanese Americans in Hawaii (see Table 3.2). All three groups had links to rival external actors that eventually fought against the United States in a major war. Despite this shared condition, the cases had divergent outcomes on peacetime exclusion, fifth-column depictions, and deportations during wars (see summary of cases in Table 3.2). In the West Coast case, Japanese Americans encountered well-entrenched anti-Asian organizations that excluded them from civil society and successfully lobbied for restrictions on further immigration as well as for

Table 3.2 Summary of Findings

|  | Japanese on the West Coast | Japanese in Hawaii | Germans in Chicago |
|---|---|---|---|
| Arrival before development of civil society | No | Yes | Yes |
| Exclusion from civil society | Yes | No | No |
| Early immigration restrictions | Yes | No | No |
| Outcomes | Limited political influence & mass internment | Significant political influence & no mass internment | Significant political influence & no mass internment |

sweeping wartime deportations. In Hawaii, Japanese American communities actively participated in the early development of important civil society organizations, which enabled them to cultivate ethnically cross-cutting relationships and challenge exclusionary immigration and wartime polices. Finally, in Chicago, German Americans utilized their previous organizational skills to directly shape the growth of the city's labor and political organizations, which they could then use to resist anti-immigration legislation and deportations.

These findings make two contributions to the literature on ethnic violence and immigration. First, these cases demonstrate that the timing of when a group has access to civil society associations has significant implications for whether a group becomes a target for mass deportations during wars. Second, while many studies on ethnic violence focus on the strategic and security aspects of wartime targeting, these three cases emphasize the importance of local interactions between migrant and native-born groups, which can significantly influence the creation and implementation of national-level security policies during wartime.

The analysis in this chapter also advances understandings about the formation of fifth-column depictions within democracies. Fifth-column threats can emerge from below, when an entire group of citizens are considered subversive, or from above, when a small group of actors are accused of colluding with a hostile state (Radnitz and Mylonas, Introduction). Our findings demonstrate how discourse on fifth-column claims can emanate from local actors at the civil society level and directly influence the formation of national policies. In a bottom-up process, civil society organizations can affect the dominant discourse and public opinion toward a specific group that places pressure on political leaders to enact repressive policies. Conversely, they can also serve as a barrier against exclusionary policies and deportations. Their role in advocating or resisting fifth-column claims essentially depends on the timing and sequence of how civil society organizations emerge and which groups such organizations include.

Both Woodrow Wilson and Franklin Roosevelt were heavily influenced by the members of their administration and local officials with strong ties to nativist groups. Both presidents and their administrations continually revisited the possibility of passing even harsher wartime policies or attempted to do so. As President Trump demonstrated and as European leaders continue to frame certain ethnic groups as threats to national security, migrant groups remain vulnerable to fifth-column depictions and civil rights abuses. The experiences of German and Japanese American communities suggest that promoting cross-cutting ethnic ties and facilitating migrants' access to positions within civil society organizations can lessen the securitization of migrant groups and prevent exclusionary policies against them.

## Notes

1. Candidate Trump made these remarks during a telephone interview on ABC News' *Good Morning America* on December 8, 2015. Meghan Keneally, "Donald Trump Cites These FDR Policies to Defend Muslim Ban." *ABC News*, December 8, 2015. https://abcnews.go.com/Politics/donald-trump-cites-fdr-policies-defend-muslim-ban/story?id=35648128.

2. Note that we use the term "migrants" instead of "Americans" here because we are referring to the arrival of the ethnic group. As we will specify, many of the migrants and their families subsequently became citizens.

3. We use "mass violence" as an inclusive term that encompasses ethnic cleansing, genocide, mass killing, and civilian victimization. The term "ethnic cleansing" includes internal and external deportations as well as killings that target ethnic groups. Since deportations are a necessary first step to internment, cases of mass internment targeting ethnic groups also count as cases of ethnic cleansing. For more detail on the definition of ethnic cleansing and the relationship between ethnic cleansing, genocide, mass killings, civilian victimization, and deportations, see Bulutgil (2015). For specific definitions of mass killings and civilian victimization, see Valentino et al. (2004) and Downes (2008).

4. Germany's submarine warfare in the Atlantic threatened US citizens and shipping and culminated in the highly publicized sinking of the *Lusitania* in 1915, which claimed over a hundred American lives. A complete rupture in diplomatic relations was engendered by the leaking of the Zimmerman Telegram in January 1917, a secret cable from the German Foreign Office to the Mexican government that proposed a military alliance against the United States (Luebke 1974).

5. The deportation and internment of German expatriates during World War I was also carried out in far larger numbers in Canada (9,000) and in the UK (32,000) and was clearly a feasible policy the United States could have pursued (Ellis and Panayi 1994).
6. According to Daniels 1977 (28, 29), the AEL officially included some 231 organizations, of which 195 were labor unions.
7. Warren would later become governor of California and chief justice of the United States.
8. The notable exception was the attorney general of Hawaii, who was a member of the American League.
9. Wilson openly fueled the idea of a German fifth column. In his 1917 Flag Day speech, he said, "the military masters of Germany have filled our unsuspecting communities with vicious spies and conspirators who have sought to corrupt the opinion of our people" (Luebke 1974: 234).
10. The idea to reinstate the Enemy Alien Act was suggested by Assistant Attorney General Charles Warren, an original founder of the Immigration Restriction League (Nagler, 1993: 196).

# References

Adamson, Fiona B. 2006. "Crossing Borders: International Migration and National Security." *International Security* 31(1): 165–199.

Azuma, Eiichiro. 1998. "Racial Struggle, Immigrant Nationalism, and Ethnic Identity: Japanese and Filipinos in the California Delta." *Pacific Historical Review* 67, no. 2 (May): 163–199.

Bigo, Didier. 2002. "Security and Immigration: Toward a Critique of the Governmentality of Unease." *Alternatives* 27: 63–92.

Bulutgil, H. Zeynep. 2015. "Social Cleavages, Wartime Experience, and Ethnic Cleansing in Europe." *Journal of Peace Research* 52, no. 5 (September): 577–590.

Bulutgil, H. Zeynep. 2016. *The Roots of Ethnic Cleansing in Europe.* New York: Cambridge University Press.

Bulutgil, H. Zeynep. 2017. "Ethnic Cleansing and Its Alternatives in Wartime: A Comparison of the Austro-Hungarian, Ottoman, and Russian Empires." *International Security* 41 (4): 169–201.

Bulutgil, H. Zeynep. 2020. "Prewar Domestic Conditions and Civilians in War." *Journal of Global Security Studies* 5 (3): 528–541.

Daniels, Roger. 1977. *The Politics of Prejudice: The Anti-Japanese Movement in California and the Struggle for Japanese Exclusion.* Berkeley: University of California Press.

Daniels, Roger. 1988. *Asian America: Chinese and Japanese in the United States since 1850.* Seattle: University of Washington Press.

Downes, Alexander B. 2008. *Targeting Civilians in War.* Ithaca, NY: Cornell University Press.

Eisenberg, Ellen. 2003. "'As Truly American as Your Son': Voicing Opposition to Internment in Three West Coast Cities." *Oregon Historical Quarterly* 104, no. 4 (Winter): 542–565.

Ellis, Mark, and Panikos Panayi. 1994. "German Minorities in World War I: A Comparative Study of Britain and the USA." *Ethnic and Racial Studies* 17 (2): 238–259.

Ensslen, Klaus, and Heinz Ickstadt. 1983. "German Working-Class Culture in Chicago: Continuity and Change in the Decade from 1900 to 1910." In *German Workers in Industrial Chicago, 1850–1910*, edited by Hartmut Keil and John Jentz. DeKalb: Northern Illinois University Press, 238–252.

Fong, Eric W., and William T. Markham. 2002. "Anti-Chinese Politics in California in the 1870s: An Intercounty Analysis." American *Sociological Review* 45 (2): 183–210.

Gems, Gerald. 2009. "The German Turners and the Taming of Radicalism in Chicago." *International Journal of the History of Sport* 26 (13): 1926–1945.

Griffith, Sarah M. 2018. *The Fight for Asian American Civil Rights: Liberal Protestant Activism, 1900–1950*. Urbana-Champaign: University of Illinois Press.

Grodzins, Morton. 1949. *Americans Betrayed: Politics and the Japanese Evacuation*. Chicago: University of Chicago Press.

Gubler, Joshua, and Joel Sawat Selway. 2012. "Horizontal Inequality, Crosscutting Cleavages, and Civil War." *Journal of Conflict Resolution* 56 (2): 206–232.

Guevarra, Rudy P. 2012. "Mabuhay Compañero: Filipinos, Mexicans, and Interethnic Labor Organizing in Hawaii and California, 1920s–1940s." In *Transnational Crossroads: Remapping the Americas and the Pacific*, edited by Camilla Fojas and Rudy P. Guevarra. Lincoln: University of Nebraska Press, 171–197.

Guistatis, Joseph. 2016. *Chicago Transformed: World War I and the Windy City*. Carbondale: Southern Illinois University Press.

Güney, Ülkü. 2015. "German Ethnic Identity in Chicago before and during the First World War." *Hacettepe University Journal of Faculty of Letters* 32: 152–162.

Hainmueller, Jens, and Daniel J. Hopkins. 2014. "Public Attitudes toward Immigration." *Annual Review of Political Science* 17: 225–249.

Hayashi, Brian Masaru. 2004. *Democratizing the Enemy: The Japanese American Internment*. Princeton, NJ: Princeton University Press.

Higham, John. 2002. *Strangers in the Land: Patterns of American Nativism, 1860–1925*. New Brunswick, NJ: Rutgers University Press.

Hopkins, Daniel J. 2010. "Politicized Places: Explaining Where and When Immigrants Provoke Local Opposition." *American Political Science Review* 104 (1): 40–60.

Huysman, Jef. 2006. *The Politics of Insecurity: Fear, Migration and Asylum in the EU*. London: Routledge.

Ichioka, Yuji. 1977. "Japanese Associations and the Japanese Government: A Special Relationship, 1909–1926." *Pacific Historical Review* 46 (3): 409–437.

Jensen, Kimberly. 2013. "From Citizens to Enemy Aliens: Oregon Women, Marriage, and the Surveillance State during the First World War." *Oregon Historical Quarterly* 114 (4): 453–473.

Jung, Moon-Kie. 1999. "No Whites, No Asians: Race, Marxism, and Hawaii's Preemergent Working Class." *Social Science History* 23 (3): 357–393.

Jung, Moon-Kie. 2003. "Interracialism: The Ideological Transformation of Hawaii's Working Class." *American Sociological Review* 68 (3): 373–400.

Kashima, Tetsuden. 2003. *Judgment without Trial: Japanese American Imprisonment during World War II*. Seattle: University of Washington Press.

Keil, Hartmut. 1985. "German Immigrant Workers in 19th Century America: Working-Class Culture and Everyday Life in an Urban Industrial Setting." In *America and the Germans, Volume 1*, edited by Frank Trommler and Joseph McVeigh. Philadelphia: University of Pennsylvania Press, 189–206.

Krammer, Arnold. 1997. *Undue Process: The Untold Story of America's German Alien Internees*. New York: Rowan & Littlefield.

Kurashige, Lon. 2016. *Winds of War: Internment and the Great Transformation, 1941–1952: Two Faces of Exclusion*. Chapel Hill: University of North Carolina Press.

Lahav, Gallya, and Marie Courtemanche. 2012. "The Ideological Effects of Framing Threat on Immigration and Civil Liberties." *Political Behavior* 34 (3): 477–505.

Levine, Bruce. 1983. "Free Soil, Free Labor, and *Freimaenner*: German Chicago in the Civil War Era." In *German Workers in Industrial Chicago, 1850–1910*, edited by Hartmut Keil and John Jentz. DeKalb: Northern Illinois University Press, 168–182.

Levine, Bruce. 1992. *The Spirit of 1848: German Immigrants, Labor Conflict, and the Coming of the Civil War*. Urbana and Chicago: University of Illinois Press.

Levine, Bruce. 2001. "Conservatism, Nativism, and Slavery: Thomas R. Whitney and the Origins of the Know-Nothing Party." *The Journal of American History* 88 (2): 455–488.

Lillquist, Karl. 2007. "Imprisoned in the Desert: The Geography of World War II–era, Japanese American Relocation Centers in the Western United States." Ellensburg: Central Washington University.

Liptak, Adam. 2018. "Travel Ban Case Is Shadowed by One of Supreme Court's Darkest Moments." *New York Times*, April 16.

Luebke, Frederick C. 1974. *Bonds of Loyalty: German Americans and World War I*. DeKalb: Northern Illinois University Press.

Maynard, Jonathan L. Forthcoming. *Ideology and Mass Killing: How Groups Justify Genocides and Other Atrocities against Civilians*. Oxford: Oxford University Press, c. 2022.

McClain, James. 2002. *Japan: A Modern History*. New York: W. W. Norton.

Mylonas, Harris. 2012. *The Politics of Nation-Building: The Making of Co-Nationals, Refugees, and Minorities*. New York: Cambridge University Press.

Nagler, J. 1993. "Victims of the Home Front: Enemy Aliens in the United States during the First World War." In *Minorities in Wartime*, edited by P. Panayi. Oxford: Oxford University Press, 191–215.

Odo, Franklin. 2004. *No Sword to Bury: Japanese Americans in Hawaii*. Philadelphia: Temple University Press.

Okamura, Jonathan Y. 2014. *From Race to Ethnicity: Interpreting Japanese American Experiences in Hawaii*. Honolulu: University of Hawaii Press.

Okihiro, Gary Y. 1991. *Cane Fires: The Anti-Japanese Movement in Hawaii, 1865–1945*. Philadelphia: Temple University Press.

Robinson, Greg. 2009. *A Tragedy of Democracy: Japanese Confinement in North America*. New York: Columbia University Press.

Shaffer, Robert. 1999. "Opposition to Internment: Defending Japanese American Rights during World War II." *The Historian* 61 (3): 597–619.

Sniderman, Paul M., Louk Hagendoorn, and Markus Prior. 2004. "Predisposing Factors and Situational Triggers: Exclusionary Reactions to Immigrant Minorities." *American Political Science Review* 98 (1): 35–49.

Spinney, Robert G. 2000. *City of Big Shoulders: A History of Chicago*. DeKalb: Northern Illinois University Press.

Straus, Scott A. 2012. "Retreating from the Brink: Theorizing Mass Violence and the Dynamics of Restraint." *Perspectives on Politics* 10 (2): 343–362.

Suhrbur, Thomas J. 1983. "Ethnicity in the Formation of the Chicago Carpenters Union: 1955–1890." In *German Workers in Industrial Chicago, 1850–1910*, edited by Hartmut Keil and John Jentz. DeKalb: Northern Illinois University Press, 86–101.

Tamura, Eileen H. 1994. *Americanization, Acculturation, and Ethnic Identity: The Nisei Generation in Hawaii*. Urbana: University of Illinois Press.

Tischauser, Leslie Vincent. 1981. "The Burden of Ethnicity: The German Question in Chicago, 1914–1941." PhD diss., University of Illinois.

United States Commission on Wartime Relocation and Internment of Civilians. 1983. *Personal Justice Denied: Report of the Commission on Wartime Relocation and Internment of Civilians*. Washington, DC: The Commission.

United States, Executive Office of the President [Donald Trump]. 2017. Executive Order 12769: "Protecting the Nation from Foreign Terrorist Entry into the United States," March 6.

United States, Executive Office of the President [Franklin Roosevelt]. 1942. Executive Order 9066: "Authorizing the Secretary of War to Prescribe Military Areas February," February 19.

Valentino, Benjamin, Paul Huth, and Dylan Balch-Lindsay. 2004. "'Draining the Sea': Mass Killing and Guerrilla Warfare." *International Organization* 58 (2): 375–407.

Varshney, Ashutosh 2003. *Ethnic Conflict and Civic Life: Hindus and Muslims in India*. New Haven, CT: Yale University Press.

Ward, Robert D. 1960. "The Origins and Activities of the National Security League, 1914–1919." *The Mississippi Valley Historical Review* 47 (1): 51–65.

Yui, Daizaburo. 1992. "From Exclusion to Integration: Asian Americans' Experiences in World War II." *Hitotsubashi Journal of Social Studies* 24 (2): 55–67.

# 4

# The Geopolitics of "Fifth-Column" Framing in Xinjiang

*Kendrick Kuo and Harris Mylonas*[1]

Fifth columns are seldom characterized by a single frame or discourse. They are unstable political constructs that can and do change over time. This chapter focuses on the Chinese government's framing of its Uyghur minority—a Muslim Turkic-speaking group living in the northwestern province of Xinjiang that has been treated as a subversive element at least since the late 1950s. Government discourse about the Uyghurs, however, has changed over time, ranging from depictions as separatists, to objects of indoctrination by Islamic extremist groups operating in countries bordering western China or the Arab world. Why would the same conflict with a particular non-core group[2]—that is, any ethnic group perceived as unassimilated by the governing elites—be portrayed by a government as a self-determination struggle at one point in time and as religious in nature later on? What accounts for the framing of a government's fifth-column claims?

In this chapter, we focus on the relationships between geopolitical environment, discourse, and policy. Geopolitical shifts can have significant consequences for how governing elites frame non-core groups as fifth columns, which in turn can shape and justify policy changes. This is not surprising when we consider that conflicts in the Cold War were described in ideological terms, only to transform into conflicts about nationalism and self-determination in the 1990s (Kalyvas and Balcells 2010). We argue that interventions—covert or overt—by external powers (state or non-state actors) can have a discernible effect on a government's framing of the real or imagined fifth column. Interventions, in turn, are grounded on a particular geopolitical context. External powers can use non-core groups within other states in order to increase their leverage or simply to destabilize an adversary. In this effort, they often target non-core groups, especially those residing in border regions, that have not completely assimilated into the dominant constitutive story of the state of residence (Mylonas 2012). This process can at times be initiated by the non-core group itself in an attempt

Kendrick Kuo and Harris Mylonas, *The Geopolitics of "Fifth-Column" Framing in Xinjiang* In: *Enemies Within.*
Edited by: Harris Mylonas and Scott Radnitz, Oxford University Press. © Oxford University Press 2022.
DOI: 10.1093/oso/9780197627938.003.0005

to secure an external patron. Regardless of the initiator, the links between the non-core group and the potential external backer have a significant constitutive impact on perceptions among ruling political elites with respect to the goals and nature of the relevant non-core group. The state's reaction to this perceived threat, in turn, is informed by the geopolitical context and can reinforce the constitutive impact. As this process unfolds, it is hard for any of the parties involved to undo the prevailing fifth-column framing until there is either a change in the international system structure or an identity reconfiguration in the constitutive story of the alleged external backer or the state.

We explore the plausibility of this framework as applied to China's Xinjiang region, focusing on how geopolitics, perceived external linkages, and fifth-column framing changed in lockstep from 1949 to the present. The case study highlights the way international configurations influence the alternation between different prevailing frames of the conflict. To be clear, we do not argue that geopolitical threat perceptions determine the government's discourse or domestic policies toward alleged fifth columns. Geopolitical changes can and are used instrumentally by governing elites to justify a particularly resonant discourse or internationally unpopular policy solution. We also contrast the government's framing to alternative narratives of events offered by human rights organizations and the Uyghur diaspora community, but adjudicating an objective truth is not the purpose of our examination. Our goal is to explore the dynamic relationships between geopolitics, perceived external linkages, and fifth-column framing.

Since the late 1950s, Uyghurs and other Turkic minorities in Xinjiang have been treated as potential or actual fifth columns. An overwhelming majority of Uyghurs are Muslim and the ethnic and religious dimensions of Uyghur identity are intertwined.[3] External linkages evolved with the geopolitical environment—characterized at various times by Soviet revisionism, pan-Turkic nationalism, and transnational jihadism. Shifts in Chinese elite discourse about the Uyghurs corresponded with geopolitical changes. Uyghurs were described variously as potential Soviet sympathizers, ethnoreligious national separatists, and eventually religious extremists. Perhaps unsurprisingly, Chinese policies in Xinjiang have also changed as Chinese authorities tried to manage and ultimately prevent political violence, most recently culminating in a mass detention system of political and ideological "re-education."

We analyze the relationship between transformations in China's geopolitical environment and the evolution of fifth-column discourse in Xinjiang's history. To do so, we trace the way the PRC framed and re-framed Uyghur identity across three phases: an ideological framing under the Sino-Soviet split, an ethnoreligious nationalist one after the 1990 Baren rebellion, and a transnational jihadist framing since September 11, 2001.

## The Rise and Fall of Sino-Soviet Relations
## (1949–1990): Uyghur "Counterrevolutionaries"

Prior to the founding of the People's Republic of China (PRC), the Soviet Union supported bids for Uyghur independence in the first and second East Turkestan Republics, which lasted during the periods 1934–1944 and 1944–1949 (Garver 1988: 154–155; Forbes 1986: 140–141). Once the Chinese Communist Party (CCP) took power in 1949, their Soviet ally no longer supported Uyghur independence, instead offering economic and technical assistance in exchange for the privilege of keeping Soviet consulates in Xinjiang (Fedyshyn 1957; Kraus 2010). Sino-Soviet relations were favorable and Uyghurs experienced a period of government accommodation (Han and Mylonas 2014). China allowed the Soviet Union access to Xinjiang's natural resources and Moscow was deeply invested in industrializing the region. During this early Sino-Soviet honeymoon period, Uyghurs enjoyed a greater degree of religious freedom, ethnic autonomy, and economic benefits; some Uyghurs were even allowed to emigrate to the Soviet Union (Bovingdon 2004: 5–6; Yuan 1990). The Uyghurs were not seen as a fifth column despite being openly backed by the USSR. The alliance between the PRC and the USSR precluded that dynamic.

After Stalin died in 1953, Mao Zedong sought to renegotiate the Soviet presence in Xinjiang and complained that Xinjiang was a Soviet semi-colony (Kraus 2019: 508–509; 2010, 153–163). In hopes of maintaining good Sino-Soviet relations, Nikita Khrushchev agreed in 1955 to abandon Soviet ownership of Xinjiang's mineral and oil industries and reduce Sino-Soviet cooperation in the region. Meanwhile, the Soviet Union started encouraging its citizens in Xinjiang to return and build the Soviet economy, not to mention escape China's land reform and collectivization policies (Wu 2015: 334). Soviet nationals included Russians, Kazakhs, and Uyghurs, among others. The Soviet population in Xinjiang shrank from 91,000 in 1953 to about 30,000 or 40,000 in 1956, then down to 3,500 by 1962.

Nonetheless, Beijing remained suspicious of the Soviets and viewed local nationalism in Xinjiang as an outgrowth of Soviet influence. During the Cold War, the Soviet Union indeed provided support and sympathy to Uyghur discontent and separatism (Bovingdon 2011: 133–135). To be sure, the exodus of human capital, not to mention uprisings, had several domestic drivers as well beyond the breakdown in Sino-Soviet relations, including China's assimilationist policies, the Great Leap Forward, and inter-ethnic tensions spurred by Han migration into Xinjiang (Kraus 2019: 510–512).

From 1958 to 1962, combating local nationalism was part of a broader political movement to "de-Sovietize" Xinjiang through the "communization" of society to improve inter-ethnic relations and culling allegedly pro-Soviet cadres from the

Xinjiang Party (Wu 2015: 324–329). Further, beginning in 1958, the Great Leap Forward's ill-conceived economic policies impoverished the country at large and *minzu* minority groups in particular.[4] Han Chinese immigration picked up and the Xinjiang Production and Construction Corps began competing with local residents for arable land and water (Dillon 2004: 56). The CCP therefore lacked legitimacy in the eyes of non-Han groups that felt economically, culturally, and politically disenfranchised (J. M. Jacobs 2016: 191–193).

The 1960s witnessed a fracturing Sino-Soviet relationship and the transformation of the Soviet Union from an ally to an enemy. With the Soviet Union's change in status, China began implementing exclusionary policies against Uyghurs, fearing that Soviet support of Uyghur independence could turn them into an instrument of subversion (Mylonas 2012; Han and Mylonas 2014). The Great Leap Forward deepened Sino-Soviet tensions and antagonized inter-ethnic relations in Xinjiang, stoking positive memories of the Soviet Union among non-Han minorities. This came to a head in the Yi-Ta (Ili-Tacheng) Incident of 1962, when in early April hundreds of Uyghurs and Kazakhs from Ili prefecture started crossing into Soviet Kazakhstan, with numbers reaching up to 60,000 before the Chinese government closed the border. At the end of May, a protest in the prefectural seat of Ghulja over the border closure turned into a larger riot. The CCP-controlled Xinjiang Production and Construction Corps violently suppressed the protesters. China accused the USSR of trying to detach Xinjiang, resulting in a deepening of the Sino-Soviet dispute (Whiting 1987: 521). The CCP Central Committee described the Yi-Ta incident as "a thorough exposure that some foreign forces have been conducting long-term subversive activities in Xinjiang" (quoted in Kraus 2019: 515).[5] The following month, China closed all Soviet consulates in Xinjiang, and in August ordered all Soviet businesses and trade offices to leave (Kraus 2019: 516).

After the 1962 exodus, the CCP revived Maoist thinking that the nationalities problem was linked to class struggles among *minzu* minorities. A "glorious patriotic movement," launched by the CCP, called Uyghurs and other *minzu* minority groups to disaffiliate from the Soviet Union and its wayward policies (Dreyer 1976: 193). Meanwhile, the Xinjiang Production and Construction Corps established military farms along the Sino-Soviet border and resettled hundreds of thousands of Han Chinese from urban centers to Xinjiang (Kraus 2019: 516). After sealing the Sino-Soviet border, Beijing increased its propaganda denouncing the Soviet Union as a revisionist power attempting to subvert China. Anti-Han and anti-CCP sympathizers were not framed as "local nationalists" and "capital nationalists," but mainly as "counterrevolutionaries" and "revisionists" (Wu 2015: 335).

Among those who left during the Yi-Ta Incident, most were farmers, workers, and herdsmen, but there were also many senior political and military officers who later formed anti-Chinese resistance groups. Perhaps the largest resistance organization in Xinjiang during this era was the East Turkestan People's Revolutionary Party (ETPRP), which was founded in 1967 or 1968. Established in Urumqi and Kashgar, the group took advantage of the Cultural Revolution's turmoil to organize an entity modeled on the CCP and the Communist Party of the Soviet Union (Dillon 2004: 57). The ETPRP, which advocated for an independent East Turkestan, was perceived as a party established by Soviet spies.

During the Cultural Revolution period (1966–1976), Han migration into Xinjiang continued to grow and Chinese authorities suppressed cultural and religious practices of *minzu* minorities (McMillen 1979: 187–195; Dreyer 1986: 724–731; Millward 2007: 265–274). The excesses of the Cultural Revolution triggered a policy reevaluation in Beijing after Mao's death. A reorientation of China's Xinjiang strategy began during the brief chairmanship of Hua Guofeng (1976–1978), but truly emerged after Deng Xiaoping took control in 1978. Deng's Open Door policy was a "double opening" strategy that simultaneously sought greater integration of Xinjiang into the Chinese polity as well as stronger economic ties to Central Asia.

In the 1980s, as Sino-Soviet relations improved, China incorporated *minzu* autonomy rights as a part of "stability maintenance" and added an economic component through Jiang Zemin's Great Western Development, which was touted as a way to mitigate economic inequality in the region. Uyghurs were re-framed once again as one of the many *minzu* minorities of China, not as a "counterrevolutionary" or "revisionist" fifth column backed by a threatening enemy. Policies such as airing programs in minority languages and affirmation action were designed to transform local elites into government stakeholders and channel minority rights advocacy into approved processes (Rudelson and Jankowiak 2004: 308). To encourage engagement with the Muslim world, the Xinjiang provincial government relaxed policies toward Islamic practices (Christoffersen 1993: 136). The government also reformed the Xinjiang Islamic Association, which reintroduced the Arabic script for the Uyghur language. Mosques were established, theological schools opened, and Qur'ans and religious literature imported (Kuo 2012: 529).

The Chinese government encouraged Uyghurs and other Xinjiang Muslim *minzu* minorities to establish relationships with their religious compatriots across the border. Thousands of Chinese Muslims undertook the *hajj*. Cross-border linkages facilitated Uyghur participation in growing Islamic movements. China allowed Xinjiang Muslims to study abroad in *madrasas* and most went to Pakistan due to its proximity. China even permitted Uyghurs to join jihadist movements fighting the Soviets in Afghanistan, where they studied side by side with Afghan Taliban and their supporters in Pakistan and were exposed to an anti-Communist, pan-Islamic political curriculum that promoted an ideology of national liberation (Kerr and Swinton 2008: 128; Roberts 2004: 229–230; Kuo 2012: 529–530).

The relaxation of assimilationist government policies in Xinjiang did not mean that non-Han minorities enjoyed a golden era of freedom. Periodic outbreaks of violence and protests against ethnic discrimination occurred throughout the 1980s. In October 1981, Kashgar experienced riots related to a Han man accused of killing a Uyghur during a fight. Thousands of college and university students demonstrated in Urumqi in December 1985 against the election of Tomur Dawamat as Xinjiang's governor, who, despite being Uyghur, was a known protégé of the region's party secretary Wang Enmao.

These policy shifts created contradictory pressures in Xinjiang: stronger ties with Central Asia and the Soviet Union, improved Sino-Soviet relations, accommodation of *minzu* minorities, and economic development in Xinjiang, yet at the same time greater interaction with foreign Islamic movements and space for Uyghurs to aspire for greater autonomy or even independence. The collapse of the USSR then created a critical juncture. The new geopolitical landscape that emerged in the 1990s changed the frames applied to the Uyghurs, who again emerged in government discourse as an alleged fifth column, albeit this time an ethnoreligious nationalist one.

## The End of the Cold War: Uyghurs as Ethnoreligious National Separatists (1990–2001)

When the Soviet Union collapsed, the Chinese government became concerned once again about cross-border insecurity in Xinjiang. With the end of the Cold War, it feared that self-determination movements by the Kazakhs, Uzbeks, and Kyrgyz could diffuse among its Uyghur population, which shared linguistic and cultural affinities with these Central Asian peoples (Macakerras 2001; J. Smith 2000). The independence of their various Central Asian neighbors raised the prospect that secession from China was possible. Newly formed Central Asian states emerging from the breakup of the Soviet Union could play an important role in Xinjiang's development, modeling independent governance, economic management, and providing a safe haven for Uyghur political entities. More directly, the Uyghur diaspora in Kazakhstan and Kyrgyzstan advocated rebellion under the banner of pan-Turkism, along with sporadic pan-Islamist appeals (Clarke 2017: 12). Beijing was also concerned with the Islamic revival in Central Asia. During the 1990s, the Taliban fought and won the civil war in Afghanistan, turning the country into a safe haven for a range of radical Islamist groups in the region. As the

Central Asian states failed to meet Uyghurs' expectations, hope diminished (Beller-Hann et al. 2007: 3–4). Then came a decade of increased violence, of bombings and assassinations targeting Uyghur government officials and co-operative religious leaders deemed traitors by Uyghur separatists (Millward 2007: 330). China blamed the outbursts of violence in the 1990s on "hostile external influences" that stoked ethnic minority opposition and a revival of political Islam (Macakerras 2001).

The Baren rebellion in April 1990 marked a turning point in government policies in Xinjiang. In protest of Han migration and new family planning regulations, about two hundred Uyghurs clashed with security forces at Baren, south of Kashgar, leaving thirty Uyghurs and fifty Kyrgyz dead (Millward 2004: 14–15, 2007: 325–327). The organizers used religious rhetoric calling for a jihad against Han "infidels" and an independent East Turkestan. The authorities rapidly quashed the rebellion and claimed that the leader, Zahideen Yusuf, led a group called the Islamic Party of East Turkestan and had connections to mujahideen groups in Afghanistan (Clarke 2007b: 50–54). China's 1980s accommodationist policies had brought about in Xinjiang what one scholar termed the "Taliban syndrome" (Ahrari 2000: 659). Meanwhile, Western commentators and international organizations described the incident as a peaceful demonstration against religious restrictions that turned violent (Rodríguez-Merino 2019: 33–34). Beyond the Baren rebellion, Uyghurs participated in a number of bus bombings in 1992 and 1993, and bombings and assassinations from 1996 to the Yining incident of February 1997 (see Millward 2004: viii–xi).

After the Baren rebellion, China initiated several changes to its domestic policies in Xinjiang. The government simultaneously increased its presence in the region, while also asserting control of religious and cultural practices. In an attempt to weaken ethnoreligious nationalist attitudes, China implemented several policies to better integrate Xinjiang into the Chinese polity, including settler colonization through Han migration to the region, exploiting the oil resources of the Tarim Basin, and building transportation infrastructure, all of which culminated in 1999 with the announcement of the campaign to Open Up the West (Becquelin 2000, 2004; Clarke 2007a). China's control of Xinjiang was represented through the Han paramilitary Xinjiang Construction and Production Corps, which controlled 48 percent of the province's land (Becquelin 2000: 68, 78). China also emphasized political, economic, and infrastructural linkages between Xinjiang and Central Asian states as key contributors to the region's development (Clarke 2003).

The CCP reversed the religious freedoms enjoyed in the 1980s, while operationalizing the new fifth-column framing of the Uyghurs. The CCP Central

Committee Document No. 7, published in 1996, described "national separatism and illegal religious activity" as the ultimate source of instability in
Xinjiang. Chinese authorities mounted a series of massive crackdowns called
"strike hard" (*yanda*) campaigns and tightened restrictions on religious
practices. These campaigns, from 1996 to 1997 and a 100-day crackdown from
January to March 1999, were officially aimed against Islamists and separatists,
but the thousands of arrests and human rights violations lent credence to a
widespread belief among Uyghurs that these campaigns were aimed at the
Uyghur identity itself. Arabs were forbidden from teaching at the Xinjiang
Theological Seminary and Uyghurs returning from Pakistan were tightly monitored. Uyghurs faced all types of religious persecution, including prohibitions
on children attending mosque and fasting during Ramadan (Human Rights
Watch 2005: 25–57).

China also embarked on a diplomatic campaign to coopt Central Asian
states, as well as Pakistan and Turkey. The Shanghai Five—China, Russia,
Kazakhstan, Kyrgyzstan, and Tajikistan—supported China's position in
Xinjiang. China pressured Central Asian members to suppress Uyghur separatist movements. Kazakhstan and Kyrgyzstan complied and even extra-
dited Uyghur dissidents back to China (Chung 2004: 999). In the 1990s, even
Turkey, which had a long history of being a safe haven for Uyghur nationalists
in exile ever since the CCP took power in 1949, only occasionally lent verbal
support to self-determination for East Turkestan, but often preferred to pla-
cate China (J. M. Jacobs 2016: 208–210, 222–223).[6] By 1999, Ankara started to
distance itself from activities promoting East Turkestan independence (Clarke
2017: 9).

These government initiatives did not completely suppress dissent. In 1996
and 1997, there were a series of attacks, including bombings in Urumqi on the
day of ceremonial mourning for Deng Xiaoping's death (Tschantret 2018: 576–
577). In February 1997, Uyghurs marched in protest of government policies in
Yining/Ghulja, carrying banners emblazoned with common Islamic mantras.
The crowds called for Uyghurs to "use the Qur'an as a weapon" and other religious slogans while they burned identification cards and residency permits
and stripped themselves of their "Han" clothes, disavowing all things Han
(Bovingdon 2011: 127). The police presence turned violent and a riot resulted,
leaving many dead.

The framing of Uyghur identity as a fifth-column threat persisted into
the 2000s, but as the geopolitical environment shifted, so did the depiction of
this threat. They were portrayed in the 1960s and 1970s as counterrevolutionaries, then re-framed as ethnoreligious nationalists in the 1990s. The attacks
of September 11, 2001 offered an opportunity to again reinterpret Uyghurs in
Xinjiang as a transnational jihadist threat.

## The Post-9/11 World: Uyghurs as a Transnational Jihadist Fifth Column (2001–Today)

Over the past two decades, the PRC has portrayed political violence in Xinjiang as the result of subversive infiltration by transnational jihadist networks (Kanat 2012). Before 9/11, China had already linked terrorism, religious extremism (*zongjiao jiduan zhuyi*), and *minzu* separatism (*minzu fenli zhuyi*) in a series of statements including the Sino-Russian joint anti-terrorism statement (1999), the Shanghai Five's Dushanbe Statement (2000), and the Shanghai Convention on Combating Terrorism, Separatism, and Extremism (2001).[7] Beginning with the Baren rebellion in 1990, the Chinese government described violent incidents in Xinjiang as premeditated attacks by separatist or religious-extremist groups inspired or directed by foreign forces (Rodríguez-Merino 2019: 37).

But what emerged after 9/11 was a new PRC fifth-column discourse centered on foreign radical Islamists recruiting Uyghurs in Xinjiang to overthrow Chinese rule. The Chinese government started emphasizing the term "terrorism" to describe all incidents of violence in Xinjiang and retroactively cast prior incidents as terrorism. Thus, in 2002, Chinese authorities re-framed the violence in Baren township in 1990 and Ghulja in 1997 as the work of "terrorist organizations" (Rodríguez-Merino 2019: 27, 31).

In the rest of this section, we present three phases of so called terrorism and counter-terrorism activity in Xinjiang. The first phase began after 9/11 and involved new policies targeting the Uyghur population in Xinjiang, justified by connecting the East Turkestan Islamic Movement to Al-Qaeda. The 2009 Urumqi riots triggered a second phase that gradually militarized law enforcement efforts in Xinjiang and selective targeting of alleged fifth-column elements. A third phase, beginning in March 2014 with an attack in Kunming, is also associated with the rise of the so-called Islamic State and the Syrian civil war. China's Xinjiang strategy in recent years has morphed into one of mass detention and a "re-education" campaign.

### The "Global War on Terror" Comes to Xinjiang (2001–2009)

The US "global war on terror" linked a state's international reputation to its counterterrorism effort. A state with "terrorists" residing within its sovereign territory could either be a "victim of terrorism" or a "sponsor of terrorism," depending on whether the state was actively seeking to deny suspected terrorists a safe haven. Counterterrorism therefore emerged as a mark of legitimate governance, at least from a United States perspective (Foot 2005; Sasikumar 2010). China responded to this new situation by framing Uyghur separatism as an international terrorism problem, securing "anti-terrorism" cooperation from various partner countries, and actively improving its counterterrorism capacity both at home and abroad (Clarke 2010).

On September 11, President Jiang Zemin offered his condolences to President George W. Bush and in a phone call the following day promised to work with the United States on counterterrorism efforts. China voted in favor of key UN resolutions to set up an international counterterrorism regime. These signals of support continued with a delegation of Chinese counterterrorism experts visiting Washington, DC, on September 25. On December 29, 2001, the National People's Congress adopted the "Anti-Terror Amendment" to the Criminal Code, which intensified the crackdown on terrorist organizations, criminalized terrorism financing and handling of dangerous substances, and suppressed money laundering.

From 2001 onward, China supported counterterrorism cooperation at the United Nations, participating in regional anti-terrorism mechanisms, and established a number of bilateral cooperation mechanisms.[8] In 2004, for example, the Shanghai Cooperation Organization established the Regional Anti-Terrorism Structure to share information and mobilize collective action against terrorist threats. Meanwhile, Beijing mounted a case that foreign-backed jihadists operated in Xinjiang. In 2002, Beijing retroactively accused the East Turkestan Islamic Movement (ETIM) as "terrorists" responsible for earlier attacks in Xinjiang, and claimed that ETIM had links with al-Qaeda:

> Over a long period of time—especially since the 1990s—the "East Turkistan" forces inside and outside Chinese territory have planned and organized a series of violent incidents in the Xinjiang Uygur Autonomous Region of China and some other countries, including explosions, assassinations, arsons, poisonings, and assaults, with the objective of founding a so-called state of "East Turkistan." These terrorist incidents have seriously jeopardized the lives and property of people of all ethnic groups as well as social stability in China, and even threatened the security and stability of related countries and regions.[9]

In a 2003 white paper, the State Council of the PRC linked religious extremism with separatism in Xinjiang:

> In the early twentieth century, and later, a small number of separatists and religious extremists in Xinjiang, influenced by the international trend of religious extremism and national chauvinism, politicized the unstandardized geographical term "East Turkistan," and fabricated an "ideological and theoretical system" on the so-called "independence of East Turkistan." . . . They incited all ethnic groups speaking Turki [sic] and believing in Islam to join hands to create

a theocratic state. They denied the history of the great motherland jointly built by all ethnic groups of China. (China State Council Information Office 2003)

The PRC's Ministry of Public Security thereafter began releasing lists of terrorist organizations. The first one was issued in December 2003, and listed the ETIM, East Turkestan Liberation Organization, World Uyghur Youth Congress, and East Turkestan Information Center, along with eleven individuals. Beijing claimed these organizations, trying to establish an independent, theocratic "East Turkestan," were associated with al-Qaeda networks and responsible for many recorded terrorist attacks. Subsequent lists prominently featured individuals linked to ETIM.

Almost immediately, China's attempts to graft its "Xinjiang problem" onto the "global war on terror" came under international scrutiny. Although Washington encouraged closer strategic cooperation between China and the United States—the United States officially recognized, for example, ETIM and the Turkestan Islamic Party as terrorist organizations—it was also vocal about China's human rights abuses (Foot 2004). China labeled Uyghur separatists as terrorists (China State Council Information Office 2002; Chung 2002: 8; Menon 2003: 187, 192, 198–199), but Washington refused to recognize the Uyghur independence movement as a terrorism issue in its entirety, preferring to view the situation in Xinjiang through the lens of ethnoreligious nationalism and human rights (Malik 2002: 27). Uyghur advocacy groups also contested China's transnational jihadist framing, highlighting the violation of human rights and religious freedom as the root cause of violence in Xinjiang.[10]

For almost two decades now, Beijing has consistently promoted a fifth-column discourse about transnational jihadism in Xinjiang. Further, while the existence of Uyghur jihadists is not disputed, their actual subversive capabilities in China remain unclear. Criticism focuses on China's claim that growing evidence of Uyghur participation justifies human rights violations in the name of counterterrorism. Critics describe China's counterterrorism strategy as a form of state terror that triggers a repression-violence-repression cycle, deepening alienation and hostility among the Uyghur population.[11] In a similar vein, Western academics criticize China for using the discourse of the "global war on terror" to transform a separatist threat into a transnational terrorist threat in order to discredit domestic opposition, legitimize state violence against Uyghur dissent, and impose increasingly draconian policies in Xinjiang under the political cover of "counterterrorism" (e.g., Dillon 2004: 162; Clarke 2018: 3; Rodríguez-Merino 2019).[12] Trust between and within the Han and Uyghur communities in Xinjiang steadily disappeared under the shadow of mass surveillance and violence (Caprioni 2011; Cliff 2016; H. Yee 2005).

## Selective Targeting: Stability and Development (2009–2014)

What became known as the 2009 Urumqi riots intensified existing trends. In July 2009, demonstrators in Urumqi demanded the Chinese government fully investigate the murder of two Uyghurs in a Guangdong factory brawl, which stemmed from the rumored rape of two Han women by six Uyghur men several days earlier. The demonstration turned violent as Uyghurs, Han, and police clashed in Xinjiang's capital city. Chinese authorities later claimed that Uyghur exile leader Rebiya Kadeer and the World Uyghur Congress, which Beijing considers a terror-sponsoring organization, orchestrated the violence (Xinhua 2009). They described the events in July 2009 through "representational tropes of the criminal, terrorist, and outside agitator" (Barbour and Jones 2013: 96). The "outside agitator" label bestowed upon the Chinese government the role of caretaker of all people within Chinese territory irrespective of ethnic affiliation, protecting them against foreign interests undermining internal harmony (Barbour and Jones 2013).

In the aftermath of the Urumqi riots, Al-Qaeda in the Islamic Maghreb called for reprisals and announced that it would target the thousands of Chinese workers and projects in northwest Africa (ANI 2009). Three months later, an Al-Qaeda militant group called on Uyghurs to "make serious preparations" for holy war against the Chinese government (quoted in People's Daily 2009). The Turkestan Islamic Party (TIP), a Uyghur jihadist group with unclear connections with ETIM, also appealed to Muslims around the world to attack Chinese interests.[13]

For CCP leaders, the rioting in Urumqi in July 2009 was the result of lax security in Xinjiang. China immediately began increasing its coercive capacity in Xinjiang and police surveillance grew significantly (Roberts 2018a; Smith Finley 2019). At the same time, a new party secretary, Zhang Chunxian, introduced a "flexible" strategy of "stability maintenance" and economic development that both increased the government's coercive capacity through a growing police presence while also facilitating a transfer of economic wealth to the region (Cliff 2012; Kuo 2014a). Zhang's appointment coincided with a renewed emphasis on Xinjiang's economic development as the solution to the region's instability.

Meanwhile, the Urumqi riots also triggered a debate among CCP leaders about the efficacy of China's *minzu* minority policy. China's policy at the time still centered on ethnic differentiation and regional autonomy, following the Soviet approach toward different nationalities. The CCP's traditional approach to *minzu* policy nominally accommodated *minzu* minorities and supported varying degrees of autonomy. After the ethnic riots in July 2009, not to mention similar violence in Tibet the year prior, CCP elites began to wonder whether it was time for a second generation of *minzu* minority policies based on a "melting-pot" model to create a unified "state-race" (Leibold 2012).

From 2009 to 2014, Xinjiang continued to experience sporadic attacks and violent clashes between police and protesters. In state media, the Chinese government consistently described the violence in terms of "terrorism" and drew a connection to foreign-backed religious extremism. At the same time, the Chinese government appeared to emphasize a religious motivation behind the violence while downplaying the role of ethnic identity or demands for national self-determination. Beijing singled out the Islamic identity of Uyghurs while trying to maintain "nationality solidarity" (*minzu tuanjie*), propping up the Chinese identity of the Uyghurs (Tobin 2013). The official line was that the terrorists, inspired by religious extremism, were the enemy of Han and Uyghur alike.

Several incidents exemplified this official line (see Dasgupta 2011; China Daily 2011; China Daily 2012). In 2013, Xinjiang experienced five major incidents of violence: three attacks in or around Kashgar (April 23, October 10, December 11), one in Turpan (June 26), and one in Hotan (June 28). Two of these incidents are representative of China's fifth-column framing strategy during this period. On April 23, three community workers in Kashgar called the police to report a suspicious group of men. These men took the workers hostage, attacked the police and officials responding to the call, then set the house on fire. World Uyghur Congress spokesman Dilxat Raxit said the violence resulted from the shooting and killing of a young Uyghur during illegal home searches (Branigan 2013). Local government officials were instead quick to label the violence jihadist in nature (Chan 2013). Hou Hanmin said that the assailants "had been training in their own house for several months. They were affected by extremism and hoped to commit themselves to jihad" (quoted in CNN 2013). Two of the alleged terrorist leaders were sentenced to death after a court investigation revealed that the group watched and listened to religious extremist materials from abroad and promoted religious extremism since March 2008 (Cui 2013). Probably to avert greater polarization, the government delayed revealing the Uyghur identity of the assailants, while immediately announcing that the ten police and officials killed were Uyghur, Han Chinese, and Mongolian.

On June 26, in Turpan's Lukqun township, a group of people wielding knives attacked a police station, government offices, and a construction site, killing nine security personnel and eight civilians before police gunfire killed ten of the assailants. Again, the government insinuated early on that this was a terror attack but delayed release of information on the Uyghur identity of the attackers. The state-run newspaper *China Daily* ran an opinion piece linking the incident with the April 23 attacks, as well as the July 2009 riots, reinforcing the narrative in which terrorists plan "to sow the seeds of hatred and fear among local residents to facilitate their own selfish goals. . . . Local people of different *minzu* minorities should continue to take a firm and united stand to crush the sinful brutality of those who seek their own gains at the cost of harmonious coexistence" (China Daily 2013). Chinese authorities attributed these two clashes, along with the other three major incidents in 2013, to "extremists and terrorists" waging "jihad" in collaboration with "hostile external forces." They also claimed that up to 100 Uyghurs had traveled to Syria for training in terrorism (Economist 2013).

There were other violent incidents extending into 2014 not discussed here, that shared these common traits. Our cursory review of political violence in Xinjiang from 2009 to 2014 reveals a pattern. First, the Chinese authorities attributed the violence to influences by international religious extremism. Second, the authorities withheld the ethnic identity of the assailants, highlighting the fact that the victims included Uyghurs, and described the perpetrators as terrorists.

## Indiscriminate Targeting: "De-extremification and Re-education" (2014–2021)

Beginning in 2014 and intensifying in 2017, the PRC developed a sophisticated security state characterized by mass internment, political and ideological "re-education," and strict limitations on the freedoms of Uyghur citizens in Xinjiang (see Greitens, Lee, and Yazici 2019/2020: 14–21). Some scholars argue that preventive repression reflects a genuine concern by Chinese authorities about the terrorist threat, while others contend that it is a thinly veiled attempt to quash all political opposition in Xinjiang to create a platform for the Belt and Road Initiative.[14] Regardless of the motivation, the fifth-column frame of transnational jihadist networks remained the same despite the change in strategy.

On March 1, 2014, in what became known as "China's 9/11," a group of Uyghurs from Xinjiang attacked the Kunming train station using foot-long knives, killing 29 and injuring 143. *Radio Free Asia* suggested that the Uyghurs were seeking political asylum before resorting to violence in desperation. In contrast, Qin Guangrong, the Yunnan Province party secretary, claimed the eight assailants tried to escape to Vietnam in order to wage jihad overseas, but failed to gain passage and so decided to stage an attack in China (A. Jacobs 2014).

Chinese authorities described the attacks as domestic manifestations of global Islamic terror. In April 2014, Xi Jinping declared a nationwide anti-terror campaign. The PRC began a religious "de-extremification" (*qu jiduanhua*) campaign that essentially equated religious piety with "extremist" behavior. Everyday Islamic practices such as fasting during Ramadan, long beards, abstaining from alcohol, owning a Qur'an, and attitudes unsupportive of the CCP were conflated with separatism and/or a predilection toward terrorism (Byler 2018; Famularo 2018; Brophy 2019).

State-sponsored surveillance and political "re-education" in Xinjiang were paired with discourse about "thought liberation" (Smith Finley n.d.; Byler 2018). Chinese authorities targeted those whose behavior and beliefs deviated from Han linguistic and cultural norms with political re-education. Conversely, they rewarded those who spoke Mandarin Chinese, entered into interethnic marriages, removed their veils, and embraced other "civilized" trappings (Leibold 2020: 56–57).[15] Chinese authorities also surveilled the population through camera-based systems in urban centers; the Internet and telecommunications; physical face-to-face supervision and monitoring by hundreds of thousands of CCP cadres; and informal policing (Leibold 2020: 59–56; Zenz and Leibold 2020). Finally, counterterrorism also took the form of a government-community partnership whereby the local population, including Uyghurs, were encouraged to share information about suspicious individuals who might be "terrorists." Village-based work teams visited households to identify deviant individuals for "educational transformation." This "mass line" strategy or "open-door counterterrorism" was designed to add "soft" methods to "hard tactics" (Xie and Liu 2019: 6–7, 10).

As time went by, pressure mounted to adopt a new approach to the "Xinjiang problem," which eventually culminated in the creation of a mass internment system beginning in 2017. These concentration camps sit at the confluence of five streams of political change that have been coursing through Xinjiang since at least 2014. First, there was a shift in official Chinese discourse that re-framed violence in Xinjiang away from blaming subversive elements in the population and instead described large swaths of the Uyghur population as infected or contaminated by extremist ideology. Following the Kunming attack in March 2014, Zhang Chunxian declared a "people's war on terror" in which Xinjiang would "promote the eradication of extremism, further expose and criticize the 'reactionary nature' of the 'three forces,' enhance schools' capacity to resist ideological infiltration by religious extremism, and resolutely win the ideological battle against separation and infiltration" (China Daily 2014).

While in 2014 local governments targeted about 1 percent of the Uyghur population in their respective jurisdictions, by 2016 there were reports that 20–30 percent of individuals in southern Xinjiang were "infected" with extremist thought and behavior (Leibold 2019: 4–5). By 2016, the discourse around extremism began to adopt medical analogies, describing it as a mental illness, an addiction, and akin to "poisonous medicine" (Samuel 2018; Grose 2019; Zenz 2019b: 121–122). Religious thought and behavior were analogized to a "malignant tumor" and a "communicable plague" that required quarantine and long-term "hospitalization" (Leibold 2019: 5; Raza 2019: 497).

Second, there was mounting evidence of Uyghur participation in transnational jihadist networks operating in Southeast Asia and the Middle East.

Beginning in 2014, reports surfaced that Uyghur militants were linking up with jihadist groups in the Philippines and Indonesia. Then, in the following two years, Uyghurs were involved in violent incidents in Indonesia and Thailand.[16] China put pressure on Southeast Asian countries such as Cambodia, Vietnam, and Thailand to extradite Uyghurs found within their borders on the charge that they are seeking to join jihadist groups.

Meanwhile, the concentration of Uyghur militant groups in Afghanistan and Pakistan shifted toward the Middle East as the Syrian civil war attracted similar groups. After Abu Bakr al-Baghdadi declared a caliphate in 2014, the Islamic State released its magazine *Dabiq*. The first issue contained a speech by al-Baghdadi in which he named China as an enemy. Later issues of *Dabiq* and its successor magazine *Rumiyah* had Uyghur-language translations. The Islamic State also released a video in 2017 in which a Uyghur pledged allegiance to the caliphate and condemned China. As the Islamic State appealed to Muslim Uyghurs, Al-Qaeda followed suit, condemning China in its magazine *Resurgence* and strengthening its ties to TIP (Botobekov 2016, 2017; Lin 2014). The extent of Uyghur migration to these conflict zones is unclear, but the number of Uyghurs collaborating with Al-Qaeda affiliates and the Islamic State has grown significantly since 2014.[17] In 2017, President Xi Jinping called for a "great wall of iron" to prevent Uyghur fighters in Syria from returning to Xinjiang and continuing their jihad there (Ruwitch, Martina, and Shepherd 2017).

Third, the debate among CCP elites between accommodationists and assimilationists ended at the Central Ethnic Work Conference in September 2014. There, Xi Jinping laid out the goal of a collective Chinese national consciousness that encouraged interethnic mingling and remolding minority cultures (Leibold 2019: 2–3). This shift in China's *minzu* minority policy arguably started in 2009, when assimilationists began to use the Urumqi riots as evidence that ethnic accommodation had failed (Leibold 2019: 2; Tobin 2020).

Fourth, Beijing gradually established a legal apparatus to support preventive counterterrorism measures. Before 2014, counterterrorism work occurred under provisions in the criminal law (see Clarke 2010: 546–550), but after the Kunming attack Xinjiang officials announced that they were considering regional counterterrorism laws. While criminal law's punishments are designed as deterrents, regional counterterrorism laws would authorize preventive measures (Kuo 2014b). China passed its first comprehensive anti-terrorism bill at the end of 2015,[18] adopting expansive definitions of "terrorism" and "extremism" and giving police the power to impose severe restrictions on suspected terrorists with little or no evidence. The law centralized China's anti-terrorism policies, but also recognized the "people's war" as a key principle and instructs authorities to mobilize grassroots organizations and civilians to support the prevention of terrorist activities. Xinjiang was the only region to enact the Implementing

Measures on the Counterterrorism Law, in effect since August 1, 2016, which included regulations on religious affairs and the prevention of juvenile crimes (Human Rights Watch 2017). In early 2017, Xinjiang adopted official regulations on de-extremification (People's Daily 2017). Then, in the summer, the Chinese State Council revised the Regulations on Religious Affairs, which expanded the Chinese state's authority to control, monitor, and punish religious practices with the goal to curb "infiltration and extremism" (Amnesty International 2018, 128).

Fifth and finally, in late August 2016, Chen Quanguo replaced Zhang Chunxian as the Xinjiang party secretary. Chen prioritized "stability maintenance" over economic development, suggesting that social stability was a prerequisite for all other Xinjiang work. Zhang's more targeted approach gave way to "concentrated transformation through education" in which China's Xinjiang strategy became an aggressive social engineering project to replace Turkic Islamic practices and modes of thought with "Chinese culture" (Byler 2019).

Authorities used a variety of indicators and information shortcuts to rank the "trustworthiness" of citizens, ranging from ethnicity, employment status, religious activity, and travel abroad, among others (Smith Finley 2019: 4–6; Leibold 2020: 56–57). Chen also established a mosque trusteeship system where state-trained imams (so-called red imams) replaced local ones; dispersed mosques were merged and placed under state-appointed clergy; and "excessive" mosques were demolished (Xie and Liu 2019: 13).

Beginning in February 2017, China began to establish "re-education" camps in southern Xinjiang. Then, in March and April, the Xinjiang Department of Justice swept staggering numbers of Muslim Uyghurs, Kazakhs, and Kyrgyz into a coercive internment camp system for "re-education" (Byler 2018; Zenz 2019b: 115–121). In 2018, the Xinjiang government revised its anti-extremism regulation to retroactively authorize these camps.

As the concentration camps garnered international attention, the Chinese government finally acknowledged their existence on August 13, 2018, but insisted that they were "vocational education and employment training centers" for the "rehabilitation and reintegration" of radicalized individuals (quoted in Leibold 2019: 7). But it has been well documented that internees are forced to study Mandarin Chinese, experience pressure to renounce Islam, conduct self-criticism, demonstrate patriotism, and in the process are subject to physical and psychological abuse (Smith Finley 2019: 6–8; Raza 2019: 492–495). By 2019, Adrian Zenz (2019a) estimated about 1.5 million detainees in such camps. Documents have also revealed a campaign in rural Uyghur regions to sterilize women with three or more children, which would translate to about 20 percent of all women of childbearing age (Zenz 2020). Justified by a fifth-column discourse of transnational jihadist threats, the PRC has over time become increasingly indiscriminate in its targeting and aggressive in its goals in Xinjiang.

# Conclusion

This chapter has identified a pattern in which covert or overt support by external powers (state or non-state actors) correlates with changes in government rhetoric and policy toward a real or imagined subversive fifth-column group. The variation in the government's fifth-column framing is inseparable from the international geopolitical context, as well as the actions of external actors that increase the salience of certain markers over others. How consciously these actors are constructing fifth-column frames is an empirical question, but external actors may have more agency than both the state of residence and the non-core group(s) in conflict.

External actors cannot dictate fifth-column frames, but when they do get involved, there is a high likelihood that their involvement will influence the resultant frame. Our argument is a sobering story about the limited agency individuals or groups have in shaping the way they are perceived and framed by their respective governments. At the same time, it also suggests that governments adapt their frames to changes in the geopolitical environment. Furthermore, elites of non-core groups have to respond to the incentives of external donors, sponsors, and patrons in their attempt to fund their struggles. The dynamic we are describing suggests that there is a joint production of fifth-column framings, which involves external actors' agendas, the policies emanating from the state of residence, and the non-core group's attempts to gain support. In this framework, it may not be as important whether the external involvement is actually taking place or merely perceived as such. But geopolitical transformations are crucial ingredients in the framing and re-framing process.

# Notes

1. We thank Clifford Bob, Audrey Davister, Madeleine Wells Goldburt, Malka Lasky, Karina Ochoa Berkley, Anum Pasha, Bairavi Sundaram, Scott Radnitz, Nadav Shelef, and Annelle Sheline for their helpful comments. This research has been supported in part by the U.S. Department of Defense, Minerva grant on 'International Order and Spheres of Influence' (grant number N00014-16-1-2334). The views expressed are those of the authors and do not necessarily represent the U.S. Department of Defense or its components.
2. For more on the term non-core group see Mylonas (2012: 26–29).
3. On Han-Uyghur relations from an interethnic boundaries approach, see Han 2010.
4. We use the term "minzu minorities" (also referred to as "minority nationalities") instead of the term "ethnic groups" to refer to China's recognized groups in order to distinguish them from the ethnic groups without official status within the PRC.
5. Charles Kraus (2019: 509–512) draws attention to the fact that in internal deliberations the CCP also recognized domestic causes of the Yi-Ta Incident, but chose to blame the Soviet Union.

6. During the Cold War, Beijing painted Uyghur nationalists as backed not only by "Soviet revisionism," but also by "reactionary" Turkey, a member of NATO (Shicor 2009: 17–19).

7. On separatism and terrorism in Xinjiang, see Kuo 2012; Clarke 2008; Evron 2007; Pokalova 2013.

8. For example, China supported the passing of Security Council Resolutions 1368 and 1373. On China's domestic anti-terror legislation, see Zhou 2014, 128–177. On China's international counterterrorism cooperation, see Raman 2004; Khan 2012/2013; Kan 2010; McNeal 2001; P. J. Smith 2009; Pollack 2003.

9. China State Council Information Office 2002. For analysis of ETIM's existence and its capabilities, see Reed and Rschke 2010.

10. By this time, China's diplomatic efforts in Central Asia in the 1990s had compelled Uyghur diasporic activism to relocate to the United States and Western Europe. These advocacy groups shifted toward nonviolent tactics and liberal political ideologies more aligned with the governments of their new states of residence. See Bovingdon 2011, 146–151; Shicor 2007, 117–125; Clarke 2017, 12–14.

11. On state terror, see Smith Finley 2019. On a repression-violence-repression cycle, see Roberts 2012, 2013, 2018b, 2018a: 243–245. On deepening alienation, see Smith Finley 2013: 235–293; Tohti 2015; Harris and Isa 2019.

12. On motivations behind China's counterterrorism regime, see Purbrick 2017: 249–252; Odgaard and Nielsen 2014; Anand 2019. On the effectiveness of China's counterterrorism strategy, see Roberts 2018b; Tschantret 2018; Trédaniel and Lee 2017.

13. There are reasons to believe that TIP is operationally weak (see Clarke 2017: 16).

14. Sheena Greitens, Myunghee Lee, and Emir Yazici (2019/2020: 17–21) recently argued that China's Xinjiang strategy reflects a genuine fear of transnational jihadist networks, even though this perception may be incorrect and its strategy ineffective. Their work has received criticism, for example, Robertson 2020. On Xinjiang's role in the Belt and Road Initiative, see Klimeš 2018; Clarke 2018.

15. As de-extremification took hold in Xinjiang, political reeducation shifted from discriminate to indiscriminate targeting (Zenz 2019b: 105–114), and a mix of crackdown and control (*daya guankong*) and prevention led to a government ban on religious activities such as Halal registration and Islamic veiling (Leibold and Grose 2016; Zenz and Leibold 2017).

16. For helpful overviews, see S. K. L. Yee 2018; Abuza 2017. See also Kuo and Springer 2014.

17. For helpful overviews, see Clarke 2017: 16–18; Pantucci 2018; Greitens, Lee, and Yazici 2019/2020, 32–36.

18. For overviews, see Bissell 2015; Zhou 2016.

# References

Abuza, Zachary. 2017. "The Uighurs and China's Regional Counter-Terrorism Efforts." *China Brief* 15 (16). https://jamestown.org/program/the-uighurs-and-chinas-regional counter-terrorism-efforts/.

Ahrari, M. Ehsan. 2000. "China, Pakistan, and the 'Taliban Syndrome.'" *Asian Survey* 40 (4): 658–671.

Amnesty International. 2018. *Amnesty International Report 2017/18: The State of the World's Human Rights*. London: Amnesty International.

Anand, Dibyesh. 2019. "Colonization with Chinese Characteristics: Politics of (In)security in Xinjiang and Tibet." *Central Asian Survey* 38 (1): 129–147.

ANI (Asian News International). 2009. "Al-Qaida Threatens China with Jihad in Retaliation for Urumqi Riots." *Indian Express*, October 8. https://indianexpress.com/article/news-archive/print/alqaeda-threatens-china-with-jihad-in-retaliation-for-urumqi-riots/.

Barbour, Brandon, and Reece Jones. 2013. "Criminals, Terrorists, and Outside Agitators: Representational Tropes of the 'Other' in the 5 July Xinjiang, China Riots." *Geopolitics* 18 (1): 95–114.

Becquelin, Nicolas. 2000. "Xinjiang in the Nineties." *China Journal* 44: 65–90.

Becquelin, Nicolas. 2004. "Staged Development in Xinjiang." *China Quarterly* 178: 358–378.

Beller-Hann, Ildiko, M. Cristina Cesaro, Rachel Harris, and Joanne Smith Finley. 2007. "Introduction." In *Situating the Uyghurs between China and Central Asia*, edited by Ildiko Beller-Hann, M. Cristina Cesaro, Rachel Harris, and Joanne Smith Finley, 1–14. Aldershot: Ashgate Press.

Bissell, Benjamin. 2015. "What China's Anti-Terrorism Legislation Actually Says." *Lawfare*, December 30. https://www.lawfareblog.com/what-chinas-anti-terrorism-legislation-actually-says.

Botobekov, Uran. 2016. "Al-Qaeda, the Turkestan Islamic Party, and the Bishkek Chinese Embassy Bombing." *The Diplomat*, September 29. https://thediplomat.com/2016/09/al-qaeda-the-turkestan-islamic-party-and-the-bishkek-chinese-embassy-bombing/.

Botobekov, Uran. 2017. "Al-Qaeda and Islamic State Take Aim at China." *The Diplomat*, March 8. https://thediplomat.com/2017/03/al-qaeda-and-islamic-state-take-aim-at-china/.

Bovingdon, Gardner. 2004. *Autonomy in Xinjiang: Han Nationalist Imperatives and Uyghur Discontent*. Washington, DC: East-West Center.

Bovingdon, Gardner. 2011. *The Uyghurs: Strangers in Their Own Land*. New York: Columbia University Press.

Branigan, Tania. 2013. "China: 21 Killed in Kashgar Clashes." *The Guardian*, April 24. http://www.theguardian.com/world/2013/apr/24/chinese-gangsters-police-shootout.

Brophy, David. 2019. "Good and Bad Muslims in Xinjiang." *Made in China Journal* 4 (3): 44–53.

Byler, Darren. 2018. "Violent Paternalism: On the Banality of Uyghur Unfreedom." *Asia-Pacific Journal* 24 (4): 1–15. https://apjjf.org/2018/24/Byler.html.

Byler, Darren. 2019. "Preventive Policing as Community Detention in Northwest China." *Made in China Journal* 4 (3): 88–94.

Caprioni, Elena. 2011. "Daily Encounters between Hans and Uyghurs in Xinjiang: Sinicization, Integration or Segregation?" *Pacific Affairs* 84 (2): 267–287.

Chan, Minnie. 2013. "Fatal Clash in Xinjiang Shows Cadres Lack Crisis Awareness, Analysts Say." *South China Morning Post*, April 29. http://www.scmp.com/news/china/article/1225526/fatal-clash-xinjiang-shows-cadres-lack-crisis-awareness-analysts-say.

China Daily. 2011. "7 Kidnappers Killed in Hostage Rescue in Xinjiang." *China Daily*, December 29. http://www.chinadaily.com.cn/china/2011-12/29/content_14349 795.htm.

China Daily. 2012. "Official: No Mercy for Terrorists in Xinjiang." *China Daily*, March 7. http://www.chinadaily.com.cn/china/2012-03/07/content_14780825.htm.

China Daily. 2013. "United against Terrorism." *China Daily*, June 27. http://www.chinada ily.com.cn/opinion/2013-06/27/content_16668159.htm.

China Daily. 2014. "Xinjiang's Party Chief Wages 'People's War' against Terrorism." *China Daily*, May 26. http://www.chinadaily.com.cn/china/2014-05/26/content_17541 318.htm.

China State Council Information Office. 2002. "'East Turkistan' Forces Cannot Get Away with Impunity." *People's Daily*, January 21. http://english.peopledaily.com.cn/200201/ 21/print20020121_89078.html.

China State Council Information Office. 2003. *White Paper on the History and Development of Xinjiang*, May 26.

Christoffersen, Gaye. 1993. "Xinjiang and the Great Islamic Circle: The Impact of Transnational Forces on Chinese Regional Economic Planning." *China Quarterly* 133: 130–151.

Chung, Chien-peng. 2002. "China's 'War on Terror': September 11 and Uighur Separatism." *Foreign Affairs* 18 (4): 8–12.

Chung, Chien-peng. 2004. "The Shanghai Co-operation Organization: China's Changing Influence in Central Asia." *China Quarterly* 180: 989–1009.

Clarke, Michael. 2003. "Xinjiang and China's Foreign Relations with Central Asia, 1991–2001: Across the 'Domestic-Foreign Frontier'?" *Asian Ethnicity* 4 (2): 207–224.

Clarke, Michael. 2007a. "China's Internal Security Dilemma and the 'Great Western Development': The Dynamics of Integration, Ethnic Nationalism and Terrorism in Xinjiang." *Asian Studies Review* 31 (3): 323–342.

Clarke, Michael. 2007b. "Xinjiang in the 'Reform' Era, 1978–1991: The Political and Economic Dynamics of Dengist Integration." *Issues & Studies* 43 (2): 39–92.

Clarke, Michael. 2008. "China's 'War on Terror' in Xinjiang: Human Security and the Causes of Violent Uighur Separatism." *Terrorism and Political Violence* 20 (2): 271–301.

Clarke, Michael. 2010. "Widening the Net: China's Anti-Terror Laws and Human Rights in the Xinjiang Uyghur Autonomous Region." *International Journal of Human Rights* 14 (4): 542–558.

Clarke, Michael. 2017. "The Impact of Ethnic Minorities on China's Foreign Policy: The Case of Xinjiang and the Uyghur." *China Report* 53 (1): 1–25.

Clarke, Michael. 2018. "China's 'War on Terrorism': Confronting the Dilemmas of the 'Internal-External' Security Nexus." In *Terrorism and Counter-Terrorism in China: Domestic and Foreign Policy Dimensions*, edited by Michael Clarke, 17–38. Oxford: Oxford University Press.

Cliff, Tom. 2012. "The Partnership of Stability in Xinjiang: State-Society Interactions Following the July 2009 Unrest." *China Journal* 68: 79–105.

Cliff, Tom. 2016. *Oil and Water: Being Han in Xinjiang*. Chicago: University of Chicago Press.

CNN. 2013. "Violence in Western Chinese Region of Xinjiang kills 21." *CNN*, April 25. http://edition.cnn.com/2013/04/24/world/asia/china-xinjiang-violence/index.html.

Cui, Jia. 2013. "Two Terrorist Leaders Sentenced to Death in Kashgar for Attacks." *China Daily*, August 13. http://usa.chinadaily.com.cn/china/2013-08/13/content_16889 009.htm.

Dasgupta, Saibal. 2011. "China Finally Ready to Admit Pak's Role in Xinjiang Violence." *Times of India*, July 20. http://timesofindia.indiatimes.com/world/china/China-fina lly-ready-to-admit-Paks-role-in-Xinjiang-violence/articleshow/9300792.cms?refer ral=PM.

Dillon, Michael. 2004. *Xinjiang: China's Muslim Far Northwest*. New York: RoutledgeCurzon.

Dreyer, June Teufel. 1976. *China's Forty Million: Minority Nationalities and National Integration in the People's Republic of China*. Cambridge, MA: Harvard University Press.

Dreyer, June Teufel. 1986. "The Xinjiang Uyghur Autonomous Region at Thirty." *Asian Survey* 26 (7): 721–744.

Economist, The. 2013. "Ethnic Unrest in Xinjiang: Unveiled Threats." *The Economist*, July 6.

Evron, Yoram. 2007. "China's Anti-Terrorism Policy." *Strategy Assessment* 10 (3): 76–83.

Famularo, Julia. 2018. "'Fighting the Enemy with Fists and Daggers': The Chinese Communist Party's Counter-Terrorism Policy in the Xinjiang Uyghur Autonomous Region." In *Terrorism and Counter-Terrorism in China: Domestic and Foreign Policy Dimensions*, edited by Michael Clarke, 39–73. Oxford: Oxford University Press.

Fedyshyn, Oleh S. 1957. "Soviet Retreat in Sinkiang?: Sino-Soviet Rivalry and Cooperation, 1950–1955." *American Slavic and East European Review* 16 (2): 127–145.

Foot, Rosemary. 2004. "China: A Third Front?" *The Adelphi Papers* 44 (363): 61–71.

Forbes, Andrew D. W. 1986. *Warlords and Muslims in Chinese Central Asia: A Political History of Republican Sinkiang*. Cambridge: Cambridge University Press.

Garver, John W. 1988. *Chinese-Soviet Relations, 1937–1945: The Diplomacy of Chinese Nationalism*. Oxford: Oxford University Press.

Greitens, Sheena Chestnut, Myunghee Lee, and Emir Yazici. 2019/2020. "Counterterrorism and Preventive Repression: China's Changing Strategy in Xinjiang." *International Security* 44 (3): 9–47.

Grose, Timothy. 2019. "'Once Their Mental State Is Healthy, They Will Be Able to Live Happily in Society': How China's Government Conflates Uighur Identity with Mental Illness." *China File*, August 2. http://www.chinafile.com/reporting-opinion/viewpoint/ once-their-mental-state-healthy-they-will-be-able-live-happily-society.

Han, Enze. 2010. "Boundaries, Discrimination, and Interethnic Conflict in Xinjiang, China." *International Journal of Conflict and Violence* 4 (2): 244–256.

Han, Enze, and Harris Mylonas. 2014. "Interstate Relations, Perceptions, and Power Balance: Explaining China's Policies toward Ethnic Groups, 1949–1965." *Security Studies* 23 (1): 148–181.

Harris, Rachel, and Aziz Isa. 2019. "Islam by Smartphone: Reading the Uyghur Islamic Revival on WeChat." *Central Asian Survey* 38 (1): 61–80.

Human Rights Watch. 2005. "Devastating Blows: Religious Repression of Uighurs in Xinjiang." *Human Rights Watch* 17 (2): 1–115.

Human Rights Watch. 2017. "China: Disclose Details of Terrorism Convictions." *Human Rights Watch*, March 16. https://www.hrw.org/news/2017/03/16/china-disclose-deta ils-terrorism-convictions.

Jacobs, Andrew. 2014. "Opposing Narratives in Piecing Together Kunming Attackers' Motives." *New York Times*, March 5.

Jacobs, Justin M. 2016. *Xinjiang and the Modern Chinese State.* Seattle: University of Washington Press.

Kalyvas, Stathis N., and Laia Balcells. 2010. "International System and Technologies of Rebellion: How the End of the Cold War Shaped Internal Conflict." *American Political Science Review* 104 (3): 415–429.

Kan, Shirley A. 2010. *U.S.-China Counterterrorism Cooperation: Issues for U.S. Policy.* Washington, DC: Office of Congressional Information and Publishing.

Kanat, K. B. 2012. "'War on Terror' as a Diversionary Strategy: Personifying Minorities as Terrorists in the People's Republic of China." *Journal of Muslim Minority Affairs* 32 (4): 507–527.

Kerr, David, and Laura C. Swinton. 2008. "China, Xinjiang, and the Transnational Security of Central Asia." *Critical Asian Studies* 40 (1): 89–112.

Khan, Rashid Ahmad. 2012/2013. "Pakistan and China: Cooperation in Counter-Terrorism." *Strategic Studies* 32 (1): 70–78.

Klimeš, Ondřej. 2018. "Advancing 'Ethnic Unity' and 'De-extremization': Ideational Governance in Xinjiang under 'New Circumstances' (2012–2017)." *Journal of Chinese Political Science* 23: 413–436.

Kraus, Charles. 2010. "Creating a Soviet 'Semi-Colony'?: Sino-Soviet Cooperation and Its Demise in Xinjiang, 1949–1955." *Chinese Historical Review* 17 (2): 129–165.

Kraus, Charles. 2019. "Laying Blame for Flight and Fight: Sino-Soviet Relations and the 'Yi–Ta' Incident in Xinjiang, 1962." *China Quarterly* 238: 504–523.

Kuo, Kendrick. 2012. "Revisiting the Salafi-Jihadist Threat in Xinjiang." *Journal of Muslim Minority Affairs* 32 (4): 528–544.

Kuo, Kendrick. 2014a. "China's Wild West: The Problem with Beijing's Xinjiang Policy." *Foreign Affairs*, January 26. https://www.foreignaffairs.com/articles/asia/2014-01-26/chinas-wild-west.

Kuo, Kendrick. 2014b. "New Rules for China's War on Terror?" *East Asia Forum*, October 25. http://www.eastasiaforum.org/2014/10/25/new-rules-for-chinas-war-on-terror/.

Kuo, Kendrick, and Kyle Springer. 2014. "Illegal Uighur Immigration in Southeast Asia." *CogitAsia*, April 24. https://www.cogitasia.com/illegal-uighur-immigration-in-southeast-asia/.

Leibold, James. 2012. "Toward a Second Generation of Ethnic Policies?" *China Brief* 12 (3). https://jamestown.org/program/toward-a-second-generation-of-ethnic-policies/.

Leibold, James. 2019. "The Spectre of Insecurity: The CCP's Mass Internment Strategy in Xinjiang." *China Leadership Monitor* 1: 1–16. https://www.prcleader.org/leibold.

Leibold, James. 2020. "Surveillance in China's Xinjiang Region: Ethnic Sorting, Coercion, and Inducement." *Journal of Contemporary China* 29 (121): 46–60.

Leibold, James, and Timothy Grose. 2016. "Islamic Veiling in Xinjiang: The Political and Societal Struggle to Define Uyghur Female Adornment." *China Journal* 76: 78–102.

Lin, Christina. 2014. "Al Qaeda and ISIS Have Declared War on China—Will Beijing Now Arm the Kurds?" *Times of Israel*, October 28. http://blogs.timesofisrael.com/al-qaeda-and-isis-have-declared-war-on-china-will-beijing-now-arm-the-kurds/.

Macakerras, Colin. 2001. "Xinjiang at the Turn of the Century: The Causes of Separatism." *Central Asian Survey* 20 (3): 289–230.

Malik, Mohan. 2002. *Dragon on Terrorism: Assessing China's Tactical Gains and Strategic Losses Post-September 11.* Carlisle, PA: US Army War College.

McMillen, Donald H. 1979. *Chinese Communist Power and Policy in Xinjiang, 1949–1977.* Boulder, CO: Westview.

McNeal, Dewardric L. 2001. *China's Relations with Central Asian States and Problems with Terrorism*. Washington, DC: Office of Congressional Information and Publishing.

Menon, Rajan. 2003. "The New Great Game in Central Asia." *Survival* 45 (2): 187–204.

Millward, James. 2004. *Violent Separatism in Xinjiang: A Critical Assessment*. Washington, DC: East-West Center.

Millward, James. 2007. *Eurasian Crossroads: A History of Xinjiang*. New York: Columbia University Press.

Mylonas, Harris. 2012. *The Politics of Nation-Building: Making Co-Nationals, Refugees, and Minorities*. New York: Cambridge University Press.

Odgaard, Liselotte, and Thomas Galasz Nielsen. 2014. "China's Counterinsurgency Strategy in Tibet and Xinjiang." *Journal of Contemporary China* 23 (87): 535–555.

Pantucci, Raffaello. 2018. "Uyghur Terrorism in a Fractured Middle East." In *Terrorism and Counter-Terrorism in China: Domestic and Foreign Policy Dimensions*, edited by Michael Clarke, 157–173. London: Hurst.

People's Daily. 2009. "Al-Qaida Threatens to Attack China." *People's Daily*, October 9. http://english.people.com.cn/90001/90776/90883/6777600.html.

People's Daily. 2017. "Regulations on Xinjiang Uyghur Autonomous Region on De-extremification" [*Xinjiang weiwuer zizhiqu qu jiduanhua tiaoli*]. *People's Daily*, March 30. http://xj.people.com.cn/n2/2017/0330/c186332-29942874.html.

Pokalova, Elena. 2013. "Authoritarian Regimes against Terrorism: Lessons from China." *Critical Studies on Terrorism* 6 (2): 279–298.

Pollack, Jonathan. 2003. "China and the United States Post-9/11." *Orbis* 47 (4): 617–627.

Purbrick, Martin. 2017. "Maintaining a Unitary State: Counter-Terrorism, Separatism, and Extremism in Xinjiang and China." *Asian Affairs* 48 (2): 236–256.

Raman, B. 2004. "Counter-Terrorism: India-China-Russia Cooperation." *China Report* 40 (2): 155–167.

Raza, Zainab. 2019. "China's 'Political Re-education' Camps of Xinjiang's Uyghur Muslims." *Asian Affairs* 50 (4): 488–501.

Reed, J. Todd, and Diana Rschke. 2010. *The ETIM: China's Islamic Militants and the Global Terrorist Threat*. Santa Barbara, CA: Praeger.

Roberts, Sean R. 2004. "A 'Land of Borderlands': Implications of Xinjiang's Transborder Interactions." In *Xinjiang: China's Muslim Borderland*, edited by Frederick S. Starr, 216–238. Armonk: M. E. Sharpe.

Roberts, Sean R. 2012. *Imaginary Terrorism: The Global War on Terror and the Uyghur Terrorist Threat*. Washington, DC: PONARS Eurasia.

Roberts, Sean R. 2018a. "The Biopolitics of China's 'War on Terror' and the Exclusion of the Uyghurs." *Critical Asian Studies* 50 (2): 232–258.

Roberts, Sean R. 2018b. "The Narrative of Uyghur Terrorism and the Self-Fulfilling Prophecy of Uyghur Militancy." In *Terrorism and Counter-Terrorism in China: Domestic and Foreign Policy Dimensions*, edited by Michael Clarke, 99–128. London: Hurst.

Robertson, Matthew P. 2020. "Counterterrorism or Cultural Genocide?: Theory and Normativity in Knowledge Production about China's 'Xinjiang Strategy.'" *Made in China Journal*, June 12. https://madeinchinajournal.com/2020/06/12/counterterrorism-or-cultural-genocide/.

Rodríguez-Merino, Pablo A. 2019. "Old 'Counter-Revolution,' New 'Terrorism': Historicizing the Framing of Violence in Xinjiang by the Chinese State." *Central Asian Survey* 38 (1): 27–45.

Rudelson, Justin, and William Jankowiak. 2004. "Acculturation and Resistance: Xinjiang Identities in Flux." In *Xinjiang: China's Muslim Borderland*, edited by S. Frederick Starr, 299–319. Armonk: M. E. Sharpe.

Ruwitch, John, Michael Martina, and Christian Shepherd. 2017. "China's Xi Calls for 'Great Wall of Iron' to Safeguard Restive Xinjiang." *Reuters*, March 9. https://www.reut ers.com/article/us-china-security-xinjiang/chinas-xi-calls-for-great-wall-of-iron-to-safeguard-restive-xinjiang-idUSKBN16H04J.

Samuel, Sigal. 2018. "China Is Treating Islam like a Mental Illness." *The Atlantic*, August 28. https://www.theatlantic.com/international/archive/2018/08/china-pathologizing-uighur-muslims-mentalillness/568525/.

Sasikumar, Karthika. 2010. "State Agency in the Time of the Global War on Terror: India and the Counter-Terrorism Regime." *Review of International Studies* 36 (3): 615–638.

Shicor, Yitzhak. 2007. "Limping on Two Legs: Uyghur Diaspora Organizations and the Prospects for Eastern Turkestan Independence." *Central Asia and the Caucasus* 48 (6): 117–125.

Shicor, Yitzhak. 2009. *Ethno-Diplomacy: The Uyghur Hitch in Sino-Turkish Relations*. Washington, DC: East-West Center.

Smith, Joanne. 2000. "Four Generations of Uyghurs: The Shift towards Ethno-political Ideologies among Xinjiang's Youth." *Inner Asia* 2 (2): 195–224.

Smith, Paul J. 2009. "China's Economic and Political Rise: Implications for Global Terrorism and U.S.-China Cooperation." *Studies in Conflict & Terrorism* 32 (7): 627–645.

Smith Finley, Joanne. 2013. *The Art of Symbolic Resistance: Uyghur Identities and Uyghur-Han Relations in Contemporary Xinjiang*. Leiden: Brill.

Smith Finley, Joanne. 2019. "Securitization, Insecurity and Conflict in Contemporary Xinjiang: Has PRC Counter-Terrorism Evolved into State Terror?" *Central Asian Survey* 38 (1): 1–26.

Smith Finley, Joanne. n.d. "Uyghur Islam and Religious 'De-extremification': On China's Discourse of 'Thought Liberation' in Xinjiang." *Oxford Islamic Studies Online*. http://www.oxfordislamicstudies.com/Public/focus.html.

Tobin, David. 2013. "Nation-Building and Ethnic Boundaries in China's Northwest." PhD diss., University of Manchester.

Tobin, David. 2020. "A 'Struggle of Life or Death': Han and Uyghur Insecurities on China's North-West Frontier." *China Quarterly* 242: 301–323.

Tohti, Ilham. 2015. "Present-day Ethnic Problems in Xinjiang Uighur Autonomous Region: Overview and Recommendations." *China Change*, April 22. https://chinacha nge.org/2015/04/22/present-day-ethnic-problems-in-xinjiang-uighur-autonomous-region-overview-and-recommendations-1/.

Trédaniel, Marie, and Pak K. Lee. 2017. "Explaining the Chinese Framing of the 'Terrorist' Violence in Xinjiang: Insights from Securitization Theory." *Nationalities Papers* 46 (1): 177–195.

Tschantret, Joshua. 2018. "Repression, Opportunity, and Innovation: The Evolution of Terrorism in Xinjiang, China." *Terrorism and Political Violence* 30 (4): 569–588.

Whiting, Allen. 1987. "The Sino-Soviet Split." In *The Cambridge History of China*, edited by Roderick MacFarquhar and John K. Fairbank, 478–538. New York: Cambridge University Press.

Wu, Zhe. 2015. "Caught between Opposing Han Chauvinism and Opposing Local Nationalism: The Drift toward Ethnic Antagonism in Xinjiang Society, 1952–1963."

In *Maoism at the Grassroots: Everyday Life in China's Era of High Socialism*, edited by Jeremy Brown and Matthew D. Johnson, 306–339. Cambridge, MA: Harvard University Press.

Xie, Guiping, and Tianyang Liu. 2019. "Navigating Securities: Rethinking (Counter-) Terrorism, Stability Maintenance, and Non-Violent Responses in the Chinese Province of Xinjiang." *Terrorism and Political Violence*, 1–19.

Xinhua. 2009. "Evidence Shows Rebiya Kadeer behind Urumqi Riot: Chinese Gov't." *Xinhua News Agency*, July 8. http://news.xinhuanet.com/english/2009-07/09/content_11676293.htm.

Yee, Herbert. 2005. "Ethnic Consciousness and Identity: A Research Report on Uygur-Han Relations in Xinjiang." *Asian Ethnicity* 6 (1): 35–50.

Yee, Stefanie Kam Li. 2018. "Uyghur Cross-Border Movement into South East Asia: Between Resistance and Survival." In *Terrorism and Counter-Terrorism in China: Domestic and Foreign Policy Dimensions*, edited by Michael Clarke, 174–184. London: Hurst.

Yuan, Qing-Li. 1990. "Population Changes in the Xinjiang Uighur Autonomous Region (1949–1984)." *Central Asian Survey* 9 (1): 49–73.

Zenz, Adrian. 2019a. "Brainwashing, Police Guards and Coercive Internment: Evidence from Chinese Government Documents about the Nature and Extent of Xinjiang's 'Vocational Training Internment Camps.'" *Journal of Political Risk* 7 (7). https://www.jpolrisk.com/brainwashing-police-guards-and-coercive-internment-evidence-from-chinese-government-documents-about-the-nature-and-extent-of-xinjiangs-vocational-training-internment-camps/.

Zenz, Adrian. 2019b. "'Thoroughly Reforming Them towards a Healthy Heart Attitude': China's Political Re-education Campaign in Xinjiang." *Central Asian Survey* 38 (1): 102–128.

Zenz, Adrian. 2020. *Sterilizations, IUDs, and Mandatory Birth Control: The CCP's Campaign to Suppress Uyghur Birthrates in Xinjiang.* Washington, DC: The Jamestown Foundation.

Zenz, Adrian, and James Leibold. 2017. "Xinjiang's Rapidly Evolving Security State." *China Brief* 17 (4). https://jamestown.org/program/xinjiangs-rapidly-evolving-security-state/.

Zenz, Adrian, and James Leibold. 2020. "Securitizing Xinjiang: Police Recruitment, Informal Policing and Ethnic Minority Co-optation." *China Quarterly* 242: 324–348.

Zhou, Zunyou. 2014. *Balancing Security and Liberty: Counter-Terrorism Legislation in Germany and China.* Berlin: Max Planck Institute for Foreign and International Criminal Law.

Zhou, Zunyou. 2016. "China's Comprehensive Counter-Terrorism Law." *The Diplomat*, January 23. https://thediplomat.com/2016/01/chinas-comprehensive-counter-terrorism-law/.

# PART II
# COLLUSIVE FIFTH-COLUMN POLITICS

# 5

# No Collusion! Or Is There?

## Presidents as Puppets in Russia and the United States

*Scott Radnitz*

In 1996, US intelligence agencies did everything to ensure Yeltsin's electoral victory. . . . Why did US intelligence help Yeltsin seize and maintain power? . . . America needs Yeltsin to defeat Russia in the cold war and to finish it off. He is diligently performing this task.
—Yuri Kachanovsky, historian, *Sovetskaya Rossiya*, July 10, 1997

We are at war. . . . We need our president to speak directly to us and tell us the truth.
—Morgan Freeman, Committee to Investigate Russia,
YouTube, September 19, 2017

When the Soviet Union collapsed, Russian president Boris Yeltsin had the unenviable task of leading his country out of economic chaos. In the midst of a massive decline in national income, runaway inflation, and widespread shortages, Yeltsin and his team of young economists implemented a series of reforms intended to shepherd the country toward a market economy.

The recovery did not materialize, however—at least during Yeltsin's presidency—and his political opponents, deprived of power along with the Communist Party, saw something worse than failed reforms: treason. Yeltsin had risen to prominence as an opponent of communism and an advocate of liberal democracy and saw policymakers in the West as his natural allies. However, taking advice from Western experts and openly engaging in bonhomie with his friend Bill Clinton did not sit well with citizens in his own country who lamented the Soviet collapse. After all, not only was Yeltsin engaged in the wholesale transformation of a system that had provided seventy years of stability and continuity, he was promoting policies championed by those who, until very recently, had represented the Soviet Union's implacable ideological enemies.

Yeltsin's position, as both the leader of a nation undergoing extreme turmoil and a conduit for ideas and policies from the West, made him the target

Scott Radnitz, *No Collusion! Or Is There?* In: *Enemies Within*. Edited by: Harris Mylonas and Scott Radnitz, Oxford University Press. © Oxford University Press 2022. DOI: 10.1093/oso/9780197627938.003.0006

of collusive fifth-column accusations. Whereas those marked as fifth columns in previous chapters in this volume have been marginalized groups, actors in a position of power—even presidents who control large nuclear arsenals—can also be the targets, as they sit at the intersection of national allegiances and global ties. For the accusers, the nature of those ties and the leader's interactions with distrusted foreign actors generate the material for charges of disloyalty and the basis for political opposition grounded in a defense of the people or the nation.

A generation after Yeltsin left the scene in Russia, on the other side of the Cold War divide, another collusive fifth-column accusation appeared. President Donald Trump's unorthodox flattery of Russian president Vladimir Putin during the 2016 campaign, coupled with evidence that Russia had meddled in the election to aid in his victory, led to accusations that, as Hillary Clinton famously put it during a presidential debate, Trump was Putin's "puppet." Throughout Trump's term in office, Democratic politicians, media pundits, "never-Trump" Republicans, and other critics alleged that Trump was a Russian patsy and publicly speculated as to the causes of his incongruous behavior.

This chapter examines these two cases in order to identify the conditions that give rise to collusive fifth-column claims and explain their resonance. At first glance, Russia in the 1990s and the United States after 2016 appear to have little in common. Yet, comparing these cases can be instructive toward a deeper understanding of the roots of fifth-column politics. During the Cold War, geopolitics and domestic political dynamics ensured that *subversive* fifth-column claims would thrive in both countries. Even after the Cold War ended, latent distrust and institutional inertia gave rise to discourses about the continued threat of the other. Fifth-column claims persisted, but in inverted form. Now, political oppositions and independent critics accused heads of state and their entourages of not only being insufficiency hostile toward a presumed adversary, but of clandestinely advancing its interests, implying or asserting willful *collusion* between the president and his foreign "masters."

As in some instances of subversive fifth-column politics, accusers built their claims on the basis of indications of friendliness between the targets and their alleged patrons. Heads of state are cordial with many foreign leaders but when the state in question is a longtime geopolitical adversary, excessive deference can appear alarming. A key factor in an opposition's willingness to advance narratives about collusive fifth columns is that a defiant stance toward the external power is unifying and mobilizing, given that public opinion toward the (former) geopolitical rival is broadly unfavorable. Such a stance was commonsensical for Russian nationalists and unrepentant Communists coping with a historic national trauma. It was less straightforward for members of the Democratic Party, who were less prone than Republicans since the Reagan era to frame foreign policy in moralistic and black-and-white terms, and were more inclined to seek common ground with Russia (Halper and Clarke 2004; Stokes 2015). The unlikely parallels between accusations in the two countries indicate how residual Cold War logics produced similar discursive dynamics in distinct political settings, and highlight how narratives about subversion can be refashioned into claims of collusion.

To be sure, there are important differences between the cases, including the scale of the supposed conspiracy. In Russia, the guilty parties represented a broad base: pro-Western liberals who championed Yeltsin's reforms and the rising constituencies that benefited. Unlike in Russia, where an entire political class was rhetorically indicted, in the United States the purported fifth column was Trump alone. Whereas Russian reformers were acting out of ideological motives to transform Russia—from their perspective, for the better—few of Trump's critics maintained that he had any intentions beyond helping himself.

## Collusive Fifth Columns

Collusive claims are weapons of the weak (Scott 2008). They are a rhetorical recourse that can mobilize elements of an opposition that is otherwise at a disadvantage in relation to elite actors with inordinate power to set agendas. Like subversive claims, they enable the affronted party to invoke a sense of patriotism and national unity to gain rhetorical leverage, but in the collusive case the claimants seek to build or buttress a movement while out of power. Because of this power differential, the implications of collusive claims differ from subversive ones. The putative threat to the national interest posed by a fifth column embedded within the government is greater than that posed from below, in ways that raise the stakes. In the imagination of the accusers, control of the levers of power enables a maliciously inclined incumbent to inflict untold damage on the country. The magnitude of this supposed threat lends itself to extreme and dire rhetoric that can crowd out more conventional political criticism and foreclose the possibility of compromise among competing parties. In rare cases, it might provoke a harsh response from the accused leader(s) or incite a rebellion from below or a coup from within the state. Yet the actual threat posed to the target of collusive accusations is less dire than that faced by actors implicated in subversive activities. By virtue of being powerful by definition, the accused fifth-column colluder is more likely to suffer reputational damage than actual harm. However, his supporters—or enablers—may lack the protection conferred by a position in the government and are vulnerable to being targeted by association.

As detailed in the Introduction to this volume, the possibilities for claims that incumbent politicians act in the interest of a foreign master emerge naturally in small states and colonial subjects, which lie on the weaker end of a major power imbalance. Where the polity is under the effective control of a foreign hegemon, the willing collaboration of indigenous rulers can generate resentment toward both the domestic government and the external power, and spark a nationalist backlash (Thomas 2008: 70). Allegations of fifth-column subservience may carry more weight when the unequal relationship is seen to violate a foundational principle or offends national or communal identities (see Crews, Chapter 6 in this volume). In either case, structural power differentials tend to exacerbate insecurities about sovereignty and have the potential to unsettle politics in the weaker state (Cash and Kinnvall 2017).

Yet neither Russia in the 1990s nor the United States in the 2010s fits the profile of small, weak, or client states. Neither was functionally subordinate to a more powerful hegemon. The fact that collusive fifth-column claims arose in both cases attests to the versatility of the trope and owes to the combination of the unique personalities involved, the relative openness of both political systems, and immediate political shocks. Yet the most important condition enabling collusive claims in both cases was the residual resonance of Cold War tropes, which ensured that mutual hostility and suspicion between the two nations would endure.

## Reforms and Reaction in Russia

The roots of fifth-column fears between the United States and Russia run deep. During the Cold War, politics in the two rival superpowers gave rise to claims that (subversive) fifth columns were intent on undermining the government from within. From its founding, the Bolshevik regime was consumed by the threat of internal enemies, whether labeled as counterrevolutionaries, "wreckers," enemies of the people, "diversionists," or spies (Goldman 2011). The geopolitical overlay of the Cold War played into Stalin's fears of capitalist encirclement and begat assertions that the regime faced coup plots sponsored from overseas. In the United States, the notion of an enemy justified the creation of a massive security and surveillance apparatus that swept up innocent people and deterred others from open dissent. Cold Warriors, backed by evidence from J. Edgar Hoover's FBI, stoked fears about Communists infiltrating government agencies, stealing secrets, and spreading anti-American propaganda in the labor movement, academia, and Hollywood (Weiner 2012).

As the Cold War wound down, Soviet hostility toward the West gave way to cooperation and a new openness to Western influences. The proliferation of cultural and youth exchanges, scientific collaboration, and a reduction of anti-American propaganda eased the climate of suspicion on both sides. Mikhail Gorbachev saw market-oriented reforms of the Soviet economy as the path to economic recovery and regime survival. Western policymakers urged him on (Matlock 2004).

This new openness was tempered by Kremlin hardliners, who viewed Gorbachev as a traitor for undermining the authority of the Communist Party and unilaterally deescalating tensions with the United States. Conservative critics of Gorbachev's reforms decried the threat that capitalism and Western ideologies posed to Marxist-Leninist principles and Soviet morality (Brudny 2000: 198–199). Forces within the military and KGB tried to hold back the tide, but they lacked broad public support for a reversal of Gorbachev's reforms and a renewed clampdown.

When Yeltsin took power in 1991, he faced a collapsing economy that he believed he had little time to stabilize (Roberts and Sherlock 1999). During that time, prices shot up, factories shut down, salaries went unpaid, pensions were cut, and organized crime increased (Treisman 2011: Chapter 6). Yeltsin's reform plan, conceived by the economist Yegor Gaidar, entailed the rapid and complete dismantling of the system of central planning, known colloquially as "shock therapy."

Even as reformists dominated within the government, opponents of reform sought to mobilize public opinion against Yeltsin's policies. They pushed a narrative that highlighted the palpable economic damage Russia was suffering and the fact that the policies behind them were developed and enthusiastically supported by representatives of the Soviet Union's Cold War adversary. Although these issues were distinct in the abstract, owing to the Western intellectual origins of Russia's economic program, in practice it did not take great effort to imagine that the West had deliberately imposed injurious policies on Russia.

President Yeltsin and his "young reformers" personified the hollowing out of Russian sovereignty. They enjoyed the vocal political support of the administrations of George H. W. Bush and, later, Bill Clinton, which encouraged Russia's transformation into a democracy and market economy. Yeltsin's advisors worked closely with economists from Harvard University, whose involvement was paid for by a noncompetitive grant from the US Agency for International Development (Wedel 2015). The technocrats operating in the Kremlin, led by Yeltsin advisor Anatoly Chubais, could be perceived to be serving foreign masters and personally benefiting even as the standard of living of most Russians collapsed (Oushakine 2010: 44–45).

Yeltsin also had a visible friendship with Clinton, one based both on personal rapport and Russia's supplicant position at a time of weakness. Yeltsin, in his memoirs, refers to Clinton as a "young, eternally smiling man who was powerful, energetic, and handsome" who "lent hope to the idea of a future without wars, without confrontation, and without the grim ideological struggles of the past" (Yeltsin 2000: 134). Yeltsin needed Clinton to authorize aid and advocate for Russia at international organizations over which the United States held sway. Yet despite the two presidents' mutual affection and Russia's dire needs, Russia received meager amounts of humanitarian assistance from the United States, and somewhat more generous loans from the IMF and World Bank, but whose repayment created new burdens for the Russian treasury (Colton 2008: 268; Treisman 2011: 315). Adding insult to injury, Yeltsin acceded to the enlargement of NATO to encompass former Warsaw Pact and Baltic states, which the Clinton administration initially equivocated on but later championed (Colton 2008: 268–269).

Russian citizens loyal to the Communist Party were reflexively opposed to Yeltsin's reforms and the top-down manner in which they were implemented, yet the economic crisis and political instability of 1992–1993 fed discontent even among people with weaker ideological leanings and provided a growing reservoir of support for opponents of reform (McFaul 2001: 173). Struggles over economic policy reached a crisis point when Yeltsin disbanded the Supreme Soviet and, when its members refused to leave, ordered the shelling of the Russian White House. Yeltsin prevailed over his opponents in the short term, but the backlash to his approach was evident in the 1993 parliamentary election, in which nationalist and restorationist forces prevailed: the Liberal-Democratic Party of Russia won 22.9 percent of the vote and the Communists 12.4 percent.[1] The two parties together performed just as well in the 1995 elections.

## A Treacherous Transition

The early years after the Soviet collapse had produced conditions—economic strife, lingering geopolitical animosity, and a freewheeling political system—that were ripe for an emboldened opposition to exploit. Yeltsin's vocal support for rapid reforms and his eagerness to cooperate with the former adversary provided an opening for nationalist politicians to tie Yeltsin to the deterioration of Russia's international stature and the country's generalized sense of chaos.

The most piercing attacks on the government came from those who had opposed Gorbachev's reforms from the beginning, and who sought to channel the widespread suffering and sense of betrayal felt by losers in the new dispensation (Oushakine 2010). Communist Party (CPRF) leader Gennady Zyuganov quickly emerged as a prominent mouthpiece for disaffected citizens. Recognizing that Leninism was no longer a popular governing ideology, he oversaw the reinvention of the CPRF to emphasize nationalism rather than class (Brudny 1997: 268). In doing so, he tapped into Russian insecurities about territory and sovereignty dating back to the nineteenth century (Tsygankov 2014: 50–51). In 1994, Zyuganov alleged that "the Western states and the transnational banking and financial corporations are the obedient relayers of the aggressive anti-Russian policy" (Allensworth 1998: 168). Numerous critical editorials were published in *Sovetskaya Rossiya*, a Communist-nationalist newspaper that was the preferred platform for conservative elements in the party in the Soviet period and continued in that vein throughout the 1990s (Brudny 2000: 10, 244). Many of these broadsides characterized market reforms as a plot by enemies of Russia to reduce the size of its population and loot the country for its natural resources (Oushakine 2010: 127; Yablokov 2018: 77). Yeltsin's opponents in the press highlighted how the West imposed its policies with the connivance of Russian elites. One editorial colorfully commented:

> Who are the authors of these criminal ideas to deplete and flood the land? The messengers of a satanic power? Insidious agents of darkness? Or the executors of someone else's methodical, purposeful plan, the disastrous projects foisted from foreign "design tables" and implemented and ongoing in Russia by a "fifth column," clearly visible only since the 80s and 90s? With all certainty, we felt their predatory but also weak, merciless and at the same time ingratiating hand, their imaginary power, captured through clever provocations, treachery, and rigging by politicians, chameleon-like in their cunning, protected by their lies of the exaggerated promises of universal prosperity of an unprecedented industrial and food paradise after the transition to a market economy. (Bondarev 1995)[2]

Vladimir Zhirinovsky, the demagogic head of the LDPR, campaigned for parliament on the same theme, tying Russia's struggles to the deliberate design of its adversaries and their local proxies:

> This is not our program to improve our lives, but that of the West and our fifth column, a program for finally finishing Russia off. The country has fallen apart. There are no frontiers. The economy is half-dead. The entire nation is wandering the streets, some selling their bodies, some selling their possessions, some selling their souls. (BBC Monitoring 1995a)

Zhirinovsky concluded another speech by linking Yeltsin's government to Russia's primary geopolitical foe:

> The president does not govern. He knows nothing. He is like Brezhnev in 1982. And the government will continue to pursue the pro-American course. There is a fifth column everywhere. There is a fifth column in the army: 200 generals have been trained at Harvard University. (BBC Monitoring 1995b)

Some attacks against Yeltsin speculated as to how he became a tool of foreign interests:

> In September 1989, the Moscow "democrats" and their overseas partners organized an unofficial trip of Boris Yeltsin to the United States. The welcoming reception by the American business elite, venerable scientists, and talk shows

messed with Boris Nikolaevich's head, and he declared the indisputable advantages of "decaying capitalism" over socialism. This "discovery" immediately opened the door to the White House for him. (Sanin 1999)

## Yeltsin's Suspicious Triumph

The way Yeltsin won re-election in 1996 exemplified his weakness and reliance on both foreigners and dubious beneficiaries of privatization. He began the campaign with an approval rating of 5 percent and was considered likely to lose to Zyuganov (McFaul 2001: 290). After weighing delaying the election, Yeltsin's campaign team sought to unify pro-market forces to support his candidacy. Seven oligarchs worked together to generate funds and provide media coverage favorable to Yeltsin, and eventually credited themselves with Yeltsin's hard-fought victory (Hoffman 2011: 328, 358). His friendship with Clinton also paid dividends in the campaign. Yeltsin's daughter hired several American advisors with (deniable) links with Dick Morris, a consultant who had worked for Clinton. The group provided polling, advertising, and strategic advice for how Yeltsin should frame his candidacy and portray his main adversary (Kramer 1996). Most observers downplay the importance of this assistance to his victory (e.g., Stanley 1996), but the episode highlighted Yeltsin's dependence on external support and his willingness to stretch democratic rules in order to prevail.

Zyuganov, for his part, premised his campaign as a battle against the West. He referred to "forces in the West who are willing to finish us off while we are almost bedridden, who have always hated Russia and everything Russian and Soviet, everything which belongs to the people, and who continue pumping this hatred, especially through our mass media, the traitors" (BBC Monitoring 1996). Yeltsin defeated Zyuganov in the second round with 54.4 percent of the vote. Yet his victory was at most a temporary respite, as the economy continued to crater.

Aggrieved commentators played on the perception that Yeltsin owed his victory to his Western benefactors:

Yeltsin was victorious because his apparatus shamelessly used the Russian media, especially television, to promote the necessity of choosing Yeltsin. His advisers prudently advised their master to ignore other rules of the game. They convinced him to take as much money as possible from the state budget and private business in the form of subsidies for his election campaign. . . . An important role was played by quiet Americans, sent to help Yeltsin by agreement with the White House. The essence of the strategy proposed by the Americans was as follows: the main emphasis was not on propaganda promising the continuation of market and democratic reforms in the event of Yeltsin's victory. These assurances were made mainly for the outside world. The strategists assumed, on the whole correctly, that most Russians do not believe Yeltsin's promises to improve their lives in in the near future. (Obeshchali-veselilis 1997)

Russia's debt default in 1998, after months of struggling to maintain the ruble, was the death knell of its shambolic economy and a seeming vindication for critics of reform. That the IMF came to the rescue with a $22.6 billion package added insult to injury. If Yeltsin had set out to destroy Russia to serve his international masters, then the deed was seemingly accomplished. According to a columnist in *Sovetskaya Rossiya*:

> Yeltsin is a henchman [*stavlennik*] of the most reactionary circles in the United States (IMF). One can only conceive of a change if there is a change in these circles, a change in their hateful attitude towards Russia into one of respect. . . . Many countries in Latin America, Africa, and Asia suffer from ruthless exploitation under the heel of Western capital suffer. Now Russia has joined them. (Belov 1998)

Another facet of the attack on Yeltsin's government was antisemitism. That the most notorious "oligarchs" to emerge from the 1990s were Jewish was not lost on nationalist critics of the government. In 1988, a notorious letter published in *Sovetskaya Rossiya* by a schoolteacher targeting perestroika alluded to Jews and the threat of "cosmopolitanism" (Korey 1995: 167).[3] It was a straightforward move to link the rising class of wealthy businessmen, who supported and sometimes advised Yeltsin, with the international financial institutions that appeared to be profiting from Russia's decline.

Zyuganov made this connection explicit in relation to the financial crisis. In an "open letter" to the oligarchs, he did not attempt to water down his invective:

> Our people are not blind. They cannot but see that Zionization of the government authorities of Russia was one of the reasons of [*sic*] the present catastrophic conditions of the country, mass impoverishment and dying out of its population. . . . They cannot turn a blind eye to the aggressive, destructive role of Zionist capital in ruining Russia's economy and plundering her property owned by all. (Hoffman 1998)

There is little evidence that antisemitic tropes resonated broadly (Gibson and Howard 2007), but the triumph of the oligarchs and the decline of the Russian economy made this perennial fifth-column allegation all but inevitable.

Collusive fifth-column claims culminated in parliamentary impeachment hearings in 1999 against Yeltsin, who was by that time in poor health and intent on identifying a successor. Five articles were lodged against him, including for his original sin: bringing about the collapse of the USSR. According to his lead accuser, Communist Duma member Viktor Ilyukhin:[4]

> The Soviet Union collapsed not because of natural processes, not as a result of the August 1991 events, but as a result of political conspiracy on the part of the "fifth column," with the connivance, and at times with the participation, of the president of the USSR M. Gorbachev and leaders of several Union ministries and agencies, and as a result of a conspiracy headed by B. Yeltsin. (Yablokov 2018: 68)

Yeltsin's support even among those who voted against impeachment was lukewarm at best, but he was acquitted on all charges (Colton 2008: 426–427). He appointed Vladimir Putin as prime minister several months later.

Putin benefited from the contrast with Yeltsin and channeled the mass frustration stemming from the effects of liberal reforms. Early on, he styled himself a technocrat and worked to rebuild the capacity of the state. However, he also focused his wrath on the winners from privatization and fed the narrative of a chaotic decade that preceded his rise, in contrast to the order he was imposing. Putin's decision to co-opt, sideline, or persecute the oligarchs was flaunted as a demonstrable strike against collusive fifth columns who had supposedly weakened Russia, and it likely paid political dividends (Rutland 2003).

As Putin consolidated power and reduced the opportunities for opposition, fifth-column claims again appeared, but they were now voiced by the Kremlin and painted a picture of subversive elements seeking to undermine the regime (Radnitz 2021: 91). After the Orange Revolution in Ukraine, the Kremlin attacked opposition activists and foreign-backed nongovernment organizations rhetorically, legally, and on the streets (Yablokov: Chapter 4). As relations with the West deteriorated, Putin and his spokespeople promoted a narrative alleging that Western countries were backing pro-democracy activists to advance their geopolitical goals. Putin's decreasingly free opponents did not see much use in claiming that Putin himself was under the sway of any foreign masters. His government had the market cornered on rhetoric about fifth-column plots against Russia.

## Collusive Fifth Column Redux: The Empire Is Struck Back

If collusive fifth-column claims were incongruous in a powerful country like Russia, even its weakened state, the notion that the US president was a "puppet" or "asset" of the Kremlin in the late 2010s was even more confounding. A bipartisan consensus in the US viewed Russia with suspicion, especially as Putin increasingly pushed back against American policies in the post-Soviet region. Of the two major parties, it was traditionally the Republicans who saw Russia as the greater threat and were more supportive of confrontational policies. As recently as 2012, Republican presidential candidate Mitt Romney called Russia America's "number one geopolitical foe." Yet it was the Republicans' 2016 nominee who wondered, "Wouldn't it be a great thing if we could get along with Russia?" (Kaczynski et al. 2017). And it was predominantly Democrats who accused him of acting as a fifth column for Russia.

Donald Trump's interest in Russia long predated his interest in politics. As a property developer seeking to expand abroad, Trump explored the possibility of building a luxury hotel in Moscow. Although no deal was signed after his visit in 1987, he maintained a desire to do business in Russia and continued to display an affinity for the country (Harding 2017). Trump's comments over the years, on and off Twitter, indicated an uncanny fixation with Russia and unabashed admiration of Putin.[5] Apparently, the affection was mutual. The Mueller Report would later determine that the "Russian government perceived it would benefit from a Trump presidency and worked to secure that outcome" (Mueller 2019).

It is not possible to assess which if any of the theories about Trump's true motives for his sympathetic, even sycophantic, attitude toward Russia is correct—blackmail over compromising material or financial problems, future business prospects, envy of Putin's authoritarian style, or some combination thereof—but it is not important for the purposes of this chapter. The case is instructive in providing insights about collusive fifth-column discourses in a powerful country, the United States, that has itself often been accused of seeking to manipulate politicians in other states (Rid 2020).

## Before Trump: The American Right and Russia

It was not unheard of for partisan critics in the United States to claim fifth-column collusion in the government. In fact, long-standing distrust of the government provided a template for ambitious politicians and ordinary citizens to "punch up" at their supposedly treacherous rulers, but in the twentieth century, such accusations came almost exclusively from the political right. In the aftermath of the Great Depression, opponents of the New Deal charged that President Franklin Roosevelt was imposing socialist policies on the public and accused him of literally or metaphorically advancing the interests of the Soviet Union. In 1938, Texas congressman Martin Dies created the House Special Committee on Un-American Activities, charged with investigating disloyalty and subversion in the government. In one of the earliest instances of the use of the term, Dies maintained that:

> fifth columnists are not always garbed in the uniform of foreign troops nor do they always speak with a foreign accent. They may be native-born American citizens. Such figurative parachutists have already landed in the federal government. They await the "zero hour" when Stalin gives the command to attack. (Goldberg 2008, 25)

The committee was charged with investigating Nazi as well as Communist infiltration, but New Deal opponents on the committee used it as a platform to fulminate against Roosevelt (MacDonnell 1995: 6). FBI director J. Edgar Hoover warned in testimony before the committee that liberals and progressives were at risk of being "hoodwinked and duped into joining hands with the communists" (Hoover 1947).

In the early years of the Cold War, Wisconsin senator Joseph McCarthy shot to prominence by alleging fifth-column infiltration at the highest levels, launching his assault with the assertion that 205 Communist Party members were working in the State Department. He lavished special attention on Secretary of Defense George Marshall, who had spearheaded the rebuilding of postwar Europe. McCarthy alleged that Marshall was behind a "great conspiracy" that had led to America's decline to the benefit of the Soviet Union:

> To what end? To the end that we shall be contained, frustrated and finally, fall victim to Soviet intrigue from within and Russian military might from without. Is that farfetched? There have been many examples in history of rich and powerful states which have been corrupted from within, enfeebled and deceived until they were unable to resist aggression. (McCarthy 1953: 307)

Republicans encouraged or humored McCarthy's charges when they implicated the administration of President Harry Truman, a Democrat. McCarthy's downfall came only after he alleged fifth-column activity in the Republican Eisenhower administration.

Accusations about collusive fifth columns also came from civil society. The John Birch Society (JBS), founded in 1958, inveighed against social changes occurring in American society and promoted conspiracy theories about the federal government. Robert Welch, its founder, echoed many Republicans in calling the New Deal a "cancer of state socialism" (Goldberg 2008: 38). He also attacked Eisenhower, calling him "a dedicated, conscious agent of the Communist conspiracy" (Mulloy 2014: 15). Although the JBS did not have widespread popular support, Welch's thesis—that the proponents of social democracy, namely, Democrats, were hastening a Communist victory—was taken up in various guises by ambitious conservatives including Barry Goldwater and Richard Nixon.

With the end of the Cold War, such rhetoric subsided not because the right became enamored of Russia, but because discourses about Russia's weakness, as an economic basket case and a potential source of loose nuclear weapons, eclipsed pronouncements about its strength. Yet elites in both political parties remained suspicious of post-Soviet Russia and wary of the possibility that it could become a revisionist power, even as it attempted a transition to democracy and embraced capitalism (Tsygankov 2009). Putin did not endear himself to Americans, as he clashed with both the Bush and Obama administrations. In 2014, only one-quarter of Americans regarded Russia as an ally or friend (Guskin 2018). Yet there was one important exception: cultural conservatives, who were attracted to Putin's self-portrayal as a champion of conservative values and a bulwark against the dual threats of European progressivism and Islamic radicalism. As erstwhile Nixon advisor and arch-conservative political commentator Pat Buchanan admiringly wrote about Putin in 2013, "He is seeking to redefine the 'Us vs. Them' world conflict of the future as one in which conservatives, traditionalists and nationalists of all continents and countries stand up against the cultural and ideological imperialism of what he sees as a decadent west" (Buchanan 2013). This narrative, of Putin as a defender of conservative values, foreshadowed the later reversal in partisan attitudes toward Russia.

## America (or Russia) First?

The myriad ways that Trump was implicated in matters relating to Russia confounded easy explanation and, for many observers, constituted proof of a conspiracy. Trump spoke fondly of Russia long before Putin began to appeal to Republicans, and Trump was not a cultural conservative. Suggestions that Trump's interest in Russia was more than as passing infatuation came from his past business dealings, positive comments on Russia and Putin, and personal visits to Russia. Donald Trump Jr., who helped run the Trump Organization, said in 2008, "Russians make up a pretty disproportionate cross-section of a lot of our assets. . . . We see a lot of money pouring in from Russia" (Dorell 2017). A substantial number of Russians had bought condos in Trump buildings and paid in cash (Kumar 2018). Trump's campaign manager had advised Ukrainian president Yanukovych, who fled to Russia after he was ousted in a revolution. And then there was Trump's relentless effort to build a tower in Moscow, which he continued to pursue even after he won the Republican nomination for president (Alexander and Behar 2019).

Some of the material for collusive fifth-column accusations about Trump came from an ostensibly nonpartisan source, the so-called Steele Dossier. A former British intelligence officer, Michael Steele, worked for a private security firm tasked initially by Trump's Republican opponents to dig up opposition research against him. Based on anonymous sources supposedly close to the Kremlin, the dossier alleged that Russia possessed compromising material on President Trump, and included salacious details involving his 2013 trip to Moscow. The dossier found its way to then-Senator John McCain, who shared it with the FBI. The allegations were fantastical and worthy of a spy novel. And yet, given Trump's peculiar background and behavior, it could not be dismissed out of hand. The evidence, though disputed, could lead to the logical (and once outlandish) conclusion that Trump was in fact being blackmailed by the Kremlin or had made a secret deal with Putin.

The reaction to the opposition research was filtered through the polarized, distrustful, and anxious American body politic. Republicans largely dismissed it, whereas Democrats saw confirmation of their suspicion that Trump had really been compromised and evidence that his unexpected election victory was tainted (Dart and Smith 2017). Trump's reaction to the report only heightened suspicions that he was concealing the truth. Rather than provide evidence that could have disproven some of the factual claims in the dossier, as President Obama had responded to false assertions about his birth certificate, Trump lashed out at the media (Mangan 2017).

Once he took office, Trump's behavior suggested to many foreign policy experts that he was bending over backward to advance Russian interests. A short list of suspicious or sycophantic actions includes one-on-one meetings with Putin with no aides present; a press conference in which Trump accepted Putin's denial that Russia had interfered in the election, despite the consensus view of the Intelligence Community; Trump's decision to congratulate Putin on his re-election in 2017 despite explicit warnings from his foreign policy staff not to do so; his reluctance to enact new sanctions after the poisoning of a former Russian spy in the UK; and his insistence that Russia should rejoin the G-7 group of nations.

What did it all mean? The evidence of Trump's admiration for Putin, and his sympathy for the Russian position on divisive foreign policy issues, was overwhelming. But the underlying causes of his behavior were more elusive. Was Trump being blackmailed to behave against his natural instincts? Did he genuinely fear the exposure of a compromising tape or documents proving financial impropriety? Could that evidence, even if it existed, be wielded effectively to force Trump to change his behavior (Radnitz 2018)? Or did Trump, for whatever reason, harbor genuine affection for Putin and Russia? Short of hard evidence exposing the nature of the connection, the possibility of collusion persisted, and the resulting ambiguity could be marshaled for political purposes.

## A Common Cause against Collusion

Trump's unexpected election victory over Hillary Clinton was jarring to Democrats, as was the way it transpired: with a minority of the popular vote and after a last-minute intervention by FBI director James Comey (Cohn 2018). Details about how Russia may have helped Trump win emerged during the campaign but the trickle gradually became a torrent after the election. When Clinton said at the third debate that Putin would "rather have a puppet as president," she set the tone for what would later follow. Democrats, stunned and demoralized after losing what appeared to be an easy victory, made Trump's subservience to Russia a focal point of their attacks. In private, prominent Republicans seemed to harbor the same suspicions. At a congressional retreat, Republican House majority leader Kevin McCarthy said, "there's two people I think Putin pays: Rohrabacher [a vocal supporter of Russia] and Trump, swear to God" (Entous 2017).

Although Democrats and the left did not appear to have coordinated a narrative strategy about Russia to the same extent that Republicans did when they pushed back, the "Russiagate" scandal became a popular discussion topic among Trump's opponents and a rejoinder to Trump's incessant lies and exaggerations. *New York Magazine* reporter Jonathan Chait laid out one of the most elaborate arguments in a piece titled, "Will Trump Be Meeting with His Counterpart—Or His Handler?" Combining episodes from Trump's past with knowledge of how KGB recruited its agents, he posited that Russia was using financial leverage and evidence of personal improprieties to blackmail Trump (Chait 2018). The Trump-as-Russian-asset theme became a staple on cable news. MSNBC anchor Rachel Maddow dissected the Steele Dossier on her nightly show to infer how Trump may have been compromised, admitting that she was "obsessed with Russia" (Hess 2019). In an indicative exchange from "Morning Joe," recurring guest Donny Deutsch said Trump was "owned by Putin because he's been laundering money, Russian money, for the last 20, 30 years. He's owned by them." Host Joe Scarborough did not agree outright, but offered his own take:

> We all will be absolutely fascinated when we finally figure out what Vladimir Putin has on Donald Trump and why Donald Trump has surrendered the Middle East, helped ISIS, helped Iran, helped Russia, helped Turkey, helped all of our enemies and betrayed all of our allies. . . . A lot of people think . . . [Putin]

has compromising pictures or something happened in a hotel in Russia years ago. No. It goes back to money. It's always about money. (Hall 2019)

A private initiative, founded by producer Rob Reiner, launched a video in which actor Morgan Freeman ominously accuses Russia of attacking American democracy and insinuates that Trump is ignoring the threat (Kozlov 2017).

Democrats in Congress pursued the same themes. Adam Schiff, the ranking Democrat on the House Intelligence Committee, warned, "The most serious risk to the country, I think, is that the Russians possess compromising information, which they call kompromat, that can influence this president's conduct of American policy" (Wright 2017). As late as May 2020, House Speaker Pelosi wondered, "What does Putin have on Trump—personally, politically, financially?" (Vogl 2020).

As the 2020 election approached, the Russia issue had subsided, as Trump's opponents had ample other material with which to attack him. Yet one group— "never Trump" Republicans—sought to appeal to national security hawks and distressed conservatives by yet again invoking Russia's insidious influence. In a two-minute ad produced by the Lincoln Project called "Fellow Traveler," a narrator speaks Russian over English subtitles and images of Trump, Putin, Hillary Clinton, and menacing Russian soldiers, in a style reminiscent of the television series "The Americans." Eliding differences between the nonexistent Soviet Union and contemporary Russia, the narrator refers to "Comrade Putin" and says, "Donald Trump received the most important endorsement in 2016 from our great leader, Vladimir Putin" and "Our special services worked overtime to elect Comrade Trump." Although its political impact was uncertain, it garnered over 1.5 million views on YouTube within two weeks.

The Mueller Report determined that, contrary to Trump's repeated refrain that his campaign had engaged in "no collusion" with Russia, there were numerous instances in which the campaign willingly sought its assistance. Among them were a Trump Tower meeting that Donald Trump Jr. expected would yield "dirt" on Hillary Clinton, and communications regarding the release of DNC emails through Wikileaks. However, no evidence surfaced that Trump acted as a collusive fifth column, acting on Russia's orders to advance Russia's interests, whether as a product of blackmail, extortion, or other forms of coercion—either during the campaign or throughout his presidency. The FBI was not able to corroborate much of the Steele Dossier, and Steele himself assessed that (only) "at least 70 percent" of it was accurate (Jurecic 2019; Wemple 2020). As far as Mueller's team was able to ascertain, Russians with varying degrees of proximity to the Kremlin seeking to influence Trump during the campaign struggled to gain access to Trump himself. It is also worth noting that Putin did not achieve his highest policy priorities during the Trump presidency: sanctions relief, freezing or rolling back NATO, or acquiescence to its interests in Ukraine. In fact, US policy in those areas remained unchanged or intensified to Russia's detriment. Instead, Russia ended up with the consolation prize of a highly polarized America, shrinking from its global commitments, and the erosion of rule of law thanks to Trump himself. Insofar as the Kremlin viewed its interests in zero-sum terms vis-à-vis the United States, the latter's loss might be seen as the former's gain.

## The Utility of Collusive Fifth-Column Discourse for Underdogs

Why were Trump's opponents so vocal about the possibility that Trump represented a collusive fifth column? Trump's behavior raised obvious red flags, but fixating on the possibility that Russia was manipulating him, at the expense of time that could be spent highlighting Trump's corruption, bigotry, or violations of constitutional norms, was a deliberate choice. The prospect of a foreign, and specifically Russian, hand behind the 2016 election result held special allure.

The most important immediate catalyst for collusive fifth-column claims may have been the magnitude of the shock of Trump's victory. Some predictive polling models forecasted that Clinton would win with 99 percent probability and even Trump himself expected to lose on election night (Wolff 2018). For the losing side, reckoning with its historic loss would require grappling with Hillary Clinton's lackluster campaign, her decision to largely forgo campaigning in Midwestern states, low African American turnout, the alienation of the white working class, and other difficult and serious challenges. By contrast, the narrative that Russia "stole" the election for Trump satisfied two psychological needs. First is the tendency for conspiracy theories to emerge and circulate during perceived crisis situations, as people seek to restore a semblance of order after losing control (Van Prooijen and Douglas 2017). Second is the psychic benefit of externalizing the causes of a loss (Crocker et al. 1999). Seeing Clinton's unexpected defeat as emanating from outside the country deflected blame from Democratic Party and campaign officials who could have conceivably prevented the outcome and diverted attention from the more than 60 million people who voted for Trump despite the candidate's obvious flaws.

A second factor behind the fifth-column claims was that despite the recent shift among some Republicans, Russia, as personified by Putin, was widely disliked.[6] Residual distrust of post-Soviet Russia for the generations that lived through the Cold War made it an expedient malefactor. Elites in both parties continued to view Russia as an adversary due to differences over issues like arms control and NATO enlargement, and Russia's invasions of Georgia and Ukraine, its annexation of Crimea, and its intervention in the Syrian civil war. Putin, thanks to his KGB background and steely visage, played into the culturally tinged stereotype of a devious and ruthless spy. Whether Trump came off as weak and naïve or selfish and scheming, Putin could be depicted as the puppet master, manipulating Trump at will.

It is not incidental that labeling Trump a national security threat helped to boost Democrats' own martial credentials. Republicans were historically more hawkish toward the Soviet Union and perceived as more supportive of the military in general. Calling out Trump's pro-Russian behavior and framing it in terms of national security enabled Trump's opponents to flip the script. They could claim to be vigilant defenders of the national interest, in contrast to congressional Republicans who had cravenly minimized the Russian threat out of fear of drawing Trump's ire, notwithstanding the occasional quip on a hot microphone. This tack might prove especially effective by highlighting the discrepancy between Trump's belligerent and hyper-masculine rhetoric and his fawning and obsequious posture toward authoritarian leaders such as Kim Jong-un of North Korea (Kurtzleben 2016).

Leftist critics of Russiagate charged that the Democrats' focus on Russia distracted from the domestic forces behind Trump's rise. They warned against the palliative of a conspiracy theory (Gessen 2017; Taibbi 2018). They pointed out that much more misinformation during the election had domestic origins than was disseminated by Russian trolls. And even if Russian propaganda succeeded, it was only because Americans were already susceptible due to partisan polarization and social media echo chambers. In short, viewing Trump as a fifth column and focusing on possible collusion, while it may have produced short-term political gains, impeded the introspection needed to understand Trump's rise and prevent his re-election.

## Mirror Images

Russia is often accused of mirroring or projecting—falsely claiming that the United States engages in dirty practices that Russia itself carries out: subverting governments, killing adversaries, spreading disinformation, and other nefarious activities. Russiagate involved a strange inversion of this pattern: the claim by some Americans that *Russia* had gained control of a proxy to influence the *American* political system. In this way, Trump's accusers mirrored those Russians who maintained that America was intent on advancing its geopolitical interests by acting through fifth columns in Russia.

Institutional continuities from the Cold War shaped both narratives. Both the United States and USSR had built massive security and intelligence bureaucracies designed to thwart the designs of their geopolitical rivals. Even after the Cold War ended and new threats emerged, menacing images of the other side lived on in popular culture and continued to shape the outlook of foreign policy professionals socialized during the Cold War. In Russia, having an external enemy was important to justify the continued funding—and existence—of the KGB, which was at risk of dissolution because its *raison d'être* of spying on Soviet citizens had been discredited (Colton 2008: 258). Western officials, ostensibly intent on expanding trade and promoting democracy, also maintained a defensive mindset, advocating for the expansion of NATO to Russia's borders and maintaining a large American troop presence in Europe. Yet, if Russian nationalists still saw the United States as its greatest foe, the threat posed by Russia had receded in the eyes of US policymakers, especially after 9/11, although it did not disappear. When Putin bridled at Russia's treatment by the West and presided over the military invasions of Georgia and Ukraine, he confirmed the suspicions of those who had distrusted Russia all along.

Beyond these parallels, fifth-column politics played out differently in these two cases as a function of their unique circumstances. In Russia, there was no interregnum between hardliners' criticism of Gorbachev and nationalists' broadsides against Yeltsin. In the face of Russia's diminished international status, there was little Yeltsin could do to conceal Russia's supplicant position in relation to the United States. Amid the shock of the Soviet collapse and Yeltsin's embrace of Western economic theories, it was to be expected that political coalitions would coalesce around attitudes toward reform. Opponents of reform could regroup politically by simultaneously mobilizing anti-Western sentiment and appealing to people's economic interests.

In the United States, claims by prominent politicians of government collusion with the Soviets waned after the Eisenhower administration; that they appeared so suddenly and vehemently in the 2010s speaks to the unique personage of President Trump. Unlike Communists who had never reconciled themselves to the Soviet collapse and had to watch their government dismantle the only system they knew—with catastrophic results—Democrats had not suffered a historic trauma and had no history of questioning the loyalties of elected officials. It was instead the exceptional, and odd, behavior by Trump, a self-proclaimed nationalist who nevertheless cavorted with suspicious foreign business partners, organized crime figures, and shady lobbyists, that stoked concerns that the presidency was vulnerable to infiltration. Whereas Yeltsin had little choice but to ingratiate himself with the American president and other officials with control over the purse strings for the sake of his country, Trump's suspicious behavior could not be explained by any foreign policy rationale. The remaining explanation—that Trump was undermining national security solely for personal gain—made his brazen behavior appear all the more inexcusable.

In both cases, tying a disliked head of state to an external threat became a rallying point for opposition forces. This strategy resembles the classic ploy of those in power, to impugn the motives of adversarial groups in order to deflate and divide them. As other chapters in this volume have shown, calling out subversive fifth columns can be cynically effective when it triggers instinctive patriotism and provokes defiance against perceived violations of sovereignty. Collusive fifth-column accusations operate according to the same logic, except they enable the building of coalitions among groups in the opposition against incumbents (Uscinski and Parent 2014). Insofar as this discursive strategy has a unifying effect, it can lead to greater political participation and resistance to overweening authority. However, by legitimizing thin accusations of disloyalty on the basis of a leader's unseemly friendliness toward a distrusted external actor, today's accusers risk becoming tomorrow's accused.

## Notes

1. Gaidar's pro-reform party won only 15.5 percent (McFaul 2001: 274).
2. All translations from Russian are the author's.
3. The letter was thought to be authored by Yegor Ligachev, a Politburo member and hardline opponent of perestroika.
4. Incidentally, Ilyukhin accused representatives of the "Jewish nation" working for Yeltsin of genocide for the heightened mortality rate of Russians. See "MP Blames Jews," 1998.
5. Before a visit to Russia in 2013, Trump tweeted, "Do you think Putin will be going to The Miss Universe Pageant in November in Moscow—if so, will he become my new best friend?" (Kaczynski et al. 2017).
6. A survey in July 2018 showed a 27-point increase among Republicans on positive views of Russia over 2015, with 59 percent saying Russia was either an ally or friendly (Guskin 2018).

## References

Alexander, Dan, and Richard Behar. 2019. "The Truth behind Trump Tower Moscow: How Trump Risked Everything for a (Relatively) Tiny Deal." *Forbes.com*, May 23. https://www.forbes.com/sites/danalexander/2019/05/23/the-truth-behind-trump-moscow-how-the-president-risked-everything-for-a-relatively-tiny-deal/#2594cdc7bc32.

Allensworth, Wayne. 1998. *The Russian Question: Nationalism, Modernization, and Postcommunist Russia*. Lanham, MD: Rowman & Littlefield.

BBC Monitoring. 1995a. "Party Political Broadcasts; Zhirinovskiy Election Broadcast: Opinion of Caucasians, Russian Should Be Only Language." Russia TV Channel, Moscow, November 22.

BBC Monitoring. 1995b. "Party Political Broadcasts and Debates; Zhirinovskiy Says Only His Party Could Rejuvenate Russia." *Russia TV*, December 14.

BBC Monitoring. 1996. "Zyuganov Calls on Electorate to Use Their Vote Carefully." *Radio Vozrozhdeniye*, Moscow, June 13.

Belov, Yuri. 2018. "Manevr sleva napravo." *Sovetskaya Rossiya*, August 18.

Bondarev, Yuri. 1995. "Stat' i idti o krizise idei, very i morali." *Sovetskaya Rossiya*, December 21.

Brudny, Yitzhak M. 1997. "In Pursuit of the Russian Presidency: Why and How Yeltsin Won the 1996 Presidential Election." *Communist and Post-Communist Studies* 30 (3): 255–275.

Brudny, Yitzhak M. 2000. *Reinventing Russia*. Cambridge, MA: Harvard University Press.

Buchanan, Patrick J. 2013. "Is Putin One of Us?" December 17. https://buchanan.org/blog/putin-one-us-6071.

Cash, John, and Catarina Kinnvall. 2017. "Postcolonial Bordering and Ontological Insecurities." *Postcolonial Studies* 20 (3): 267–274.

Chait, Jonathan. 2018. "Will Trump Be Meeting with His Counterpart—Or His Handler?" *New York Magazine*, July 9. https://nymag.com/intelligencer/2018/07/trump-putin-russia-collusion.html.

Cohn, Nate. 2018. "Did Comey Cost Clinton the Election?: Why We'll Never Know." *New York Times*, June 14. https://www.nytimes.com/2018/06/14/upshot/did-comey-cost-clinton-the-election-why-well-never-know.html.

Colton, Timothy J. 2008. *Yeltsin: A Life*. New York: Basic Books.

Crocker, Jennifer, Riia Luhtanen, Stephanie Broadnax, and Bruce Evan Blaine. 1999. "Belief in US Government Conspiracies against Blacks among Black and White College Students: Powerlessness or System Blame?" *Personality and Social Psychology Bulletin* 25 (8): 941–953.

Dart, Tom, and David Smith. 2017. "The Trump Dossier Doesn't Faze His Voters: 'I Haven't Been Following That.'" *The Guardian*, January 13. https://www.theguardian.com/us-news/2017/jan/13/trump-supporters-reaction-russia-dossier.

Dorell, Oren. 2017. "Donald Trump's Ties to Russia Go Back 30 Years." *USA Today*, February 15. https://www.usatoday.com/story/news/world/2017/02/15/donald-trumps-ties-russia-go-back-30-years/97949746/.

Entous, Adam. 2017. "House Majority Leader to Colleagues in 2016: 'I Think Putin Pays' Trump." *Washington Post*, May 17. https://www.washingtonpost.com/world/national-security/house-majority-leader-to-colleagues-in-2016-i-think-putin-pays-trump/2017/05/17/515f6f8a-3aff-11e7-8854-21f359183e8c_story.html.

Gessen, Masha. 2017. "Russia: The Conspiracy Trap." *New York Review of Books*, March 6. https://www.nybooks.com/daily/2017/03/06/trump-russia-conspiracy-trap/.

Gibson, James L., and Marc Morjé Howard. 2007. "Russian Anti-Semitism and the Scapegoating of Jews." *British Journal of Political Science* 37: 193–223.

Goldberg, Robert A. 2008. *Enemies Within: The Culture of Conspiracy in Modern America*. New Haven, CT: Yale University Press.

Goldman, Wendy Z. 2011. *Inventing the Enemy: Denunciation and Terror in Stalin's Russia*. Cambridge: Cambridge University Press.

Guskin, Emily. 2018. "How Much Will Republicans Follow Trump on Russia and Putin?" *Washington Post*, July 20. https://www.washingtonpost.com/news/the-fix/wp/2018/07/20/how-closely-will-republicans-follow-trump-on-russia-and-putin/.

Hall, Coby. 2019. "Donny Deutsch: 'Trump Is Owned by Putin Because He's Been Laundering Money for 20, 30 Years.'" *Mediaite.org*, October 24. https://www.media ite.com/tv/donny-deutsch-trump-is-owned-by-putin-because-hes-been-laundering-money-for-20-30-years/.

Hess, Amanda. 2019. "This Is the Moment Rachel Maddow Has Been Waiting For." *New York Times Magazine*, October 1. https://www.nytimes.com/2019/10/01/magaz ine/rachel-maddow-trump.html.

Halper, Stefan, and Jonathan Clarke. 2004. *America Alone: The Neo-Conservatives and the Global Order*. Cambridge: Cambridge University Press,

Harding, Luke. 2017. *Collusion: Secret Meetings, Dirty Money, and How Russia Helped Donald Trump Win*. New York: Vintage Books.

Hoffman, David. 1998. "Communist Party Chief Joins Attack on Zionism." *Washington Post*, December 25. https://www.washingtonpost.com/archive/politics/1998/12/25/ communist-party-chief-joins-attack-on-zionism/fc60bb53-7863-40df-a765-c4c04 25a8890/.

Hoffman, David E. 2011. *The Oligarchs: Wealth and Power in the New Russia*. New York: Public Affairs.

Hoover, J. Edgar. 1947. "Speech before the House Committee on Un-American Activities." March 26. https://voicesofdemocracy.umd.edu/hoover-speech-before-the-house-committee-speech-text/.

Jurecic, Quinta. 2019. "Explaining the Steele Dossier—and How Information Flows in Washington." *Washington Post*, November 27. https://www.washingtonpost.com/ outlook/explaining-the-steele-dossier--and-how-information-flows-in-washington/ 2019/11/27/3e684dca-0d34-11ea-8397-a955cd542d00_story.html.

Kachanovsky, Yuri. 1997. "Rezhim izmeny." *Sovetskaya Rossiya*, July 10.

Kaczynski, Andrew, Chris Massie, and Nathan McDermott. 2017. "80 Times Trump Talked about Putin." *Washington Post*, March 3. https://www.cnn.com/interactive/ 2017/03/politics/trump-putin-russia-timeline/.

Korey, William. 1995. *Russian Antisemitism, Pamyat, and the Demonology of Zionism*. Vol. 2. Chur, Switzerland: Harwood Academic.

Kozlov, Vladimir. 2017. "Rob Reiner, Morgan Freeman Help Launch Committee to Investigate Russia." *Hollywood Reporter*, September 20. https://www.hollywoodrepor ter.com/news/general-news/rob-reiner-morgan-freeman-help-launch-committee-investigate-russia-1041397/.

Kramer, Michael. 1996. "Rescuing Boris." *Time* 148 (4): 28–37.

Kumar, Anita. 2018. "Buyers Tied to Russia, Former Soviet Republics Paid $109 Million Cash for Trump Properties." *McClatchyDC.com*, June 19. https://www.mcclatchydc. com/news/politics-government/white-house/article210477439.html.

Kurtzleben, Danielle. 2016. "Trump and the Testosterone Takeover of 2016." *NPR.org*, October 1. https://www.npr.org/2016/10/01/494249104/trump-and-the-testosterone-takeover-of-2016.

MacDonnell, Francis. 1995. *Insidious Foes: The Axis Fifth Column and the American Home Front*. New York: Oxford University Press.

Mangan, Dan. 2017. "Donald Trump Says He's a 'Germaphobe' as He Dismisses Salacious Allegations." *CNBC.com*, January 11. https://www.cnbc.com/2017/01/11/donald-trump-says-hes-a-germaphobe-as-he-dismisses-salacious-allegations.html.

Matlock, Jack. 2004. *Reagan and Gorbachev: How the Cold War Ended*. New York: Random House.

McCarthy, J. 1953. *Major Speeches and Debates of Senator Joe McCarthy Delivered in the United States Senate, 1950–1951*. Washington, DC: US Government Printing Office.

McFaul, Michael. 2001. *Russia's Unfinished Revolution: Political Change from Gorbachev to Putin*. Ithaca, NY: Cornell University Press.

"MP Blames Jews for Russian Deaths." 1998. *BBC.com*, December 15, http://news.bbc.co.uk/2/hi/europe/235877.stm.

Mueller, Robert S. 2019. *The Mueller Report: Report on the Investigation into Russian Interference in the 2016 Presidential Election*. Washington, DC: US Department of Justice.

Mulloy, D. J. 2014. *The World of the John Birch Society: Conspiracy, Conservatism, and the Cold War*. Nashville, TN: Vanderbilt University Press.

Obeshchali-veselilis. 1997. *Sovetskaya Rossiya*, July 3.

Oushakine, Serguei Alex. 2010. *The Patriotism of Despair*. Ithaca, NY: Cornell University Press.

Radnitz, Scott. 2018. "If Putin Has Kompromat on Trump, How Might He Use It?" *Washington Post*, April 24. https://www.washingtonpost.com/news/monkey-cage/wp/2018/04/24/if-putin-has-kompromat-on-trump-how-might-he-use-it/.

Radnitz, Scott. 2021. *Revealing Schemes: The Politics of Conspiracy in Russia and the Post-Soviet Region*. New York: Oxford University Press.

Rid, Thomas. 2020. *Active Measures: The Secret History of Disinformation and Political Warfare*. New York: Farrar, Straus & Giroux.

Roberts, Cynthia, and Thomas Sherlock. 1999. "Bringing the Russian State Back In: Explanations of the Derailed Transition to Market Democracy." *Comparative Politics* 31 (4): 477–498.

Rutland, Peter. 2003. "Putin and the Oligarchs." In *Putin's Russia*, edited by Stephen Wegren, 133–152. Lanham, MD: Rowman & Littlefield.

Sanin, A. 1999. "Meru stolitsy i senatoram. Ne stoite na puti slavyanskogo bratstva! Uvazhaemyi Yuriy Mikhailovich!" *Sovetskaya Rossiya*, February 13.

Scott, James C. 2008. *Weapons of the Weak: Everyday forms of Peasant Resistance*. New Haven, CT: Yale University Press.

Stanley, Alessandra. 1996. "Moscow Journal: The Americans Who Saved Yeltsin (Or Did They?)." *New York Times*, July 9. https://www.nytimes.com/1996/07/09/world/moscow-journal-the-americans-who-saved-yeltsin-or-did-they.html.

Stokes, Bruce. 2015. "Republicans and Democrats Sharply Divided on How Tough to Be with Russia." *Pew Research Center*, June 15. https://www.pewresearch.org/fact-tank/2015/06/15/republicans-and-democrats-sharply-divided-on-how-tough-to-be-with-russia/.

Taibbi, Matt. 2018. "The New Blacklist." *Rolling Stone*, March 5. https://www.rollingstone.com/politics/politics-news/the-new-blacklist-202612/.

Thomas, Martin. 2008. *Empires of Intelligence: Security Services and Colonial Disorder after 1914*. Berkeley: University of California Press.

Treisman, Daniel. 2011. *The Return: Russia's Journey from Gorbachev to Medvedev*. New York: Simon & Schuster.

Tsygankov, Andrei. 2009. *Russophobia: Anti-Russian Lobby and American Foreign Policy*. New York: Palgrave Macmillan.

Tsygankov, Andrei P. 2014. *The Strong State in Russia: Development and Crisis*. New York: Oxford University Press.

Uscinski, Joseph E., and Joseph M. Parent. 2014. *American Conspiracy Theories*. New York: Oxford University Press.

Van Prooijen, Jan-Willem, and Karen M. Douglas. 2017. "Conspiracy Theories as Part of History: The Role of Societal Crisis Situations." *Memory Studies* 10 (3): 323–333.

Vogl, Frank. 2020. "What Does Putin Have on Trump?" *Salon*, May 2. https://www.salon.com/2020/05/02/what-does-putin-have-on-trump_partner-2/.

Wedel, Janine R. 2015. *Collision and Collusion: The Strange Case of Western Aid to Eastern Europe*. New York: St. Martin's Press.

Weiner, Tim. 2012. *Enemies: A History of the FBI*. New York: Random House.

Wemple, Eric. 2020. "The Steele Dossier Just Sustained Another Body Blow: What Do CNN and MSNBC Have to Say?" *Washington Post*, April 18. https://www.washingtonpost.com/opinions/2020/04/18/steele-dossier-just-sustained-another-body-blow-wheres-media/.

Wolff, Michael. 2018. *Fire and Fury*. New York: Henry Holt.

Wright, Austin. 2017. "Schiff Concerned Russia Has Compromising Material It Could Hold over Trump." *Politico*, July 11. https://www.politico.com/story/2017/07/11/schiff-trump-russia-conspiracy-blackmail-240422.

Yablokov, Ilya. 2018. *Fortress Russia: Conspiracy Theories in the Post-Soviet World*. Cambridge: Polity Press.

Yeltsin, Boris. 2000. *Midnight Diaries*. Translated by Catherine A. Fitzpatrick. New York: Public Affairs.

# 6

# "Sellers of the Homeland"

## Narratives of Treason and Fidelity in Afghanistan*

*Robert D. Crews*

Bandit
For he who is a bandit of Islam and homeland
How would the earth give him a burial place?
The angels of the sky will curse him,
Hell will be ready for the country-seller and treacherous one.
For he whose hands are red with the blood of the oppressed nation,
God's torment will fall on the infidel.
As I would have dedicated my martyred spirit,
O God, I made him the red flower of a tulip for the garden.

(Gumnam 2013: 90)

Observers of Afghan politics have long privileged the role of tribe, ethnicity, and a weak state in understanding the history of the contest for power in Afghan society. In addition to reproducing colonial representations of Afghanistan, this perspective has neglected the history of Afghan political thought, overlooking the emergence and evolution of a dynamic political language that has placed treason and foreign interference at the center of narratives of political legitimacy and national belonging. Fifth-column claims in Afghanistan are as old as the modern Afghan state—and indeed are central to the history of its formation within a context of intense geopolitical competition and extended periods of incomplete sovereignty. Afghan elites and non-elites alike have resorted to this language of treason and fidelity to make sense of the interplay of internal and external political challenges and of the recurrence of horrific violence and the breakdown of social trust. Yet, as with other aspects of Afghan political thought,

* I thank Sabauon Nasseri, the editors of this volume, and our fellow University of Washington workshop participants, especially Cabeiri Robinson, for their very incisive comments and valuable suggestions for revising this chapter.

Robert D. Crews, *"Sellers of the Homeland"* In: *Enemies Within.* Edited by: Harris Mylonas and Scott Radnitz, Oxford University Press. © Oxford University Press 2022. DOI: 10.1093/oso/9780197627938.003.0007

fifth-column politics has varied over time. These charges have ranged across a spectrum encompassing what Radnitz and Mylonas in this volume call "subversive" and "collusive" claims, targeting state elites and their rivals as well as accusing whole communities of collaborating with outsiders. In the Afghan context, accusations aimed at elites who have "sold the nation" to more powerful neighbors or who have become "puppets" of a great power have been the most pervasive. I follow the editors' understanding of a "collusive" claim as one that points to "hostile activity from above" by a small number of actors "whose secretive actions appear to serve malign foreign interests" (9).

In the Afghan setting, such allegations should be understood as a highly emotive mode of conspiracy theorizing. This does not mean that such thinking is pathological or even, in many cases, inaccurate. Nor does it suggest that Afghans are uniquely or exclusively drawn to conspiracy thinking or emotion. Rather, as Mark Fenster notes, we might think of the conspiracy theory as a highly reasonable "political strategy." It is at once "a cultural practice of interpretation" and "a narrative form" that circulates in a variety of political settings (Fenster 2008: 9–13). What is specific to the Afghan variety of this global practice is that, in modern times, Afghans have generated various kinds of fifth-column claims to make sense of a realm of security politics contested, largely in secret, by a very small Afghan elite interacting with a lengthy list of international actors. At the same time, competing elites have utilized this language with the aim of discrediting rival political projects. Thus conspiracy narratives about fifth columns in the Afghan context have been both an "interpretive practice" and "political strategy," to use Fenster's productive language. In a country whose past has been stamped by repeated foreign interventions and, since 1978, decades of traumatic violence and physical displacement, the precarious nature of security, trust, and survival has made resort to such mechanisms crucial in making sense of the world (Mills 2013: 240–241).

Such strategies have gone beyond linking opponents' ideologies with foreign interests and have evolved to advance claims aimed at evoking particular emotions. As we will see, khiyānat—"treason" or "treachery"—has meant more than political disloyalty: it has also indicated a moral offense. It has signified betrayal of one's compatriots and often, of Islam, frequently symbolized by the traitor's supposed penchant for alcohol, drugs, illicit sex, and pornography (Pahwal 2004). But in such narratives the identity of the traitor has not always been visible on the surface: traitors have allegedly hidden behind masks. Thus it has been the duty of the righteous and patriotic vanguard to reveal the true face of villainy to the nation.

Long before the image of a fifth column gained traction in the early twenty-first century, the charge of "selling the homeland" (watan furushi) to foreigners operated as a keyword of political disputation. Used in Dari, Pashto, and Arabic, the term watan in the Afghan context, as Bernt Glatzer has argued, "has an emotional quality close to the German Heimat, a geographical and social area where I feel at home, where I belong, where my family and my relatives live, where I can rely on the people, where I feel security and social warmth" and also "has the connotation of something treasured and vulnerable, something which has to be defended like the female members of a man's family" (2001: 382). Thus "selling the *watan*" reflected base transactional and material interests and epitomized betrayal of the nation, and in some versions, of Islam. For all the variability of this discursive strategy, various actors have utilized it to target competing political movements. They have sought to energize an Afghan public whose impassioned outrage they hoped to mobilize to demonstrate fidelity to the nation (and to Islam) by supporting the political agenda of those willing to engage in the self-sacrificing act of unmasking the "sellers of the homeland" (Blom and Lama-Rewal 2020).

## Imperial Legacies

The distinctive history of the making of the modern state has left Afghans with a formative legacy: at crucial moments over the course of the nineteenth century, intervention by imperial powers determined the outcome of power struggles within the country. More than any other episode, the British invasion that restored Shah Shuja-ul-Mulk to the throne in 1839 (in what would become known as the First Anglo-Afghan War) haunted the consciences of later Afghan rulers. In the shadow of Russian imperial expansion in Asia, another confrontation with the British empire in the Second Anglo-Afghan War (1878–1880) facilitated the victory of 'Abd al-Rahman Khan (r. 1880–1901) over his Afghan rivals. In each of these conflicts, European backing played a central role in the contest for the throne. In the autobiography attributed to 'Abd al-Rahman, the architect of the modern Afghan state sought to deflect collusive claims against him and point instead to would-be Afghan fifth columns seeking power by cooperating with foreigners. "I cannot help being suspicious when I call to mind past historical events in Afghanistan," he warned, "kings have been murdered, unjustly dethroned and treacherously taken prisoners by their internal and external *friends*, as I may call them." "It is natural," he continued, "that a person should be suspicious when he finds himself surrounded by selfish people who are longing to snatch the first opportunity to take a slice of Afghanistan, just like robbers keenly watching the housekeeper" (Sultan 1900, vol. 2: 241–242).

'Abd al-Rahman cautioned against allowing foreign influence in his kingdom. He warned that such a presence might risk "foreign intrigues in the country, for the purpose of causing disruption among the tribes, and so dividing up the country." He also attributed the supposedly disloyal inclinations of his subjects to the fact that they "are not sufficiently patriotic to understand the value of having their own ruler," adding that "the Amir is obliged, for his own safety, to be on his guard against the plottings and machinations of a people who have killed their kings and chiefs in the past, and who are always intriguing with the amir's enemies both inside and outside his dominions" (Sultan 1900, vol. 2: 145; vol. 1: 259). Like his spy networks, his policy of cultivating dissidents from neighboring states, and his admonitions lauding suspicion, this notion that Afghans were perennially lacking in loyalty and national feeling long outlived the amir. The imperial context of Afghan state-building made the specter of dissident collaboration with foreign empires a fixture of monarchical statecraft.

But looking at the Afghan case more closely also reveals how varied and adaptive such narratives might be. 'Abd al-Rahman himself was branded an infidel in 1881 by mullahs in Kandahar because of his collusion with the British (Kakar 1979: 153). At the same time, the amir engaged in fifth-column politics by singling out groups that might have a particular affinity for foreign protectors because of their religion and ethnicity. He cast doubt on the loyalties of groups who presented a challenge to state control, who occupied a kind of "outsider" status vis-à-vis the monarchy's attempts to create religious and cultural markers of a loyal, national community, *and* who, it could plausibly be charged, maintained loyalties beyond the borders of the kingdom. In this scenario, Shi'i Muslim communities figured as exemplars of disloyalty to the Afghan state and nation. Moreover, their ostensibly heretical treachery was compounded by their legibility as simultaneously ethnic others. These included the Qizilbash and, especially, the Hazaras (Ibrahimi 2017). Critics faulted their supposed vulnerability not just to the charms of the British but also to Qajar Iran, whose king and religious scholars sometimes positioned themselves as the protectors of their Shi'i co-religionists. For Afghan ruling elites, fifth-column politics demonizing Shi'i Hazaras, often accompanied by sanctioned violence against them, has again and again operated as a strategy aimed at rallying the support of ethnically diverse communities under the banner of Sunni Islam and a more exclusive nationalism.

During military campaigns aimed at subjugating the Hazaras, Amir 'Abd al-Rahman raised doubts about their capacity to become loyal subjects. In addition to Hazara ties to Shi'i Islam, their supposedly foreign origin and mythic ties to Genghis Khan would endure as the distinguishing features of the Hazaras that accounted for their status on the margins of Afghan society. Indeed the amir's autobiography maintained that:

> The general belief in Afghanistan is that a great many of the western invaders of India were in the habit of giving houses and lands to their own people all along the road to India, in order to safeguard their rear; and that this is the reason why the Mongols planted the Hazaras from one end of Afghanistan to the other, from west to east, just as Alexander the Great did in the case of the so-called Kafirs from Kokand and Badakshan to Chitral and the Panjab borders. These people are all Shias.

Similarly, he observed that the inhabitants of Kafiristan (an area inhabited by non-Muslims and thus disparaged in official sources as "The Land of the Infidels") were a "warlike nation on the whole north-western border of Afghanistan, from east to west," who "would be the cause of great anxiety from the rear, at a time when my Government might be occupied in a war with any other country." The amir was also wary of the vulnerability of this population to Christian proselytization:

> It is the habit of some Christian missionaries to interfere wherever they have an opportunity, and I thought that they would make unnecessary trouble about my conquering Kafiristan; it was therefore necessary to lose no time in getting the fighting over and annex the country before the news could be spread abroad. (Sultan 1900, vol. 1: 276–277, 288–290)

Nonetheless, "traitors" could also be found among the subjects who were ostensibly closest to the monarchy. Pashtun tribal groups were a constant source of anxiety about anti-state subversion. They repeatedly threatened the central government and its local representatives with armed rebellion.

At court, too, dissidents could turn on the sovereign, as Amir Habibullah (r. 1901–1919) learned in February 1919, when members of his retinue who opposed his conciliatory approach to the British assassinated him during a hunting trip. An Afghan Hazara scribe who served Habibullah and his son Amanullah (r. 1919–1929) left an account of the seditious conduct of an Afghan notable that shifted suspicion, in a somewhat coy fashion, from the Hazaras to the Pashtuns. Raising doubts about the fealty of the Pashtuns to crown, nation, and Islam and highlighting their vulnerability to British manipulation, the Hazara scholar observed that

> every political provocation which has occurred and still occurs in Afghanistan, on its eastern and northern borders, is not connected so much with the ignorance and savagery of the people or the corruption and oppression of government, as it is with the subversive activities of the English government. Because of its rivalry with Russia, England gives the Pushtuns significant aid and support, for it sees in them an iron shield to repel any attack on Afghanistan by her rival. If these rumors correspond in any way to reality, then, despite Pushtun assurances of their devotion to Islam, one must consider even their leaders to be godless ignorant people. If they truly believed in the Shari'ah of the Prophet or if they had spent any time reading the Koran or listening to mullas recite the verse which says: "O believers! Take not Jews and Christians as friends; they are friends of each other. Whoso of you makes them his friends is one of them. God

guides not the people of the evildoers!" then they would never submit to the provocations of England, which professes Christianity.

Besides invoking the truth of the Qur'an, the author added that, in addition to having grown too close to the English, a key Pashtun leader stood accused of "having an affair with Aman Allah's elder sister"—an offense the amir initially opted to punish with the death penalty (McChesney 1999: 51–52). As these examples illustrate, fidelity to Islam was a crucial marker of political allegiance, one that subversive ties to the British or others could taint.

When a rebellious figure who took on the title Habibullah II forced Amanullah from the throne in 1929, the upstart issued a decree castigating his main military rival (the future Nadir Shah, r. 1929–1933), claiming that Nadir stood at the head of a group of "traitors" who collaborated with the British, offering as proof the assertion that he had "behaved like a kafir [infidel] while in Europe and ate a lot of pork, which has blackened his bones and marrow." "But thanks to the grace of God," the decree continued, "these devils will not be able to lead the faithful into error but will only corrupt themselves," declaring "the blood of these traitors is lawful according to the Shari'ah" (McChesney 1999: 98).

## Servants of Aliens and Betrayers of the Country

During the 1930s, with the establishment of a growing foreign presence, Germany, Italy, the USSR, Japan, France, and Great Britain competed for influence among elite circles in Kabul and other major towns, making fifth-column claims of collusion with foreigners a dynamic instrument of elite contestation. The Afghan government carefully sought to find a balance among these powers, warily watching Amanullah (who had fled to Italy) and his followers in the diaspora and within the country. Despite Afghan neutrality during World War II, a number of Afghan notables gravitated toward the Third Reich, some doing so with an eye to gaining German support to return Amanullah to power. With the conclusion of the war, though, critics of the authoritarian politics of the Afghan state aspired to introduce republican and democratic ideas. Some looked to the left in the neighboring Soviet Union, a development that the government used to repress its challengers, arresting those accused of being in league with the Soviets. In subsequent decades, the authorities would face challenges from various socialist-oriented groups as well as from Islamists who looked abroad to organizations such as the Muslim Brotherhood.

Following a coup d'état in July 1973, government repression of perceived opponents intensified. Ending the reign of Muhammad Zahir Shah (r. 1933–1973), his cousin, Muhammad Daoud, declared a republic. Having drawn upon leftist forces, Daoud moved against the Islamists. He also went after rival politicians and military officers, accusing a number of them of treason. Arrests and forced confessions followed. The most high-profile case targeted a former prime minister, Muhammad Hashim Maiwandwal. On October 3, 1973, the official *Kabul Times* published what it claimed was a confession written by Maiwandwal, who, it charged, had been apprehended plotting a coup against the new order. The paper claimed that he belonged to the ranks of reactionaries who regularly appear after patriotic revolutions, to "jeopardize the destiny of their nation, and millions of innocent human beings are mass murdered." The attribution of ruthless and extreme forms of violence via radio and print media would become a hallmark of Afghan revolutionary rhetoric, as would the tendency to refer to the ambiguous, even anonymous, identity of the foreign countries that supposedly orchestrated such conspiracies:

> These servants of aliens and betrayers of the country not only are not ashamed but actually enjoy the slaughter of innocent children, liquidation of thousands of patriotic students and making homeless millions of their compatriots under the orders and instructions of their masters. ("From Confession to Suicide" 1973)

Given Daoud's lengthy history of fifth-column politics in the form of support for the Pashtun and Baloch movements against Pakistan (and the Afghan government's provision of sanctuary for dissidents from among these communities), the Afghan public likely understood this formulation as an accusation aimed at Islamabad. However, the lack of specificity appears to have been a careful choice, one made to avoid direct confrontation with Afghanistan's neighbor and, perhaps more important, to enhance calls for vigilance against an enemy whose identity could not be known with certainty.

Daoud ultimately extended his campaign of repression to the left in a move that exacerbated his isolation—and provoked a coup d'état in April 1978 by the People's Democratic Party of Afghanistan. Dominated by the Khalq ("Masses") branch of the party, the new government proclaimed a "Democratic Republic of Afghanistan" and unleashed a brutal campaign against the "feudal" elements of society, murdering Daoud and his relatives, together with other elites, including major religious families. While they introduced many of the symbols and keywords of the Soviet political system, the PDPA had many local, Afghan political resources to draw upon. These included Daoud's personality cult and his pursuit of shadowy "traitors" who stood in the way of his revolution. The revolutionaries' hunt for furtive enemies quickly extended to the party itself, deepening older divisions between its Khalq and Parcham ("Banner") factions.

The language of the Khalqi campaign against subversive counterrevolutionaries differed from Daoud's in that it relied more on Marxist categories, while also seeking to evoke emotional responses. Official media portrayed the coup of April 1978 as the triumph of the "toiling masses" over the "feudal" forces of counterrevolution. In October 1978, the *Kabul Times* featured highly stylized interviews, supposedly with common Afghans, to illustrate popular and even joyous enthusiasm for the revolutionary purge of members of the Parcham faction:

> Nek Mohammad, a vegetable grower from Bib Mehro speaking to the reporter of the daily Kabul Times regarding the arrest of the Babrak Karmal clique said that he was overjoyed when he heard [of] the arrest of the traitors. He said that the traitors were a selfish lot who wanted to stage a coup d'etat in the interest of feudals and their henchmen. . . . He said that he was sure that they were instigated from beyond the borders of Afghanistan and were working for a foreign country and its lackeys in Afghanistan. They do not want peace and tranquility to prevail in Afghanistan. ("People Join State in Condemning Traitors" 1978)

Self-interest, combined with action on behalf of class foes and unnamed foreigners, was the hallmark of a traitor. Thus a welder added that, "like Daoud, the hangman," these traitors "were thinking of their own comfort," whereas "[T]he patriotic duty of every compatriot is to work selflessly for the welfare of the people and for the consolidation of the people's regime." The enemies of progress and development—and the external forces who controlled them—were the only ones who stood to gain from violence and upheaval.

Such language employed class categories, but it was also infused with emotive, moral, and religious arguments. A statement attributed to one of the peasants featured in the article asserted that the purged leaders had committed a "grave sin against humanity and their own country." Moreover, in a gesture aimed at countering the long-standing charge that Afghan socialists were "atheists," another peasant "wished the God Almighty to bestow his blessings upon the Democratic Republic of Afghanistan" ("People Join State in Condemning Traitors" 1978).

Khalqi rhetoric made another crucial adaptation of Afghan narratives of treason in the 1970s by highlighting the role of the subversive enemy who was unseen and concealed behind "masks." The history of the Soviet Communist Party offered important lessons here, of course, and a number of key Afghan leftists seem to have revered Stalin as a model revolutionary leader. However, it is also possible that anxiety about the "hypocrite" (*munafiq*), the subject not only of warnings in the Qur'an but of Sufi poetry and other literary forms with which Afghans would have been familiar, may have played a role in popularizing this image. The same interview series mentioned above reveals how such deception might be dramatized:

Mohammad Ali, a peasant from Joi Bagh said that the recent plot uncovered by our patriotic Khalqi elements was lead [sic] by the great demagoge [sic] Babrak Karmal who was in fact a wolf in sheep's clothing. Whoever works or intends to work against our Khalqi state and the interest of the majority of the people of Afghanistan will be annihilated. ("People Join State in Condemning Traitors" 1978)

The figure of a wolf-like Karmal donning the mask of a sheep and this peasant's fervor for the idea that Karmal would be killed as an expression of the popular will suggest how the authors of such pieces sought to craft narratives that would simultaneously ascribe a penchant for brutal violence to "traitors" and justify violent action in defense of a betrayed and wounded people.

## Puppets of the Foreigners

The Soviet invasion of December 1979 presented the leadership of the Democratic Republic with a strategic burden. Cries by the opposition—who gathered under the banner of *jihad*—that the "communists" had become "the puppets of the foreigners" proved difficult to refute in a convincing way at home or abroad (Gumnam 2013: 84–85). However, the anti-Soviet mujahideen came to suffer from an analogous liability. From the summer of 1979, opposition groups started to accept aid from the United States. Meanwhile, their leaders established offices in Pakistan and Iran, where the Inter-Services Intelligence (ISI) and Ayatollah Ruhollah Khomeini's revolutionaries sought to manage the activities of these groups. Soon, at least a dozen nations and hundreds of NGOs had become engaged in a remarkably internationalized Cold War theater (Crews 2015).

Given the dependence of both sides on foreign support, it is striking that the mujahideen enjoyed such success in wielding the accusation of collusive activity to undermine the Kabul government. Their approach carried such force because, of course, it was very challenging to refute given the realities of occupation by the Soviet Union, a polity broadly associated with atheism. Moreover, the mujahideen made extensive use of Islamic linguistic and legal repertoires. Nonetheless, paradoxically, both sides of this struggle developed rhetoric in common, accusing their rivals of falsity and deceitfulness in the betrayal of Afghanistan through service to foreign powers.

A collection of accounts about the jihad under the title *Kandahar Assassins* reveals how both sides came to portray the figure of the traitor as "hypocrite." Committed to exposing such villains, a narrator from among the ranks of the mujahideen explained how:

During the time of the Khalqis there were people who would stand in the first line of prayer at the mosque together with them and would claim to be doing jihad. However, these Khalqis in Muslim forms would bring many innocent humans to the hands of the oppressor executioners. They would themselves appear Muslim and thus they would not be under suspicion but behind the scenes they would be devils and would carry out great treacheries against the Islamic community. Great *Allah* revealed these hypocrites in the holy lap of *jihad*. (Gumnam 2013: 87)

The mujahideen focused on actors who had the capacity to unify or divide the nation, to create consensus or dissent. For the leftists, teachers had served as a vanguard entrusted with bringing literacy and other aspirations of the revolution to the countryside, often by force. The figure of this "ambitious deviant" then became a target of indignation for the mujahideen:

A teacher is meant to be a guide but most of the teachers in our community hang devious titles on their chests. Our homeland expects a lot from the teachers; the foundation of society is connected to teachers. Any role that teachers play also has an impact on society, so when a teacher is devious or has gone astray, then society will be spoiled and such a teacher's death is legitimate and should be carried out just with a dull knife blade. Most of our misfortunes have been due to these sinister and country-selling teachers. (Gumnam 2013: 80)

Another archetypal villain was the government spy attached to the main intelligence organ, KhAD (State Information Services):

As each day passes the KhAD secret police take their place among us and cause disunity. Our major complaint is that people have turned their backs on us. If our old employees, commanders, media people, soldiers, officers, writers, and enlightened and knowledgeable people were with us, we would be able to create a great system. . . . [O]ur jihad would have improved; there would be no place for the spies of the enemy. (Gumnam 2013: 99)

Echoing the regime's calls for unity and self-sacrifice, the mujahideen faulted infiltrators for sowing ethnic and political discord while at the same time lamenting their abandonment, even betrayal, by other social groups who retreated from the struggle to enjoy safety and comfort. All of this was the cause of "other types of disunity in the form of ethnic or party divisions." "We can say, sadly," the narrator continued, "that our power brokers and businessmen have been unfaithful to us." Referring to Afghan people of means who opted to flee, the mujahideen charged that,

[T]hey left this soil of champions and, in their words, "they saved themselves from death." They sold the love of the homeland for the luxury of foreign countries. They left us vagrants. (Gumnam 2013: 99)

As a means to delegitimize opponents, the charge of "selling the homeland" was an elastic and adaptive way of making sense of the complex and dynamic political reasoning that shaped the choices Afghans made during the jihad. Crucially, during a period of violent dislocation, anxiety, and social and political upheaval, it was the ubiquity of international resources—and Afghans' own sense of their past and precarious struggle for survival in an uncertain present—that gave fifth-column rhetoric such force. Nonetheless, it remains difficult to assess the extent to which various actors found such allegations credible, especially when the revolutionary government accused the opposition of distorting the true meaning of Islam and selling out to the Americans, Pakistanis, Chinese, and other "capitalists" and "imperialists." Clearly, the spokesmen of the Democratic Republic sought to popularize the idea, while the media arm of the mujahideen parties as well as much of the international press outside the Soviet bloc refuted it (Ahmadi 2008: 119–139). At the same time, Afghan politicians who engaged in this discourse or who were its targets seem to have recognized that the charge of "selling the homeland" was to be answered by returning the allegation.

The regime and its foes traded accusations about infidelity to the nation and Islam, but when the Soviets withdrew in 1989, and as opposition forces closed in on the capital in 1992, the mujahideen factions who clamored for power resorted to adapting these narratives to discredit other opposition groups. Although all parties had received foreign support, once again the Hazaras became a target. But this was not simply the product of a "return" to ethnic or sectarian scapegoating. Under the leadership of the head of the Hizb-i Wahdat (Unity Party), 'Abdul 'Ali Mazari, Hazaras sought to leverage their role in the jihad to claim a political share in the new order. Mazari campaigned for an Islamic state structured along federal lines that would afford equal rights to the country's diverse ethnic and religious groups, including recognition of his (mostly) Shi'i community (Ghaffārī La'lī 2018: 24–25, 43–44, 279–280, 353–354). Rival parties largely opposed his efforts, though, and the struggle for control of the new government in Kabul descended into fighting among mujahideen factions that frequently resorted to strategies of ethnic mobilization (Ibrahimi 2017).

Many opponents of the Hazaras simultaneously tried to cast their demands for political inclusion as an Iranian plot. Throughout the 1980s, Tehran had attempted to coordinate the resistance among Afghan Shi'i groups (among others), offering arms and other forms of support. Iranian authorities also sought to build upon clerical ties that preceded the Soviet invasion to propagate Khomeini's revolutionary Shi'i project, all while hosting several million Afghan refugees. Some Hazara activists were initially enthusiastic about the revolution in Iran and took inspiration from its model of clerical leadership. Iran proved to be a problematic patron, however. Discrimination against Afghan migrants in Iran presented one impediment. Another arose from the conviction in many mujahideen circles that Iran should not enjoy a monopoly on Islamic interpretation.

For Mazari, the challenge was to gain access to a share of the national government on behalf of the Hazaras while diffusing the incendiary charge that he served Iran's interests. As an intellectual with extensive training in Shi'i seminaries and ties to clerical networks in Iran and Iraq, his religious credentials were a key asset. Yet he also positioned himself as an advocate of Hazara inclusion within a more pluralistic Afghan national context. In interviews with Afghan and international journalists, Mazari repeatedly deflected the implication that he was dependent upon Iran. At the same time, he carefully disavowed representing ethnic or sectarian differences. In fact, he rejected the idea that the disputes among the mujahideen were essentially about such claims. This strategy also led Mazari to adopt the practice of criticizing rival leaders individually, without highlighting their association with particular parties or, more important, the ethnic groups they came to represent more explicitly in the early 1990s (Ghaffāri La'li 2018: 49, 349, 395, 406). However, he was not above suggesting some of his opponents were engaged in treasonous politics. In January 1995, for instance, he somewhat obliquely suggested that his rival Ahmad Shah Massoud was acting as a "betrayer of the nation" (khā'in-i milli), claiming that he was stoking conflict by insisting on the ethnic apportionment of power and thus "does not honor the territorial integrity of Afghanistan or respect its independence" (Ghaffāri La'li 2018: 439).

In 1994, a new force, the Taliban, appeared on the political scene with the aim of sweeping away such internecine conflict and inaugurating a truer form of Islamic governance; however, this movement, too, faced the challenge of overcoming rejection by those who condemned it as a fifth column of Pakistan. A local history of the southwestern province of Nimroz, written by a local Baloch intellectual, offers a window onto how many Afghans saw the rise of the Taliban. The author, Abdurrahman Pahwal, was a staunch opponent of the Taliban. In narrating his critique, he drew extensively on many of the tropes of treason that recent establishment politicians—and the Taliban themselves—had used to delegitimize their enemies, even accusing the latter of drug-smuggling and the making of pornographic films (Pahwahl 2004: 96). Invoking a canonical figure associated with "selling of the homeland," Pahwal likened the Taliban's first appointed representative (wali)—and his "great stature, long and full beard, turban, and the way he sat on his chair"—to Shah Shuja, "who in his time [the First Anglo-Afghan War] came to power in Kabul with the help of the English" (Pahwal 2004: 17). In this scenario, Pakistan had become like England: "Just as India was like the 'Golden Calf' for the English colonial masters, and the English conquered India in order to seize its riches, the Taliban, at the instruction of Pakistan, have tried to conquer Nimroz, which due to its geographic position and brisk trade, seemed like a 'Kuwait' to them" (Pahwal 2004: 22). Even the anti-Baloch attitude of the Taliban was rooted in Pakistan, whose leader, Benazir Bhutto, Pahwal maintained, was anti-Baloch. As more and more Taliban, whom Pahwal castigated as "English slaves," arrived in Nimroz from Pakistan, "[G]radually, Zaranj [the capital of the province] became a Pakistani city" (Pahwal 2004: 22–26).

Shifting from fifth-column politics focused on elites, Pahwal asserted that the Taliban were assisted by "local helpers" of all kinds in Nimroz. These "various dubious elements" came forward to gain access to government positions, thereby getting their hands on wealth and power. Echoing one of the major themes of mujahideen propaganda about the dangers of the hypocrite and the figure of the schoolteacher, Pahwal pointed out that

These were the same people who during the time of the communists' rule had worked as teachers in the secular schools of communism and who acted at that time as if they had previously devoted their entire lives exclusively to the struggle for the liberation of oppressed people. In reality, it was now about how to get as many new privileges as possible or to keep from losing the ones they already had.

Denouncing others who would be sent to jail or the gallows, "these devilish traitors" also now pretended to be loyal to Islam but, in truth, Pahwal insisted, "through informers or through the skillful manipulation of tribal, religious, and linguistic differences," they pursued their "ominous ambitions." They succeeded, as long as "their sinister faces were not recognized, and they themselves were not exposed" (Pahwal 2004: 29–30).

Despite such critiques, the Taliban repeatedly disavowed or downplayed receiving support from Pakistan (as well as from Saudi Arabia and the Gulf states). And, in a more contradictory fashion, they sought to deflect allegations that they were in any way dependent upon Osama bin Laden, al-Qaeda, or other terrorist groups—or that they in any way subordinated Afghan interests to those of foreigners. Following the attacks of September 11, 2001 on US soil and the American-led invasion of Afghanistan, the dispersal of the Taliban leadership and their flight to Pakistan deepened their reliance upon Pakistan. But the government that succeeded them in Kabul was also dependent on foreign backing. From their sanctuary in Pakistan, Taliban intellectuals capitalized on this reality to undermine the legitimacy of Hamid Karzai (r. 2001–2014), whom the United States recruited, along with other figures from the Afghan diaspora, to return to the country to rule. It was not terribly difficult for skeptics of various stripes to cast Karzai as the latest "Shah Shuja" placed on the throne by a foreign army.

The Taliban launched a relentless campaign to demonize the "puppets" and "dog washers" of the Americans who joined the new political system. They focused ire on the foreign intervention and "occupation" and then sought to discredit the post-Taliban government. "The Bonn conference," the former Taliban official Mullah Abdul Salam Zaeef wrote in his memoirs, "imposed American ideas and inflicted certain Afghans on the Afghan people." According to Zaeef, "[T]he alliances they made, cooperating with known warlords and war-criminals and helping the same people back to power who had once before ruined the country and bled its people—all of this was a policy failure" (Zaeef 2010: 239). Similarly, in his memoirs, another Taliban figure, Sayyed Mohammad Akbar Agha, underscored continuity between the Soviet-backed government and the post-2001 order, arguing that "the remnants of that puppet regime now have power in all parts of this government." "It is obvious," he concluded, "that foreigners are using them to reach their own strategic goals here" (Akbar Agha 2013: 160–161).

The Taliban claim of collusive fifth-column activity on the part of the Kabul government and its US backers evolved to blend the invocation of international legal norms and human rights language with claims of ethnic solidarity. "The Bonn Conference," Zaeef wrote, "through which America imposed its will by bringing together a small group of Afghans, often hailed as a groundbreaking moment, was a bigger violation of Afghanistan's independence than the American invasion." "America," he charged, "gave power to the Northern Alliance in order to strengthen its own position, and suppressed Pashtuns while calling them 'Taliban.'" Zaeef maintained that "there were no real representatives of Afghanistan at Bonn," adding that "[T]he decisions made were illegal in any sense," and that Pashtuns "were underrepresented, even though President Karzai is Pashtun" (Zaeef 2010: 240). Instead of relying on Pashtuns, presumed here to be Afghanistan's natural rulers,

> America made an irreversible mistake in their choice of friends, ignoring their history with Afghanistan. The Afghan allies they chose were often warlords who had returned to Afghanistan in the wake of battle, using America and damaging the very foundations of the new Afghanistan they planned to create. (Zaeef 2010: 242–243)

Zaeef appears to have composed his memoirs with multiple audiences (including international readers) in mind. However, in Taliban publications meant more explicitly for Afghans there has been a similar focus on the illegitimacy of Afghans serving in the post-2001 government and especially on members of the Islamic clerical establishment who have sided with the new authorities against the Taliban. For example, in 2004 a satirical article in the Taliban publication *Voice of the Mujahed* featured an interview with Satan, who had taken on the physical appearance of the Afghan enemies of the Taliban, including rival clerics and politicians, all of whom were under his power. Not coincidentally, he had "the eyes of Bush, the cap and gown of Karzai, the waistcoat of Mr. Qanuni, the beard of Sayyaf and the nose and trousers of the Father of the Nation [King Zahir Shah]." When asked who his "loyal servants in Afghanistan" were, Satan replied that

> Not only in Afghanistan, but around the whole world I have my allies. In Afghanistan I have very loyal friends. People know them as Pirs and Mullahs but in fact they are my people. They include Pir Sayyid Ahmad Gailiani, Fazal Hadi Shinwari and Fayaz, the head of Kandahar's Ulema Shura. Besides this: Khalilzad, the current president Karzai and the ministers of his cabinet are my closest allies. (Strick Van Linschoten and Kuehn 2018: 241–242)

Elsewhere the Taliban blended Qur'anic and cosmopolitan registers in condemning the Karzai government as "Satan's party" and a "fifth column" (Obama 2009). Delivered through an increasingly sophisticated media campaign, and reinforced via targeted assassinations and violence aimed at creating mass casualties among government officials and civilians alike, such messages succeeded in undercutting the legitimacy of Karzai and, later, Ashraf Ghani (president, 2014–2021) in the eyes of many Afghans. Authorities in Kabul responded, in turn, by representing the Taliban as the "stooges" of Pakistan. But this designation proved to be divisive. Karzai dropped the formulation in 2008 in hopes of reconciling with what he called the "native sons" of Afghanistan (*Spiegel International* 2008).

As Karzai's own career shows, narratives of fifth-column treason in Afghanistan proliferated and became even more variegated since 2001. While calling for reconciliation with the Taliban, Karzai simultaneously chastised members of the US-led international coalition who were, he claimed, "strongly connected to corrupt elements" in Afghanistan who received "land and money" from foreigners for their "loyalty" (*Spiegel International* 2008). The Taliban, in turn, adapted this charge to discredit notables who worked with the foreign-backed Karzai government, declaring in their Arabic-language journal *Al-Somood* (2001) that "the agents gathering in the Afghan tribal council are not actually Afghans, rather the sellers of their conscience and religion in exchange for a few dollars, which they perceive as the highest price in their lowly lives." Meanwhile, such framings traveled beyond elite discourse. In 2015, a mob of men in central Kabul responded to rumors of a woman, Farkhunda Malikzada, burning a Qur'an at the Shah-i Du Shamshira shrine by beating her to death. As Niamatullah Ibrahimi has shown, some Afghans were convinced that she must have been in the service of a foreign embassy who was bent on conspiring against Islam and Muslim religious scholars (Ibrahimi 2019).

It was also during this period that many Afghans began to adopt more explicit and widespread use of "fifth-column" language (*sutūn-i panjum* in Dari), though the term has not entirely displaced the emotive use of "selling the homeland." Numerous Afghan political commentators have adopted the phrase, often explicating its origins in the Spanish Civil War. For example, in 2015 the veteran politician and mujahideen leader Abdul Rasul Sayyaf charged that the growing numbers of Taliban suicide bombers who managed to access targets in Kabul, seemingly at will, must have relied on the assistance of a "fifth column" of collaborators ("Hushdār-i Sayyāf" 2015). In September 2017 Sayyaf again called on the government to "destroy the fifth column" that plagued Afghanistan. Warning of the formation of a "nest" (*lānah*) of traitors working within the political system, Sayyaf identified this "fifth column" operating "in the service of [our] enemies" as the chief obstacle to "security" and "prosperity" in the country ("Sayyāf" 2017).

Sayyaf's invocation of treason lent substantial political weight to a narrative that was gaining currency among many Afghans struggling to understand how, in the face of so much apparent investment in security by the Afghan government and American-led international forces, attacks by the Taliban and others on the capital were becoming more frequent and lethal (Ibrahimi 2019). Beyond Kabul, the deterioration of security in many parts of the north and the mysterious appearance, from 2014, of fighters pledging loyalty to a new political entity, "Islamic State–Khorasan Province" (ISKP), seemed to beg for explanations that went beyond the official accounts of these developments offered by Kabul and Washington. Who benefited from this instability? Was it the work of the Afghan government and its American backers? Why were Afghan government helicopters allegedly ferrying ISKP fighters around northern Afghanistan? Was it true, as some Iranian and Russian authorities and media claimed, that US helicopters were even transporting ISKP fighters from Syria and Iraq to northern Afghanistan to exacerbate the Afghan quagmire and give the Americans even greater leverage to manipulate the future of the region? In 2017, even former president Karzai alleged that ISKP was a "tool" of the United States (Sarwan and Zahid 2017). Abdul Hamid, a civilian who was displaced, along with 10,000 others, by ISKP violence, told *New York Times* reporters in August 2018 that he was incensed by television reports showing how government authorities treated ISKP prisoners: "We lost everything to Daesh, and now the government sends helicopters for them from Kabul and brings them here and gives them rice and meat and mineral water, and provides them with security, and we are not even able to find food" (Rahim and Nordland 2018).

In seeking to interpret this complex political landscape, such fifth-column rhetoric has thus served as an adaptable, if not entirely conclusive, explanatory framework. But one should also understand it in this setting as a framework for action. Here the case of the Hazaras again points to the multifaceted uses of this narrative. From the 1990s, the Hazara campaign for civic equality and inclusion in Afghanistan, which enjoyed substantial international backing since 2001, has unleashed a fierce contest over the definition of national belonging and loyalty among a highly politicized and intellectually resourceful Afghan public. Anxieties about Shi'i Hazaras abound as a conduit of Iranian (and sometimes US) power. The recent recruitment, sometimes coerced, of Afghan Shi'i migrants to fight in the Syrian war has further intensified debates about their status. Hazara access to international institutions within Afghanistan and beyond, especially in the educational sphere, also presents a challenge to established ideas about class and ethnicity; that is, their advancement in employment and education is likely to undermine social hierarchies, many of which four decades of war have put into flux. The Islamic State and perhaps other stealthy actors have, it seems, attempted to capitalize on a violent legacy of anti-Hazara and anti-Shi'i discrimination and exclusion by targeting these communities in a series of spectacular bomb attacks in Kabul and elsewhere in recent years. Besides seeking domestic and international security and legal protections, Hazara activists and intellectuals have responded by seeking to reframe their collective identities as national, non-sectarian, and, perhaps most strikingly, secular. Facing increasingly lethal demands that they prove their fidelity to the Afghan nation, they have labored to give their opponents no opportunity to claim that they must be "unmasked" to expose the face of the enemy of the nation and of Islam.

The Taliban offensive of summer 2021 that resulted in the return of the movement to the seat of power in Kabul reinvigorated the Afghan search for fifth columns. President Ashraf Ghani's furtive flight from the presidential palace on August 15 provoked intense speculation about his motives. Rumors asserting that he left with enormous sums of cash deepened the shock for Afghans who saw his departure as an act of "betrayal." For many Afghans, his escape (possibly with vast riches) was proof that he had "sold" Afghanistan to the Taliban. For such critics, the Taliban advance across the west and north of the country in 2021 was nothing more than a "an invasion by Pakistan," referring to the fact that Islamabad had offered support for the movement since its appearance in 1994. To critics, it was not just Ghani who was a "national traitor" and "country-seller" ("Afghan Activist" 2021). US Special Representative Zalmay Khalilzad also became the target of the same accusations. Opponents of the Taliban pointed to his lobbying for the recognition of the Taliban in 1996 and his pivotal role in negotiating an agreement with the Taliban in February 2020 in Doha, Qatar, which arranged for the final withdrawal of American forces. Anti-Taliban observers tried to puzzle through the web of relationships linking Ghani, Khalilzad, and the Taliban, castigating all of them as parties working on behalf of Pakistan.

The sudden collapse of a state that had received two decades of international backing and the swift return of armed Taliban patrols to the streets of Kabul lent new weight to fifth-column politics in Afghanistan. Resort to this language has helped make sense of a stunning political reversal and a broad sense of betrayal, by Afghan elites such as Ghani and his circle, by the United States, and by the world. This language has presented a means to comprehend dozens of influential actors large and small who have operated with impunity in the shadows and who have repeatedly denied responsibility for acts of violence against seemingly random civilian targets. Electoral fraud, opaque governance, and mass-scale and persistent violence against civilians, often inflicted by ostensible allies wielding high-tech weaponry such as drones and other aircraft, have been essential parts of this meaning-making environment. Alongside persistent poverty, despite billions of dollars in aid, the widespread perception of corruption and of ethnic, regional, and linguistic bias, as well as perceptions of foreign favoritism for groups on the basis of gender, ethnicity, or proximity to American or other foreign institutions, have only strengthened the credibility of such analyses. Shaped by state secrecy, widespread social anxiety, and a burgeoning mediascape, fifth-column politics has offered a vocabulary to understand a chaotic civil war, a multifaceted international intervention, and the obscure and erratic decision-making of national and foreign political elites who have dominated Afghan politics while excluding a wider public.

# References

"Afghan Activist Says Ashraf Ghani and Joe Biden Caused Misery and Chaos." 2021. *Deutsche Welle*, August 27. https://www.dw.com/en/afghan-activist-says-ashraf-ghani-and-joe-biden-caused-misery-and-chaos/a-58998732.

Ahmadi, Wali. 2008. *Modern Persian Literature in Afghanistan: Anomalous Visions of History and Form*. London and New York: Routledge.

Akbar Agha, Sayyed Mohammad. 2013. *I Am Akbar Agha: Memories of the Afghan Jihad and the Taliban*. Edited by Alex Strick van Linschoten and Felix Kuehn. Berlin: First Draft Publishing.

Blom, Amélie, and Stéphanie Tawa Lama-Rewal, eds. 2020. *Emotions, Mobilisations and South Asian Politics*. London and New York: Routledge.

Crews, Robert D. 2015. *Afghan Modern: The History of a Global Nation*, Cambridge, MA, and London: Belknap Press of Harvard University Press.

Fenster, Mark. 2008. *Conspiracy Theories: Secrecy and Power in American Culture.* Minneapolis and London: University of Minnesota Press.

"From Confession to Suicide." 1973. *Kabul Times*, October 3, 1.

Ghaffārī Laʿlī, ʿAbd Allāh, ed. 1396 [2018]. Faryād-i ʿadālat: majmūʿah-i muṣāḥabahʾhā-yi shahīd-i vaḥdat-i millī Ustād ʿAbd al-ʿAlī Mazārī. Kābul: Intishārāt-i Bunyād-i Andīshah.

Glatzer, Bernt. 2001. "War and Boundaries in Afghanistan: Significance and Relativity of Local and Social Boundaries." *Die Welt des Islams* 41 (3): 379–399.

Gumnam, Mohammad Tahir Aziz. 2013. *Kandahar Assassins: Stories from the Afghan-Soviet War.* Edited by Alex Strick van Linschoten and Felix Kuehn. Berlin: First Draft Publishing.

"Hushdār-i Sayyāf dar sālgard-i tirūr-i Rabbānī: Muvāẓib-i Nufūẕīhāy-i dushman bāshīd." 2015. *BBC Persian*, Shahrīvar 29, 1394/September 20, 2015. https://www.bbc.com/pers ian/afghanistan/2015/09/150920_k02-sayyaf-warning.

Ibrahimi, Niamatullah. 2017. *The Hazaras and the State: Rebellion, Exclusion and the Struggle for Recognition.* London: Hurst and Co.

Ibrahimi, Niamatullah. 2019. "Rumor and Collective Action Frames: An Assessment of How Competing Conceptions of Gender, Culture, and Rule of Law Shaped Responses to Rumor and Violence in Afghanistan." *Studies in Conflict & Terrorism*, DOI: 10.1080/ 1057610X.2019.1647678.

Kakar, M. Hasan. 1979. *Government and Society in Afghanistan: The Reign of Amir ʾabd Al-Rahman Khan.* Austin: University of Texas Press.

McChesney, R. D., ed. and trans. 1999. *Kabul under Siege: Fayz Muhammad's Account of the 1929 Uprising.* Princeton, NJ: Markus Wiener Publishers.

Mills, Margaret A. 2013. "Gnomics: Proverbs, Aphorisms, Metaphors, Key Words and Epithets in Afghan Discourses of War and Instability." In *Afghanistan in Ink: Literature between Diaspora and Nation*, edited by Nile Green and Nushin Arbabzadah, 229–251. New York: Columbia University Press.

"Obama Arrogantly Assumes Power as the Pharaoh of His Time after Bush Suffered Humiliating Collapse." 2009. *Al-Somood*, January 27, accessed via Taliban Sources Project at FFI/UiO, https://tsp.hf.uio.no.

Pahwal, Abdurrahman. 2004. *Die Taliban im Land der Mittagssonne: Geschichten aus der Afghanischen Provinz*, ed. and trans. Lutz Rzehak. Wiesbaden: Reichert.

"People Join State in Condemning Traitors: Babrak-Qader Clique Cursed." 1978. *Kabul Times*, October 1, 3.

Rahim, Najim, and Rod Nordland. 2018. "Are ISIS Fighters Prisoners or Honored Guests of the Afghan Government?" *New York Times*, August 4, https://www.nytimes.com/ 2018/08/04/world/asia/islamic-state-prisoners-afghanistan.html.

Sarwan, Rahim Gul, and Noor Zahid. 2017. "Former Afghan President Karzai Calls Islamic State 'Tool of US.'" *VOA News*, April 19, https://www.voanews.com/extrem ism-watch/former-afghan-president-karzai-calls-islamic-state-tool-us.

"Sayyāf: ḥukūmat bāyad sutūn-i panjum rā az nābūd kunad." 2017. *Voice of America Dari*, Mīzān 6, 1396/September 28, 2017. https://www.darivoa.com/a/sayaf-jihadi-leader-ask-government-to-destroy-fifth-pillar/4047812.html.

"Spiegel Interview with Afghan President Hamid Karzai: 'I Wish I Had the Taliban as My Soldiers.'" 2008. *Spiegel International*, June 2. https://www.spiegel.de/international/ world/spiegel-interview-with-afghan-president-hamid-karzai-i-wish-i-had-the-tali ban-as-my-soldiers-a-557188.html.

Strick Van Linschoten, Alex, and Felix Kuehn, eds. 2018. *The Taliban Reader: War, Islam and Politics*. Oxford and New York: Oxford University Press.

Sultan Mahomed Khan, Mir Munshi. 1900. *The Life of Abdur Rahman: Amir of Afghanistan*. London: John Murray.

"What Lies behind the Afghan Tribal Council Session?" 2011. *Al-Somood*, October 28, accessed via Taliban Sources Project at FFI/UiO: https://tsp.hf.uio.no.

Zaeef, Abdul Salam. 2010. *My Life with the Taliban*. London: Hurst and Co.

# PART III
# CHALLENGES TO FIFTH-COLUMN POLITICS

# 7

# Security Threats or Citizens?

## Fifth-Column Rhetoric in Jordan

*Lillian Frost*

*When and why do political elites publicly frame groups as fifth columns?* While some argue that geopolitical factors explain when states identify groups as fifth columns, others emphasize domestic political concerns. This chapter finds that, although geopolitical dynamics are central in defining groups as fifth columns, the decision to frame groups in this way tends to reflect domestic political interests. This chapter unpacks these domestic political interests and argues that political elites, such as executive rulers and their advisors, can frame groups, whether citizens or not, strategically as fifth columns to *avoid* introducing new policies. When political elites label a group as a fifth column, they can divert constituents' attention away from the content of a new policy to the security concerns surrounding it. However, when political elites favor a new policy, they will avoid engaging fifth-column frames and securitizing it—even when these frames easily are accessible.

This chapter assesses these arguments by examining variations in the Jordanian government's portrayal of citizen and noncitizen Palestinians as fifth columns over time. The chapter looks at two cases of nationality law reform in Jordan that cannot be accounted for by geopolitical fifth-column theories (Mylonas 2012). One case demonstrates how political elites could refrain from engaging fifth-column frames despite geopolitical incentives to do so, to facilitate the adoption of a nationality law. This case involves the successful adoption of the 1954 nationality law, which extended Jordanian nationality to all Palestinians living on the East and West Banks. Jordan originally consisted of only the East Bank (or "Transjordan"), but Jordan annexed the Palestinian West Bank during the 1948 Arab-Israeli War. Jordan described the policy as supporting Palestinians until they could establish their own state. This accommodating approach is surprising given the geopolitical conditions at the time, when Jordan was aiming to expand its borders and Palestinian groups were linked to enemy Egyptian and Syrian regimes as well as Palestinian leaders who challenged the king's legitimacy.

The second case demonstrates how political elites used fifth-column rhetoric to avoid changes to the nationality law to satisfy domestic, rather than

Lillian Frost, *Security Threats or Citizens?* In: *Enemies Within.* Edited by: Harris Mylonas and Scott Radnitz, Oxford University Press. © Oxford University Press 2022. DOI: 10.1093/oso/9780197627938.003.0008

geopolitical, interests. This case focuses on Jordan's failed nationality law reform in 2014, in which leaders decided not to revise the law to enable women,
like men, to pass their nationality on to their children and spouses. Jordanian
officials claimed this decision stemmed from security concerns about the threat
these spouses and children, who they claimed were predominantly noncitizen Palestinians, would pose to the state's sovereignty because of their links to
Israeli plans to make Jordan the alternative homeland of the Palestinians. This
exclusionary approach is surprising in geopolitical terms because Jordan had
maintained peaceful, cordial relations with Israel and the Palestinian Authority.
However, this approach makes sense from a domestic political perspective because Transjordanian political elites used this fifth-column framing to block the
nationality law reform and preserve their existing privileges.

This chapter uses process tracing to unpack the events leading up to each
proposed nationality law and highlights the key role of domestic politics in each
case. It also demonstrates how noncitizen insiders can be subject to fifth-column
frames, challenging neat distinctions between noncitizen outsiders and citizen
insiders. My narrative builds on data from two hundred interviews I conducted
between 2016 and 2019 in Jordan with former and current government officials,
journalists, lawyers, activists, development staff, and Jordanian women married to noncitizens.[1] All of my interviewees remain anonymous for their protection and are referenced using alphanumeric identifiers.[2] In addition, I draw
from hundreds of files collected on Jordan's internal affairs from 1946 to 1954
from the US National Archives at College Park and British National Archives
at Kew. Overall, this chapter highlights the prominent role of domestic political interests in ruling elite decisions to frame groups as fifth columns, while
challenging the binary insider-outsider distinction in this volume's definition
of fifth columns.

## Fifth Columns, Geopolitics, and Elite Interests

Although few studies in political science address fifth columns by name, much of
the related literature emphasizes geopolitics or domestic politics as driving forces
behind when and why political elites describe a group as a fifth column. In terms
of geopolitics, some argue that when a state's regime associates a group with a
threatening country's regime, then the state will repress and exclude that group
to shore up the regime's strength (Gause 2010; Jenne 2007; Kuo and Mylonas,
this volume). This decision to exclude the group may stem from traditional security threats, for example if members of the group have undertaken violent attacks
against the state, or it may reflect fears that the group will commit such attacks in
the future.

Others contend that a state's decision to repress and exclude a group is driven by the state's foreign policy goals (i.e., whether the state is revisionist or status quo) and its relationship with the external power perceived as supporting the group (i.e., rivalry or alliance). In this case, a state is more likely to exclude and potentially frame a group as a fifth column when an enemy external power supports the group and when the state of residence has revisionist ambitions (Mylonas 2012). This argument assumes that ruling elites are attempting to homogenize their societies along a particular constitutive story. Building on this logic, Kuo and Mylonas (this volume) suggest that geopolitical shifts can have significant consequences for how governing elites frame non-core groups as fifth columns, which in turn can shape and justify policy changes.

These accounts are logical because fifth columns, by definition, must have a real or imagined connection to a foreign enemy and pose an internal threat. Thus, as a baseline, the state must have a foreign enemy for political elites to connect with a group in order to frame that group as a fifth column. In addition, a history of this connection—between the group and a foreign enemy—helps legitimize these fifth-column claims. Similarly, clear evidence of the group's involvement in attacks against the state or support for the foreign enemy also helps strengthen fifth-column frames.

However, elite decisions about *when* to frame a group as a fifth column do not necessarily stem from geopolitical concerns or traditional security threats. Instead, domestic political interests can influence these framing decisions, and at times, these framings can falsely accuse a group of receiving support from an external power (Radnitz and Mylonas, Introduction). Political elites, such as executive rulers and their advisers, can wield fifth-column frames and apply these identities strategically, including to foment violence (Brass 1997; Gagnon 1994/ 1995; Harff 2003), deconstruct collective identities (Lane 2011), or create particular forms of nationalism (Snyder 2000).

This chapter builds on domestic politics explanations of fifth-column frames by employing securitization theory to propose one mechanism that accounts for domestic political elites' decisions to portray a group publicly as a fifth column. In doing so, this chapter argues that political elites can frame groups as fifth columns to avoid introducing new policies by diverting constituents' attention away from a new policy's content to the security concerns surrounding it. The securitization of the new policy enables political elites to remove the issue from the public agenda and avoid reform (Lori 2017). Securitization is a process that involves the social construction of security issues through discourses and practices that represent and define security threats (Buzan et al. 1998; Williams 2003; Balzacq 2005). These security threats can encapsulate different types of threats beyond war and force, including threats to a state's economy, sovereignty, identity, and environment (Buzan et al. 1998).

The securitization process involves using a security threat to justify casting an issue as an existential threat that calls for extraordinary measures beyond the routines and norms of everyday politics (e.g., allowing the limitation of otherwise inviolable rights). This process lifts the issue "above politics" (Williams 2003). Thus, political elites can frame a policy reform as empowering an alleged fifth-column group, and as such, the government can claim that it must handle the securitized policy outside of everyday politics and without legal constraints.

## Fifth-Column Threats and Nationality Law

Nationality law is one clear policy area where political elites can deploy fifth-column frames to avoid reform. Opportunities for nationality law reform provoke formal discussions about whom the state should include as citizens and whom the state should exclude (Howard 2009). In more authoritarian states, the ruling regime often dominates these discussions by pushing forward its preferences concerning which groups to incorporate (Shevel 2011). These preferences often boil down to whether incorporating the group bolsters regime survival and regime interests, particularly in the short term. Specifically, if incorporating a group brings the regime more votes or otherwise solidifies the regime's hold on power, then the regime will include that group, regardless of its links to foreign enemies and threats.

However, if including the group increases votes for the regime's opposition or otherwise weakens the regime's hold on power, then the regime will exclude that group. One useful strategy for ensuring the group's exclusion is to frame that group as a fifth column. Drawing from securitization approaches, I suggest that a regime can connect that group to external security threats based on the group's identity, past behavior, or apparent interests.

This strategy can exclude the group, while also dividing potential cross-cutting oppositions (Brumberg 2014; Charnysh, Chapter 1, this volume). For example, following the Arab Uprisings of 2011, the Saudi regime promoted sectarian tensions between Sunni and Shia Muslims to inhibit the creation of joint platforms for political mobilization and prevent the development of national non-sectarian politics (Al-Rasheed 2011). Other examples from the same period and region include state-fomented divides between Jordanians of Transjordanian and Palestinian descent, Sunni and Shia Bahrainis, and leftists and Islamists in Syria and Libya (Lynch 2012).

Likewise, this securitization approach can support regime survival by placating important constituencies that are against introducing more inclusive nationality laws, such as far-right political parties or movements and autochthonous, sons of the soil groups (Hammerstadt 2014). At the same time, this approach can provide an apparently legitimate excuse, based on security threats, to liberal international actors for adopting or maintaining an exclusionary policy. Regimes can tap into Western concerns with threats, such as communism or terrorism, to justify their decisions to delay inclusive reforms in favor of strengthening security (Forester 2019).

This analysis finds that domestic political concerns, rooted in regime survival interests, tend to explain political elites' decisions to frame a group publicly as a fifth column. Geopolitical dynamics may play a role in these decisions, but political elites ignore these dynamics when they contradict their domestic political interests. Therefore, how a group relates to incumbent elite interests is a critical factor to consider in assessing whether political elites will frame that group as a fifth column.

## Jordan's Successful Nationality Law Reform in 1954

### Case Background

In 1921, the British established the Emirate of Transjordan as a protectorate and recognized Hashemite leader, Emir Abdullah, as its leader.[3] Abdullah was the second son of Sharif Hussein bin Ali, the Sharif and Emir of Mecca, who aimed to establish an independent Arab state across the Arabian Peninsula, Greater Syria,[4] and Iraq under his and his descendants' rule. Sharif Hussein failed to achieve this goal, but ultimately Hashemites were placed on thrones in Iraq and Transjordan, in addition to their existing throne in the Hejaz (in present-day western Saudi Arabia).[5]

After twenty-five years, Transjordan received independence from the British in 1946 as the Hashemite Kingdom of Transjordan, changing Abdullah's title from emir to king (Salibi 1998: 153). Soon after its independence, Transjordan faced a massive shock when the 1948 Arab-Israeli War broke out in May of that year. Transjordan fought on the Arab side of the war and successfully annexed much of the territory that became known as the "West Bank" during the first phase of fighting (Massad 2001: 229).[6] This move enabled King Abdullah to expand his rule beyond Transjordan. However, it also involved a tripling of his population. In 1945, Transjordan had a population of 476,000, while the West Bank recorded 433,000 residents in 1947.[7] The 1948 war brought over half a million refugees to both Banks, with 101,000 refugees registered on the East Bank and 430,000 refugees registered on the West Bank in 1949.[8]

With this greatly expanded population, King Abdullah needed to gain the support of Palestinians for his vision of a union between the East and West Banks. He viewed this union as a first major step on his path toward his dream of establishing and ruling Greater Syria (Abu Odeh 1999: 31–33; Brand 1988: 149). Abdullah also desperately wanted access to the Mediterranean through a Hebron-Gaza corridor.[9] Negotiations between Transjordan and Israel highlighted that "Jordan considered access to the Mediterranean of vital importance."[10]

However, gaining Palestinian support would not come easily, particularly with rival power bases in Gaza, Cairo, and Damascus. Amid the war, in September 1948, the Egyptian regime helped to establish, through the Arab League, the All-Palestine Government, which declared its independence after fleeing Gaza for Cairo in December 1948 (Robins 2004: 71). King Abdullah's longtime rival, Palestinian Arab nationalist leader Hajj Amin al-Husseini, was appointed as the government's president.

Given these rival developments, King Abdullah wanted to act quickly to secure a direct appeal from the Palestinians in the West Bank "demanding their annexation" to Jordan (Nevo 1996: 166). This would frame the planned union as a "magnanimous gesture from the throne under pressure from an expectant populace" rather than as "the acquisitive move of an ambitious dynast" (Robins 2004: 71). As such, King Abdullah organized, from behind the scenes, two gatherings of Palestinians in October and December 1948 to demonstrate popular Palestinian interest in the union and to counter the All-Palestine Government as well as its claims to represent all Palestinians (Robins 2004: 71).

These efforts toward gaining Palestinian support also were critical for King Abdullah's physical survival, with Syria and Egypt supporting Palestinian political and physical attacks against him. For example, in March 1949, the government uncovered a plot to assassinate King Abdullah by several Palestinians, who "had been sent by persons acting on behalf of [Syrian president] Shukri Kuwatli and Haj Amin el Husseini."[11] Likewise, in May 1949, Palestinians were caught in an unsuccessful attempt to assassinate the king and confessed to receiving orders "expressly from Damascus to assassinate Abdullah" and that "Syrian President Kuwwatly [sic] urged that [the] plot should be carried out."[12]

In this context, the Transjordanian government quickly took steps to include Palestinians in the political community to bolster King Abdullah's legitimacy in ruling over the West Bank. In May 1949, the Kingdom released an order to change the name of the country to "the Hashemite Kingdom of the Jordan" (rather than the Hashemite Kingdom of Transjordan), implying that the Kingdom included both sides of the Jordan River.[13] In addition, throughout 1949, the Jordanian government undertook measures to grant Palestinians on the East and West Banks Jordanian passports and nationality. These efforts culminated in an additional nationality law passed in December 1949 that formalized Palestinian access to nationality.[14] This nationalization in turn enabled plans to move forward to hold elections on both Banks in 1950, where all Palestinians and Transjordanians (i.e., East Bankers, or those living on the East Bank in 1946), as citizens, could vote and run for office in the House of Deputies.[15] The ultimate goal of holding elections was for Palestinians and Transjordanians to vote in favor of a union between the East and West Banks.

On April 24, 1950, the Jordanian Parliament convened for the first time to confirm the union between the East and West Banks of the Jordan River under the rule of the Hashemite Kingdom of Jordan, led by King Abdullah. This union institutionalized the complete equality of rights of all people living on both Banks as citizens of Jordan, while acknowledging that this union would not influence the final settlement of the Palestine question (e.g., Palestinians maintained their right of return and claims to an independent Palestinian state).[16]

However, King Abdullah did not rule over the two Banks for long. On the morning of July 20, 1951, a Palestinian gunman assassinated King Abdullah at the entrance of the Al Aqsa Mosque in Jerusalem.[17] King Abdullah's second son, Naif, served as regent until his first son, Talal, could assume the throne, as required by the Jordanian constitution. King Talal had been plagued with mental health issues, which quickly brought his reign to an end. Talal's first son, Hussein, took over the throne at age seventeen in August 1952, under the guidance of a regency council, until he formally acceded to the throne in May 1953 on his eighteenth birthday (according to the Islamic lunar calendar).

King Hussein oversaw a flurry of revised laws in January 1954, which passed through the parliament without much public attention or fanfare. One of these laws was the 1954 Nationality Law, which extended Jordanian nationality to Palestinians who arrived in the country after the 1949 additional nationality law (Massad 2001; Frost 2022).

## Fifth-Column Links

King Hussein's decision to incorporate more Palestinians into the Jordanian citizenry is surprising, considering the clear links at the time between Palestinians and the hostile Egyptian and Syrian states, as well as Palestinian responsibility for the assassination of King Abdullah. The details of the assassination case revealed that supporters of Hajj Amin al-Husseini (i.e., the ex-Mufti) hatched the plot in Cairo, Egypt, and recruited assassins in Syria and Jerusalem.[18] The trial did not find any evidence of the ex-Mufti's direct involvement, and it remains unclear who paid for and requested the operation.[19]

In British correspondence with the Israelis just after the assassination, the Israeli government indicated that they saw "a real possibility that the Mufti or his supporters might gain power in Jordan."[20] Further, a few weeks later, British General Glubb[21] wrote a letter to the Foreign Office describing the political situation surrounding King Abdullah's assassination. He noted that for "nearly two years, we have been hearing these stories that the Mufti's supporters will murder King Abdulla [sic] and then the West Bank (Arab Palestine) will rebel against Jordan. Why it took so long I am not sure."[22] He then noted that "there is no doubt that the Egyptians knew about it and that the Egyptian Legation in Amman was cooperating in the plot."[23]

General Glubb adds that after the murder "there was an unconcealed outburst of joy in the Egyptian Press" and the "Egyptian Government alone sent no representative to the funeral." He suggests that there is "no doubt that the plan put into execution was the original Mufti's idea to kill the King and raise rebellion and civil was [sic] between 'Trans-Jordan' and 'Arab Palestine.' "[24] Although the Egyptians, Saudis, Syrians, and ex-Mufti supporters were trying to organize "civil war between Trans-Jordan and Arab Palestine," these efforts failed and instead, "the King's murder caused a remarkable demonstration of solidarity."[25] These observations, from one of the most prominent men in Jordanian politics at the time, highlight that fifth-column frames were at least accessible at this time, if not well founded and based on real threats.

## Domestic Reactions

There were many social, political, and economic conditions that were conducive to excluding the Palestinians in Jordan. First, after King Abdullah's assassination, there were outbreaks of anti-Palestinian feelings among Transjordanians as well as Palestinian celebrations on the West Bank. Just two days after the assassination, Sir Alec Kirkbride[26] reported to London from Amman that "anti-Palestinian feeling is mounting and is present within the Arab Legion," with "small incidents in Amman[,] Salt[,] Swelleh[, and] Mafrak," taking the form of "stoning cars, attacks on refugees and in Akaba demonstrations against the continuance of work by the Palestinians."[27] Later that day, Sir Kirkbride specified that "there has been ill feeling evident between Palestinians and Transjordanians over the murder of the King."[28] A British officer in Jerusalem also wrote London that day stating that "the citizens of Es Salt, on the East Bank, have been throwing rocks at motor-cars coming from the West Bank," and "the police could or would do nothing about it."[29]

Two weeks after the assassination, American officials reported that "anti-Palestinian feeling still runs strong among indigenous Jordanians," and the "open jubilation of many Palestinians at the King's death, allegedly carried to the extent of public veneration of the assassin's photograph, is not calculated to attenuate this ill-feeling."[30] The report also speculated that "If accompanied by U.N. and Arab League pressure for internationalization of the Jerusalem area," as the Egyptians already had advocated, "Jordan leaders, supported by public opinion in East Jordan, might well feel that Palestine could be jettisoned as more of a liability than an asset."[31]

Second, the British may have supported the Jordanian state in marginalizing and undermining the Palestinians because British officers often noted that the Palestinians were a more difficult population to manage. In their annual review of Jordan in 1951, the British reported that "the once peaceful Amirate of Transjordan has been largely taken over by the Palestinians."[32] In another report in November 1952, the British noted that "the East Bankers are comparatively easy to rule,"[33] and in December 1952, they stated again "the Jordan people, and especially the East Bankers, are relatively easy to govern."[34]

Third, Palestinian refugees often fared better economically than Transjordanians, offering another fault line for the government to take advantage of to exclude Palestinians. The British noted in 1951 that "the general standard of living of large sections of the indigenous population has often been worse than that of the refugees themselves whose basic needs are provided by UNRWA."[35] The Americans also pointed out that the economic situation in Jordan was terrible in November 1952 and that economic conditions have fostered discontent among West Bankers; although "econ[omic] conditions on [the] West Bank are no (rpt no) worse than econ[omic] conditions on [the] East Bank," the "population of [the] East Bank has never been accustomed to the high econ[omic] standards formerly enjoyed in Pal[estine]."[36] Thus, the West Bankers not only fared better economically but they also protested for better economic conditions, making them a more problematic constituency for the regime to manage.

## Domestic Motivations for Fifth-Column Framing

Despite these conditions, the Jordanian regime refrained from publicly framing the Palestinians as a fifth column. Instead, the government described King Abdullah's assassin as a "terrorist" and "criminal," rather than as a Palestinian. In the correspondences immediately following the assassination, the British described the assassin as someone "known to be a terrorist"[37] and "employed by the ex-Mufti of Jerusalem."[38] Jordan News and the BBC also repeated these descriptions and did not refer to the assassin as a Palestinian.[39] Further, the British noted in their annual report on Jordan in 1951 that the Jordanian government controlled news-messaging about the assassination, stating: "The murder of King Abdullah was treated in a restrained manner owing to Government supervision."[40]

Why did the Jordanian regime decide not to frame Palestinians publicly as a fifth column and prevent more Palestinians from receiving Jordanian nationality in the early 1950s? Despite the connections of some Palestinians to external rivals and their participation in attacks against the king, the regime decided to maintain Palestinians as equal citizens and provide more Palestinians with access to nationality through legal reforms in 1954. This decision cannot be accounted for by geopolitical arguments predicting the exclusion of Palestinians, and it instead reflected Hashemite goals for Jordan to serve as a pan-Arab nation ruling over Greater Syria.

Before King Abdullah's death, he expressed his desire to incorporate the Arab parts of Palestine into his kingdom in order to achieve this goal of Hashemites ruling Greater Syria (Abu Odeh 1999: 31–33). A former minister noted that, in 1950, King Abdullah "had dreams of Greater Syria, though this dream was not necessarily shared by the indigenous population."[41] King Abdullah passed much of this vision onto his grandson, King Hussein, who was with King Abdullah when he was murdered. Another former minister noted, "King Hussein was close to his grandfather and inherited the duty to honor the legacy of his grandfather."[42] King Hussein himself wrote in his autobiography that his grandfather "of all men, had the most profound influence on my life" (Hussein bin Talal 1962: 10). Days before King Abdullah's assassination, Hussein promised his grandfather that he would not let his work go unfinished and to pass on his legacy (Shlaim 2007: 46–47).

Although King Hussein tried to stay on good terms with all Arab states (not just fellow Hashemite Iraq, as his grandfather did), he maintained his grandfather's policy of extending equal Jordanian nationality rights to the Palestinians, despite Arab condemnation and disapproval of this policy (Shlaim 2007: 67–78; Plascov 1981). A Jordanian historian and former diplomat noted that as children in Jordan, they were taught that "Greater Syria is our land, not just Jordan," and it was not until the 1960s that "Jordanian identity became clear" as Palestinian nationalism grew.[43]

In August 1952, Jordan's prime minister, Tawfik Abu Al-Huda, castigated a proposal by Stanley Morrison, a British missionary, for Arab Palestine to secede from Jordan as an effort that would appeal "to Jewish interests and certain other foreign powers," and the Jordanian government expelled Morrison from Jordan.[44] This incident is interesting because, as the next case study demonstrates, Jordanian leaders later would frame the naturalization of more Palestinians and the reacquisition of the West Bank as an Israeli plot to turn Jordan into Palestine, often referred to as the "alternative homeland (الوطن البديل)." However, in the 1950s, the opposite rang true, with proposals to separate the West from the East Bank reflecting "Jewish interests" to divide and weaken Arabs.

In addition, an American foreign officer report in 1954 describes discussions with King Hussein about his views on Palestine and the Palestinian refugees. The foreign officer notes: "His Majesty said that something should be done for the refugees. They should have the right to return to their homes, but that he felt that in making a new and happier existence, most of them would wish to stay in Jordan." In discussing the political challenges in undertaking refugee resettlement programs (which the Palestinians and other Arab states opposed), the king "intimated that he personally would be interested in endeavoring to break down this political impediment."[45]

Overall, King Hussein did not want to frame Palestinians as fifth columns publicly because at the time, he and his advisors were focused on continuing the Hashemite legacy of striving to rule Greater Syria, including Arab Palestine. These efforts required including the Palestinians as part of Jordan to strengthen Hashemite legitimacy in ruling the West Bank.

## Jordan's Failed Nationality Law Reform in 2014

### Case Background

On June 5, 1967, war broke out again between Arab and Israeli forces.[46] This brief war resulted in the Israeli occupation of the Gaza Strip and West Bank, among other Arab territories. The Israeli occupation created a new wave of refugees flooding into Jordan's East Bank. By February 1968, there were 241,000 Palestinians displaced from the West Bank and 20,000 displaced from Gaza.[47] Those who arrived from the West Bank maintained their access to Jordanian nationality and full citizen rights. However, those arriving from Gaza, which had been under Egyptian rule, were treated as foreigners under the law, but they received access to Jordanian passports and other citizen-like rights in practice, including access to public education and healthcare.

Despite King Hussein's efforts to cultivate a Jordanian identity that encompassed citizens of Palestinian and Transjordanian (i.e., East Bank) origin, a distinct Palestinian national identity had persisted. As of 1964, the main representative of this identity was the Palestine Liberation Organization (PLO), which was founded during a summit that year in Egypt. Tensions between the PLO and Jordanian government, which had hosted the PLO since the 1967 war, broke out in September 1970 (i.e., Black September). That month the Jordanian army attacked PLO fighters (i.e., the fedayeen) following the PLO's increasingly intransigent behavior in Jordan, which included acting like an autonomous government and challenging King Hussein's authority and, in two assassination attempts, his life. Although many Palestinian-Jordanians fought dutifully in the Jordanian army and some Transjordanians joined the PLO, this conflict shook Palestinian-Transjordanian relations and reignited questions about Palestinian loyalty to King Hussein and Jordan.

After 1970, several developments amplified this shift toward the Jordanian government viewing the Palestinians in Jordan as a threat rather than an asset, particularly as it became increasingly unlikely that Jordan would regain the West Bank. To start, in 1974, the Arab League voted to recognize the PLO as the sole legitimate representative of the Palestinians. In response, King Hussein suspended parliament (which, since 1950, represented the East and West Banks) and formed a new cabinet, 80 percent of whom were East Bankers (Abu Odeh 1999: 211). In addition, Israeli claims that "Jordan is Palestine" became louder in the late 1970s and early 1980s, with the slogan's main proponent, Ariel Sharon, serving in successive cabinets and supporting the expansion of Israeli settlements on the West Bank. Further, a stream of failed peace negotiations between Jordan and the PLO as well as Jordan and Israel in the 1970s and 1980s amplified fears in the Jordanian government that not only would Jordan lose the West Bank permanently, but also that the East Bank would become the "alternative homeland (الوطن البديل)" of the Palestinians (Salibi 1998: 261).

These frustrations culminated in King Hussein's decision on July 31, 1988, to dismantle "the legal and administrative links" between the East and West Banks, which he had maintained after Israel occupied the West Bank in 1967.[48] This announcement reduced Jordan to the East Bank alone and ended the 1950 Unity of the Banks. One major outcome of this decision was the removal of Jordanian nationality from the one million Palestinians residing on the West Bank at the time of the decision (Robins 1989: 168–170).

Although the disengagement denationalized all residents on the West Bank in July 1988, the king made clear in the disengagement announcement that these measures did not apply to any Jordanians of Palestinian descent residing on the East Bank. However, many Palestinians who were not in the West Bank in July 1988 have been subject to nationality revocation based on unclear and varying terms of the disengagement. Although there are no clear figures on how many Palestinian-Jordanians have been affected, a former minister suggested that there were about 1,600–1,700 cases a year,[49] and Human Rights Watch (HRW) suggested that there were spikes in the number of cases in the mid-1990s and 2010s. The same HRW report indicated that government officials defended these withdrawals because they reflected "opposition to Israeli expansionism and a further uprooting of Palestinians in the West Bank," a concern with "Jordan's poverty of resources," and "the need for a 'demographic balance'" in Jordan (HRW 2010: 11).

These events generated concerns about nationalizing additional Palestinians in Jordan and set the context for government resistance to calls to reform Jordan's nationality law to enable women to transmit their nationality to their children and spouses. Like most states before 1979, Jordan does not allow citizen women, unlike citizen men,[50] to pass their nationality on to their children or spouses.[51]

However, diverging from most states, Jordan has not removed this discrimination toward women (DTW) from its nationality law since the United Nations (UN) General Assembly's adoption of the Convention on the Elimination of all Forms of Discrimination Against Women (CEDAW) in 1979 (Cole 2013). Instead, Jordan, like twenty-one other states, signed CEDAW with a reservation to Article 9 in 1984,[52] which it confirmed when ratifying CEDAW in 1992.

Concerns about the alternative homeland and nationalizing Palestinians in Jordan persisted in the 1990s and 2000s. Paramount among these concerns was the failure of the Palestinian-Israeli peace process after the Oslo Accords in 1993 and 1995, which did not create a Palestinian state. In addition, the second Palestinian intifada broke out in 2000 in the still-Israeli-occupied West Bank and Gaza. Although Jordan aimed to protect its sovereignty by signing its own peace treaty with Israel in 1994, anxieties persisted about Israel expelling more Palestinians to Jordan (Ryan 2011: 370).

After assuming the Jordanian throne following his father's death in 1999, King Abdullah II undertook several initiatives to promote a more distinctly Jordanian identity. In 2002, he announced the "Jordan First (الأردن أولا)" initiative. This initiative stressed that "Jordanian interests are above any other interests,"[53] which seemed to include the ongoing second Palestinian intifada (Bani Salameh and El-Edwan 2016: 995). In 2005 and 2006, King Abdullah launched the "National Agenda (الأجندة الوطنية)"[54] and "We Are All Jordan (كلنا الأردن)"[55] initiatives, which focused on Jordan's political and economic reform process. Although national identity was not at the center of these initiatives, debates about the reform process reflected a broader division in Jordanian society between those who wanted democratic reforms and a national comprehensive identity versus those who wanted to privilege security over reforms and entrench tribal, East Bank identities in the state (Bani Salameh and El-Edwan 2016: 995).

These opposing visions for Jordan became more outspoken after 2010, with public letters to the king highlighting their interests. On May 1, 2010, the National Committee of Retired Military Servicemen (اللجنة الوطنية للمتقاعدين العسكريين) released a statement to King Abdullah calling for the passage of laws to implement Jordan's administrative and legal disengagement with the West Bank and to halt the nationalization of any Palestinians from the West Bank or Gaza. The statement also warned about a "Zionist scheme for liquidating the Palestinian Question at the expense of the Jordanian People" (quoted in Tell 2015: 7). In response, the king personally met with the movement's leaders and decided in March 2012 to increase the pensions of military retirees (Tell 2015: 10). Although these efforts coopted some of the movement's members, the movement itself helped establish the predominance of Transjordanians in the roughly eight thousand protests, marches, and strikes that occurred in Jordan between January 2011 and August 2013 (Tell 2015: 9).

However, there were efforts by Palestinian-Jordanians to highlight their grievances. On August 2, 2012, a coalition of Palestinian-Jordanian elites, referred to as the Jordanian Initiative for Equal Citizenship (المبادرة الأردنية لمواطنة متساوية), wrote a letter to the king that documented the public attitudes and official policies that discriminated against Palestinians (Bani Salameh and El-Edwan 2016: 997). The letter also strongly affirmed that Palestinian-Jordanians were committed to the right of return to their original homeland and rejected the notion that "Jordan is Palestine" (Rantawi and el-Abed 2012). The king responded to this initiative by increasing the representation of Palestinian-Jordanians in parliament through a new elections law in 2012 (Bani Salameh and El-Edwan 2016: 997).

These competing demands and the predominance of Transjordanian protest formed the context in which Jordanian women started campaigning to obtain equal nationality rights in 2009. At that time, Nimah Habashneh founded the campaign and Facebook page "My Mother Is Jordanian, and Her Nationality Is a Right for Me (أمي أردنية وجنسيتها حق لي)." This campaign emerged out of her posts describing the problems she faced in caring for her children, who could not receive Jordanian nationality because their father was Moroccan. Habashneh tirelessly campaigned for nationality law reform and discussed the issue with Jordanian parliamentarians, international organizations, and the media.

Her efforts cultivated a group of women married to non-Jordanians (including husbands from Palestine, Egypt, Syria, Iraq, Pakistan, Britain, and the Netherlands) who were facing the same problems in raising their children. This interest led to Habashneh's campaign joining other civil society organizations to form the coalition "My Nationality Is the Right of My Family (جنسيتي حق لعائلتي)" in February 2013. The campaign and coalition efforts culminated in 2014 when a member of parliament (MP), Mustafa Hamarneh, and his Mubadara (كتلة المبادرة الوطنية)[56] added the issue to the legislative agenda and advocated forcefully for nationality law reform (Frost 2022).

The legislative debates generated a firestorm of public criticism from other MPs and East Bank opinion leaders. Prominent among these critiques was the claim that the nationality reform would ensure "Israel's ultimate plan of creating a substitute homeland for Palestinians in Jordan" (Husseini 2014). This increasingly powerful opposition, with the backing of many in the government, framed women's equal nationality as an existential threat to Jordan's identity that would depopulate Palestine and make Jordan the alternative homeland.

Ultimately the government, led by Prime Minister Abdullah Ensour, decided not to reform the nationality law, noting publicly that the reform "might affect the demographic balance in Jordan and might lead to empty[ing] Palestine from its people" (Husseini 2014). Instead, he granted the children of Jordanian women married to non-Jordanians (and residing in Jordan for at least five years) a set of privileges (referred to colloquially as "مزايا"), including free public education and health services, access to more work professions with fewer work permit fees, and access to driver's licenses.[57] However, Prime Minister Ensour emphasized publicly that the government had no intention "now or in the future" of granting these children nationality or of "accepting them in the army" (Husseini 2014).

## Fifth-Column Links

The main group highlighted as a threat linked to the alternative homeland through this nationality law reform were the Gazan refugees who had fled to Jordan following the 1967 war, whom I refer to as the Gazans (Kvittingen et al. 2019: 16). Although they never received Jordanian nationality, they had been treated more like quasi-citizens than foreigners in practice since their arrival. A prominent example of this quasi-citizen status is the temporary passports Jordan has issued to Gazans since 1968 (Frost 2020). These passports look identical to citizen passports on the outside, and they have enabled Gazans not only to travel internationally but also to register their children in schools, obtain work, and prove their legal residency (El-Abed et al. 2014).

Although Gazan access to these passports is not guaranteed in law, Jordan typically has granted Gazans more rights than other noncitizens in practice. Many former ministers commented on the extent of Gazan rights in Jordan. One former minister noted, Gazans "have all rights but political rights; they can do everything: own, work, etc."[58] Another former minister stated: "those with temporary passports have all rights but political rights."[59] While these comments minimize and ignore many areas where Gazans are not treated like full citizens (e.g., property ownership, access to tertiary education, etc.), they also highlight that government officials recognize that Gazans have a special status between citizen and foreigner. Thus, by 2014, Gazans were de facto insiders in Jordan.

This insider status enabled Jordanian officials and MPs to frame Gazans publicly as an externally backed threat to Jordan's sovereignty and identity operating from within, by describing them as pawns in Israeli plots to make Jordan Palestine in 2014. This framing occurred despite Jordan's largely cooperative relations with Israel since their peace treaty in 1994. In addition, the Obama administration's stalled peace efforts at the time had not triggered major concerns about the alternative homeland because these efforts focused on a two-state solution and, by the end of 2014, the Palestinian Authority (in collaboration with Jordan) joined UN bodies as a distinct state.

Although Jordanian officials have had real concerns about the alternative homeland since its founding and particularly since the 1970s, these concerns did not reflect an imminent threat in 2014. This comes across clearly when comparing Obama's peace efforts from 2010 to 2014, which involved negotiations between Palestinian and Israeli representatives toward a two-state solution, and the Trump administration's peace plans. Trump's peace efforts largely excluded the Palestinians and favored Israeli goals to an unprecedented extent, including by moving the US embassy to Jerusalem and allowing additional Israeli settlements in the West Bank. Trump's peace efforts, unlike Obama's, raised substantive fears about the alternative homeland in Jordan.

Regardless, in 2014, Jordanian officials publicly equated nationality law reform with Israeli plans to make Jordan Palestine because the reform would nationalize more Palestinians (Husseini 2014). Prominent among these noncitizen Palestinians were the Gazans, who make up a large, visible proportion of noncitizen Palestinians in Jordan and of those married to Jordanian women. One former minister remarked that most of the roughly 600,000 Palestinians living in Jordan without nationality are Gazan.[60] Another former minister explained "80–90 percent of women married to non-Jordanians are married to non-Jordanian Palestinians," and "most of these men are from Gaza; this would nationalize 300,000 at once."[61] A third former minister explained that women's nationality reform would give "hundreds of thousands of children Jordanian citizenship, and most of them are Gazan;" he added that "with a resolution to the Palestinian situation, then Jordan can reform women's citizenship."[62]

Jordanian officials also directly referenced noncitizen Palestinians when speaking to Jordanian women married to foreigners. For example, a Jordanian woman married to an Egyptian explained that when she demands rights for her children, "the government brings up the [Palestinian] right of return, [Palestinian] resettlement ("توطين"), the alternative homeland, and emptying Palestine . . . they talk about demographic changes and act like the children are security threats."[63] Although this woman is not married to a Palestinian, the government explicitly linked nationality law reforms that would grant more rights to her children to concerns about the alternative homeland.

However, when providing details on why the demographic issue was so important to blocking this reform, Jordanian officials privately emphasized threats to the privileges of Transjordanian leaders rather than fears about Israeli connections to Gazans and their combined efforts to make Jordan Palestine. A former minister explained that "the women's citizenship issue is really about women married to Gazans; even though it is the same policy for all women, the government is really worried about the Gazans."[64] He added that the "deep state . . . refused to give these kids citizenship or political rights." He described the deep state as including Transjordanian officials in the royal court, intelligence, and bureaucracy, who can "use the pretext of security" when they want to block a reform.[65]

Other ministers also highlighted the link between alleged security concerns and Transjordanian leaders in the security apparatus. One former minister noted that in policymaking, "security dimensions and measures are most important," even though "Jordan is secure." He added that the security apparatus "says security is the most important thing over everything because they want to stay in their jobs."[66] Another former minister described the reform's failure as "explicitly tied to Palestinian citizenship," and said this link is the source of the "vicious" rhetoric against reform because people are worried that "Jordan will become Palestine."[67] Another former minister echoed this point, commenting, "Unlike other women's policies, citizenship law stems from the security apparatus . . . it is not cultural or religious, it is political."[68] These comments highlight that, although Jordanian officials framed Gazans as fifth columns, the primary motivation for this framing reflected domestic elite interests rather than serious concerns about Gazan links to external threats.

## Domestic Motivations for Fifth-Column Framing

The 2014 debate over nationality law reform highlights how political elites can portray groups as fifth columns and security threats to avoid introducing domestic reforms that hurt their interests. Specifically, influential leaders from the General Intelligence Directorate (GID), referred to as the *mukhabarat* (المخابرات), as well as conservative Transjordanian advisors to the king opposed the reform because it would disturb the status quo in terms of demographics and, potentially, in terms of the privileges that Transjordanians have in the state, including special access to jobs and state benefits.

Like the king's advisors, GID leaders have substantial influence over policymaking in Jordan, particularly over nationality policies. Curtis Ryan and Jillian Schwedler (2004: 143) quote a former prime minister as stating: "Nothing, nothing is decided on any topic without the *mukhabarat*. No policy, political or otherwise, is uninfluenced by them." In addition, the GID has close links to the Ministry of Interior (MOI)—which handles nationality issues—with leaders circulating between these institutions (Tell 2015: 2; Moore 2019: 251). A former government official emphasized that "There's high coordination between the MOI and GID . . . some ministers of interior are ex-GID, including former directors of the GID."[69]

These vested interests refer in part to the Transjordanian composition of the GID. As a former minister observed, "the GID has major influence over any policy that affects the national interest," and "most of the GID is made up of East

Bank Jordanians."[70] Another former minister highlighted this linkage, describing the security apparatus as the "bastion" of East Bank Jordanian feelings.[71] A third former minister explained that in the security apparatus, "there is a strong feeling to have pure Jordanian national tendencies."[72]

The linkage between the security apparatus and Transjordanian communities is not new and reflects several historic policies. To start, it reflects the policy, initiated by the British in the 1930s, of recruiting loyal Transjordanian Bedouins into the army (Abu Odeh 1999: 17–18). It also reflects decisions to purge Palestinian-Jordanians unofficially from sensitive positions due to their questioned loyalty to the regime, following Black September (Baylouny 2008: 289; Frost 2020: 199–218). More recently, it reflects Transjordanian nationalists' efforts to establish Jordan as a Transjordanian state, where Palestinian-Jordanians will return to Palestine after a final peace settlement or decide to exchange their Palestinian identity for a more Transjordanian one (Abu Odeh 1999: 214–215, 228).

Transjordanians also populate most government positions, both military and civilian, through nepotism. Many Jordanians, regardless of their descent, highlight the importance of "wasta (واسطة)," or social connections to access government jobs and services. A Jordanian development worker noted that Jordan "needs reforms to wasta, to the 'system of clans (نظام العشائر)', because "wasta is everywhere."[73] The reference to clans (al-'ashaa'ir العشائر) highlights that most of the Transjordanians benefiting from wasta are those hailing from prominent East Bank clans and tribes. Even in 1968, Jordanian MOI leaders candidly told US officials that they would not provide Gazans with nationality because the Gazans would "compete with Jordanian workers." US officials were surprised that Jordanian leaders "frankly admitted" that in practice the "best qualified are often passed over . . . to make room for relatives of job holders and for those hailing from the East Bank."[74]

Transjordanians, particularly from prominent tribes, benefit not only from special access to state jobs but also from the benefits these jobs entail, especially in the military. These include access to healthcare and insurance, subsidized goods, social security and retirement benefits; access to loans, subsidized housing, and higher education; and university quotas for veterans and their family (Baylouny 2008: 280, 288–289). As a former minister highlighted, Transjordanians view these benefits as part of their "social contract" with the state; he explained "the Hashemites rule by consensus" and "this consensus includes giving [Trans]Jordanians more than other groups."[75] Further, when asked about Gazan exclusion, an Islamist leader pointed out that "the deep state does not want the Gazans; it is conservative and does not want to give them more jobs or influence (تأثير)."[76]

Overall, the GID and other influential Transjordanian leaders opposed the nationality reform because it threatened their privileges and promoted equal citizenship, not because it represented a substantive threat to Jordan's sovereignty. For example, the nationality law reform would increase the number of Palestinian-Jordanians, which in turn could increase Palestinian pressures on the Jordanian government to provide them with greater, more proportionate representation in government. A Jordanian political analyst observed that the nationality law reform "touches on their [GID] territory and they do not want to lose any territory."[77] A Jordanian woman married to a non-Jordanian similarly stated that the blocked reform reflected a "competition over resources; the government system wants to secure its special access to benefits and services . . . they do not want competition from these children."[78]

The limited number of noncitizen Palestinians married to Jordanian women further indicates the domestic rather than geopolitical impetus for framing the Gazans as fifth columns. Even using official state figures, which may be inflated, only 62–64 percent of cases involve women married to noncitizen Palestinians (Abuqudairi 2014; Husseini 2014). This figure is much lower than the frequently stated "90 percent" of cases. As one pro-reform activist described, the "government did not listen when they said that most of these women are not married to Palestinians . . . it is a problem in the north and south because of Syrian and Saudi fathers due to [cross-border] tribal links." She added that, instead, the government publicly "enlarges the issue, as if all the husbands are Gazan." She also observed that, when it comes to the alternative homeland, "the core of the problem with Palestinians is with the West Bank," not the Gaza Strip.[79] This comment highlights the disjuncture between the geopolitical framing of Gazans in public as fifth columns representing an Israeli plot and the reality of the threats the government claims they represent.

The framing of Gazans as a fifth column in public debates about the nationality law reform was not based on hostile relations with Israel or on imminent threats posed by the Gazans. None of the Jordanian officials interviewed provided clear connections between Gazans and the Israelis. Instead, as the interview quotes demonstrate, these officials identified these links as false or referred loosely to these links, often relying on preconceived notions and historic examples. In addition, since the 1970–1971 conflict, Gazans have not perpetrated any violence against the state or regime.[80] Thus, the use of fifth-column frames to block the 2014 nationality law reform did not stem from geopolitical interests or well-founded security threats. Instead, political elites, particularly GID leaders, seemed to use these frames to support their own interests.

Ultimately, King Abdullah supported the securitization of the policy and did not pressure the government to reform the nationality law. A former minister summarized the king's response stating, "the royal court did not fully support it [the reform], but the intelligence was against it . . . the intelligence's interference is heavy-handed, and the royal court always discusses with the intelligence first."[81] Given the GID's influence in Jordan, as well as the long-held view of Transjordanian tribes as the backbone of the Hashemite monarchy in Jordan, King Abdullah seemed to decide that this issue was not worth the political capital needed to oppose the GID. He also may have considered it expedient to frame this decision, which diminished women's rights, as linked to external security threats to justify the lack of reform both domestically and internationally. Overall, domestic political elite interests drove the fifth-column framing and securitization of the reform.

## Conclusion

*When and why do political elites publicly frame groups as fifth columns?* This chapter finds that political elites, namely a state's executive leader and his or her advisors, frame groups as fifth columns only when it strengthens their hold on power and suits their professional and personal interests. Contrasting with theories based on geopolitical factors, this chapter finds that Jordan's use of fifth-column frames on Palestinians has not aligned with Jordan's relations with external powers or with moments when Palestinians were linked clearly to external threats. Instead, this chapter provides support for the prominent role of domestic political interests in explaining the manifestation of fifth-column politics.

These findings suggest that political elites can use fifth-column frames to justify policies that might otherwise be unappealing to domestic and international audiences. These frames do not have to reflect current threats and can draw upon a group's historic links to an external threat. In addition, these frames can portray a group as indirectly linked to an external threat, where that group, perhaps unknowingly, is part of a foreign plot against the state. Jordanian officials used these strategies to connect Gazans to Israeli plans to take over the West Bank and Gaza and designate Jordan as the Palestinian state. Jordanian officials viewed Gazans as threatening—even though Gazans also condemn the alternative homeland—because, for Jordanian officials, the act of nationalizing Gazans in Jordan would provide support to Israeli claims that Jordan is Palestine. However, as my analysis of interview and archival data demonstrated, applying these claims to the women's nationality law reform made little sense in geopolitical terms, particularly because Gazans did not make up the vast majority of foreign husbands and the reform would not have nationalized all Gazans.

The second case study also highlights that government officials can use fifth-column frames on groups that are not legal members of the state. Although the Gazans have lived in Jordan for generations, carry Jordanian passports, and enjoy more rights than other noncitizens, they never received Jordanian nationality. As such, they have operated as quasi-citizens of Jordan, exercising their membership in practice rather than in law. This dynamic does not appear unique to Jordan, with states around the world fluctuating in how they portray long-term, noncitizen migrant groups, such as Germany's treatment of Turkish migrants or the United States' treatment of Central American migrants. This case highlights that states can construct who is an insider as well as who is a fifth column, and it challenges the insider-outsider distinction often used when studying modern states.

Overall, this chapter describes two critical junctures where government officials could have used subversive fifth-column frames. In both cases, these frames concerned a group defined by ethnic descent with potential links to external threats. However, the key difference between the decision not to use the fifth-column frame in the first case but to use it in the second case was the political and personal interests of domestic ruling elites.

# Notes

1. I recruited participants using a purposive snowball sampling technique. I conducted these interviews in English or Arabic, depending on the interviewee's preference. I typically conducted these interviews alone, but in some cases, I brought a research assistant to increase my safety and to help in interpreting different Arabic dialects.
2. The majority of ministers interviewed were prime ministers, ministers of interior, or ministers of foreign affairs. They also were predominantly Transjordanian men.
3. This section draws from Frost 2020: 99–109.
4. Greater Syria refers to current-day Lebanon, Syria, Israel, Palestine, and Jordan.
5. Hashemites ruled the Hejaz until 1925 when Ibn Saud successfully annexed it. The Hashemite dynasty lasted in Iraq until a coup d'état removed them in 1958. The Hashemites remain in power in Jordan today.
6. Until the end of 1949, officials did not refer to this territory as the West Bank. British officials referred to it as "Arab Palestine," "Eastern Palestine," or "the Arab areas of Palestine." Transjordanian officials described it as "west Jordan," "the western territory," or "the western territories" (Massad 2001: 229). For simplicity, I use the term "West Bank" throughout this chapter.
7. Memorandum from Foreign Office Research Department, October 27, 1949, Foreign Office File 371/75287, File E13020/1081/80, The National Archives of the UK at Kew.
8. Ibid.
9. Telegram to the Secretary of State from Fritzlan, December 12, 1949, 1945–49 Central Decimal File, File 767n.90i/12-1249, The US National Archives College Park.
10. Memorandum of Conversation to Department of State, November 30, 1949, 1945–49 Central Decimal File, File 767n.90i/11-3049, The US National Archives College Park.
11. Situation Report from Sir A. Kirkbride, Amman, April 4, 1949, Foreign Office File FO 371/75273, File E4646/1013/80, 1–2, The National Archives of the UK at Kew.

12. Incoming Telegram to the Secretary of State from Stabler, March 30, 1949, 1945–49 Central Decimal File, File 890i.00/3-3049, The US National Archives College Park.

13. Jordan: Annual Review for 1949 from Sir A. Kirkbride in Amman, January 13, 1950, Foreign Office File FO 371/82702, File ET1011/1, The National Archives of the UK at Kew. This is the specific name Kirkbride instructed the Foreign Office to use. However, as Joseph Massad (2001) points out, a more accurate translation of the Arabic name "المملكة الأردنية الهاشمية" is the Hashemite Jordanian Kingdom. The accepted official name today is the Hashemite Kingdom of Jordan.

14. "Law Number 56 of 1949— Additional Law for the Nationality Law" Issue Date "December 20, 1949" Issue Number 1004.

15. Amman Despatch No. 91 from Sir A. Kirkbride, December 12, 1949, Foreign Office File FO 371/75277, File E15075/10111/80, The National Archives of the UK at Kew.

16. Incoming Telegraph to the Secretary of State from Gerald Drew, April 25, 1950, 1950–54 Central Decimal File, File 684.85/4-2550, The US National Archives College Park.

17. Outward Telegram from Commonwealth Relations Office, July 20, 1951, Foreign Office File 371/91838, File ET 1942/9(c), The National Archives of the UK at Kew.

18. Incoming Despatch to the Minister of State Kenneth Younger, September 14, 1951, Foreign Office File 371/91839, File ET 1942/66, The National Archives of the UK at Kew.

19. Ibid.

20. Incoming Telegram to G. W. Furlonge in the Eastern Department Foreign Office, July 24, 1951, Foreign Office File 371/91839, File ET 1942/34, The National Archives of the UK at Kew.

21. Glubb had led Jordan's army during the mandate period (since 1930) and remained in that position until 1956.

22. Incoming Letter to Lt. Col. R.K. Melville, August 16, 1951, Foreign Office File 371/91839, File ET 1942/60, The National Archives of the UK at Kew.

23. Ibid.

24. Ibid.

25. Ibid.

26. Kirkbride served as a British advisor to King Abdullah since Abdullah's arrival in Transjordan in 1920 in various official British positions.

27. Incoming Telegram from Sir A. Kirkbride, July 22, 1951, Foreign Office File 371/91838, File ET 1942/20, The National Archives of the UK at Kew.

28. Incoming Telegram from Sir A. Kirkbride, July 22, 1951, Foreign Office File 371/91836, File ET 1941/22, The National Archives of the UK at Kew.

29. Incoming Telegram from A. R. Walmsley, July 22, 1951, Foreign Office File 371/91839, File ET 1942/41, The National Archives of the UK at Kew.

30. Foreign Service Despatch No. 32 to the Department of State, Washington from Gerald A. Drew, August 9, 1951, 1950–54 Central Decimal File, File 785.00/8-951, The US National Archives College Park.

31. Ibid.

32. Annual Report Sent to the Foreign Office, December 26, 1951, Foreign Office File 371/98856, File ET 1011/1, The National Archives of the UK at Kew.

33. Incoming Despatch to Principal Secretary of State for Foreign Affairs Anthony Eden, November 6, 1952, Foreign Office File 371/98859, File ET 1015/19, The National Archives of the UK at Kew.

34. Incoming Dispatch to the Secretary of State for Foreign Affairs Anthony Eden, December 24, 1952, Foreign Office File 371/98857, File ET 1013/14, The National Archives of the UK at Kew.

35. Annual Report Sent to the Foreign Office, December 26, 1951, Foreign Office File 371/98856, File ET 1011/1, The National Archives of the UK at Kew. UNRWA stands for the United Nations Relief and Works Agency for Palestine Refugees in the Near East. UNRWA precedes UNHCR (i.e., the Office of the United Nations High Commissioner for Refugees) and focuses solely on Palestinian refugees from the 1948 Arab-Israeli War residing in Jordan, Lebanon, Syria, the West Bank, and Gaza.

36. Incoming Telegram to the Secretary of State from Green, November 28, 1952, 1950–54 Central Decimal File, File 785.00/11-2852, The US National Archives College Park.

37. Outward Telegram from Commonwealth Relations Office, July 20, 1951, Foreign Office File 371/91838, File ET 1942/9(c), The National Archives of the UK at Kew.

38. Incoming Telegram to Foreign Office, July 20, 1951, Foreign Office File 371/91838, File ET 1942/10, The National Archives of the UK at Kew.

39. Press News Summaries, July 20, 1951, Foreign Office File 371/91838, File ET 1942/26, The National Archives of the UK at Kew.

40. Annual Report Sent to the Foreign Office, December 26, 1951, Foreign Office File 371/98856, File ET 1011/1, The National Archives of the UK at Kew.

41. Author interview with former minister (EU87), January 2016.

42. Author interview with former minister (LG53), December 2016.

43. Author interview with Jordanian historian and former diplomat (PW98), May 2017.

44. August 1952 Situation Report to Foreign Office, September 4, 1952, Foreign Office File 371/98857, File ET 1013/10, The National Archives of the UK at Kew.

45. Despatch No. 144 to the Department of State, Washington from the Amman Embassy, October 23, 1954, 1950–54 Central Decimal File, File 785.11/10-2354, The US National Archives College Park.

46. This section draws from Frost 2020: 230–237 and 317–325.

47. Telegram from American Embassy in Amman to the Secretary of State/Department of State, February 2, 1968, 1967–69 Subject Numeric File, File REF 3 UNRWA, The US National Archives College Park.

48. King Hussein Bin Talal, "Address to the Nation," Amman, July 31, 1988. http://www.kinghussein.gov.jo/88_july31.html.

49. Author interview with former minister (MO43), October 2019.

50. This discrimination reflects the diffusion of European laws regarding citizenship and nationality to the region, rather than a direct application of Islamic or tribal law (Sonbol 2003). Much of Jordan's 1954 nationality law matched British law verbatim (Massad 2001). Although tribal and religious heritage, including the traditional notion of blood passing through the father (Charrad 2001), likely made policymakers

more amenable to including this discrimination in nationality laws, they were not the source of these laws.

51. The sole exception legally is for children who would otherwise become stateless. However, the government often ignores this policy in the case of stateless Palestinian children.

52. Article 9 of CEDAW stipulates that "States Parties shall grant women equal rights with men with respect to the nationality of their children."

53. King Abdullah II Website, "Jordan First." https://rhc.jo/en/hm-king-abdullah-ii/jordan-first.

54. King Abdullah II Website, "National Agenda." https://rhc.jo/en/hm-king-abdullah-ii/national-agenda.

55. King Abdullah II Website, "We Are All Jordan." https://rhc.jo/en/hm-king-abdullah-ii/we-are-all-jordan.

56. *Mubadara* translates to "initiative." The translated full name is the National Initiative Bloc.

57. However, in practice, these privileges often have not been enforced, and during numerous author interviews, women have complained about spending time and money trying to access them with no success.

58. Author interview with former minister (MI39), May 2017.

59. Author interview with former minister (MO43), October 2019.

60. Author interview with former minister (EU87), January 2016.

61. Author interview with former minister (ZE98), June 2017.

62. Author interview with former minister (MO43), October 2019.

63. Author interview with Jordanian woman married to a non-Jordanian (LY44), February 2017.

64. Author interview with former minister (EA89), January 2016.

65. Ibid.

66. Author interview with former minister (TV54), January 2019.

67. Author interview with former minister (EU87), October 2017.

68. Author interview with former minister (MH78), January 2016.

69. Author interview with former government official (TK26), April 2017.

70. Author interview with former minister (EU87), January 2016.

71. Author interview with former minister (LG53), December 2016.

72. Author interview with former minister (EA89), January 2016.

73. Author interview with Jordanian development worker (RI20), October 2019.

74. Airgram from American Consulate in Dhahran to the Department of State, June 3, 1968, 1967–69 Subject Numeric File, File POL 7 JORDAN A-129, The US National Archives College Park.

75. Author interview with former minister (EU87), October 2019.

76. Author interview with Islamist leader (WJ61), October 2019.

77. Author interview with Jordanian political analyst (FF23), January 2019.

78. Author interview with Jordanian woman married to a non-Jordanian (BF75), May 2017.

79. Author interview with Jordanian women's activist (JD27), October 2019.

80. Some Gazans likely supported the PLO in the 1970–1971 conflict. However, Gazans as a group did not necessarily join the conflict, and they did not necessarily join the PLO's side in it if they did.

81. Author interview with former minister (EU87), January 2019.

# References

Abu Odeh, Adnan. 1999. *Jordanians, Palestinians, and the Hashemite Kingdom in the Middle East Peace Process*. Washington, DC: United States Institute of Peace Press.

Abuqudairi, Areej. 2014. "Women 'Punished' for Marrying Non-Jordanians." *Al-Jazeera*, December 20. http://www.aljazeera.com/news/middleeast/2014/12/women-punished-marrying-non-jordanians-20141215121425528481.html.

Al-Rasheed, Madawi. 2011. "Sectarianism as Counter-Revolution: Saudi Responses to the Arab Spring." *Studies in Ethnicity and Nationalism* 11, no. 3 (December): 513–526.

Balzacq, Thierry. 2005. "The Three Faces of Securitization: Political Agency, Audience and Context." *European Journal of International Relations* 11 (2): 171–201.

Bani Salameh, Mohammed Torki, and Khalid Issa El-Edwan. 2016. "The Identity Crisis in Jordan: Historical Pathways and Contemporary Debates." *Nationalities Papers* 44 (6): 985–1002.

Baylouny, Anne Marie. 2008. "Militarizing Welfare: Neo-liberalism and Jordanian Policy." *Middle East Journal* 62 (2): 277–303.

Brand, Laurie. 1988. *Palestinians in the Arab World: Institution-Building and the Search for State*. New York: Columbia University Press.

Brass, Paul. 1997. *Theft of an Idol*. Princeton, NJ: Princeton University Press.

Brumberg, Daniel. 2014. "Theories of Transition." In *The Arab Uprisings Explained: New Contentious Politics in the Middle East*, edited by Marc Lynch, 29–54. New York: Columbia University Press.

Buzan, Barry, Ole Wæver, and Jaap de Wilde. 1998. *Security: A New Framework for Analysis*, Boulder, CO: Lynne Rienner.

Charrad, Mounira. 2001. *States and Women's Rights: The Making of Postcolonial Tunisia, Algeria, and Morocco*. Berkeley: University of California Press.

Cole, Wade M. 2013. "Government Respect for Gendered Rights: The Effect of the Convention on the Elimination of Discrimination against Women on Women's Rights Outcomes, 1981–2004." *International Studies Quarterly* 57: 233–249.

El-Abed, Oroub, Jalal Husseini, and Oraib Al-Rantawi. 2014. "Listening to Palestinian Refugees/Displaced Persons in Jordan: Perceptions of Their Political and Socio-economic Status." Al Quds Center for Political Studies-Amman, Jordan (January).

Forester, Summer. 2019. "Protecting Women, Protecting the State: Militarism, Security Threats, and Government Action on Violence against Women in Jordan." *Security Dialogue* 50 (6): 475–492.

Frost, Lillian. 2020. "Ambiguous Citizenship: Protracted Refugees and the State in Jordan." PhD diss., George Washington University. https://www.proquest.com/openview/cc137bdaaee818a84851468e755bac58/1?pq-origsite=gscholar&cbl=18750&diss=y.

Gagnon, V. P. 1994/1995. "Ethnic Nationalism and International Conflict: The Case of Serbia." *International Security* 19, no. 3 (Winter): 130–166.

Gause, F. Gregory. 2010. *The International Relations of the Persian Gulf.* New York: Cambridge University Press.

Hammerstadt, Anne. 2014. "The Securitization of Forced Migration." In *The Oxford Handbook of Refugee and Forced Migration Studies*, edited by Elena Fiddian-Qasmiyeh, Gil Loescher, Katy Long, and Nando Sigona, 265–277. Oxford: Oxford University Press.

Harff, Barbara. 2003. "No Lessons Learned from the Holocaust?: Assessing Risks of Genocide and Political Mass Murder since 1955." *American Political Science Review* 97, no. 1 (February): 57–73.

Howard, Marc Morjé. 2009. *The Politics of Citizenship in Europe.* New York: Cambridge University Press.

Human Rights Watch. 2010. "Stateless Again: Palestinian-Origin Jordanians Deprived of their Nationality." (February 2).

Hussein bin Talal. 1962. *Uneasy Lies the Head: An Autobiography.* London: Heinemann.

Husseini, Rana. 2014. "Gov't Announces Privileges for Children of Jordanian Women Married to Foreigners." *Jordan Vista*, November 9. http://vista.sahafi.jo/art.php?id=dcd832e583bcddbd74a3b00cf3f96d765394697b.

Jenne, Erin K. 2007. *Ethnic Bargaining: The Paradox of Minority Empowerment.* Ithaca, NY: Cornell University Press.

Kvittingen, Anna, Åge A. Tiltnes, Ronia Salman, Hana Asfour, and Dina Baslan. 2019. "'Just Getting By': Ex-Gazans in Jerash and Other Refugee Camps in Jordan." *Fafo Report*, 34.

Lane, David. 2011. "Identity Formation and Political Elites in the Post-Socialist States." *Europe-Asia Studies* 63 (6): 925–934.

Lori, Noora. 2017. "Statelessness, 'In-Between' Statuses, and Precarious Citizenship." In *Oxford Handbook of Citizenship*, edited by Ayelet Shachar, Rainer Bauböck, Irene Bloemraad, and Maarten Vink, 743–766. Oxford: Oxford University Press.

Lynch, Marc. 2012. *The Arab Uprising: The Unfinished Revolutions of the New Middle East.* New York: PublicAffairs.

Massad, Joseph. 2001. *Colonial Effects: The Making of National Identity in Jordan.* New York: Columbia University Press.

Moore, Pete W. 2019. "A Political-Economic History of Jordan's General Intelligence Directorate: Authoritarian State-Building and Fiscal Crisis." *Middle East Journal* 73 (2): 242–262.

Mylonas, Harris. 2012. *The Politics of Nation-building: Making Co-nationals, Refugees, and Minorities.* New York: Cambridge University Press.

Nevo, Joseph. 1996. *King Abdullah and Palestine: A Territorial Ambition.* Basingstoke, UK: Macmillan.

Plascov, Avi. 1981. *The Palestinian Refugees in Jordan 1948–1957.* London: Frank Cass.

Rantawi, Oraib, and Oroub el-Abed. 2012. "Modest but Powerful Activism for Palestinian-Origin Jordanian Rights." Al-Shabaka, The Palestinian Policy Network, October 1. https://al-shabaka.org/commentaries/modest-but-powerful-activism-for-palestinian-origin-jordanian-rights/.

Robins, Philip. 1989. "Shedding Half a Kingdom: Jordan's Dismantling of Ties with the West Bank." *Bulletin (British Society for Middle Eastern Studies)* 16 (2): 162–175.

Robins, Philip. 2004. *A History of Jordan.* Cambridge: Cambridge University Press.

Ryan, Curtis. 2011. "Political Opposition and Reform Coalitions in Jordan." *British Journal of Middle Eastern Studies* 38, no. 3 (December): 367–390.

Ryan, Curtis, and Jillian Schwedler. 2004. "Return to Democratization or New Hybrid Regime?: The 2003 Elections in Jordan." *Middle East Policy* 11 (2): 138–151.

Salibi, Kamal. 1998. *The Modern History of Jordan.* New York: I.B. Tauris.

Shevel, Oxana. 2011. *Migration, Refugee Policy, and State Building in Postcommunist Europe.* New York: Cambridge University Press.

Shlaim, Avi. 2007. *Lion of Jordan: The Life of King Hussein in War and Peace.* New York: Alfred A. Knopf.

Snyder, Jack. 2000. *From Voting to Violence: Democratization and Nationalist Conflict.* New York: W. W. Norton.

Sonbol, Amira El-Azhary. 2003. *Women of Jordan: Islam, Labor, & The Law.* Syracuse, NY: Syracuse University Press.

Tell, Tariq. 2015. "Early Spring in Jordan: The Revolt of the Military Veterans." Civil-Military Relations in Arab States Series. Carnegie Middle East Center, November 4.

Williams, Michael C. 2003. "Words, Images, Enemies: Securitization and International Politics." *International Studies Quarterly* 47: 511–531.

# 8

# External or Internal Enemies?

## Polish Citizens in Interwar France and the Ethnic Politics of Citizenship*

*Kathryn Ciancia*

Between 1918 and 1939, members of Poland's political class feared that their state, which had emerged on the map of eastern Europe following the collapse of the continental empires at the end of World War I, was being undermined by enemies within. In this nominal nation-state, where less than 70 percent of the population was identified as ethnically Polish, much of their anxiety was tied to the so-called national minorities, including Germans, Jews, Ukrainians, Lithuanians, and Belarusians (Tomaszewski 1985). While the vast majority of these people were Polish citizens (singular: *obywatel*), the Polish state categorized them as separate nationalities (singular: *narodowość*) who were granted certain rights as such by Poland's 1921 constitution. Yet despite Poland's ostensible protection of minority rights—or, as Eric Weitz argued, precisely because of the entangled histories of minority protection and discrimination against minorities (Weitz 2008)—Polish officials assumed that national minorities were more likely to be disloyal to the state. This was true for both those nationalities that could be linked to external states, like ethnic Germans, who were perceived to be working on behalf of Germany, particularly after Hitler's ascent to power in 1933 (Chu 2012), and those, such as Ukrainians and Jews, who did not possess their "own" nation-states (Brykczynski 2016; Charnysh, this volume; Schenke 2004). Indeed, the latter two groups were frequently viewed as threats to the state because of their alleged allegiance to a "foreign" ideology—Bolshevism—which Poles associated with both a supra-national external state, in the shape of the Soviet Union, and international networks of agitators that transcended state borders.

Although Poland's status as a new, geopolitically vulnerable, and post-imperial state with a large percentage of national minorities contributed to heightened anxieties about enemies within, Polish concerns were not unusual during this

* I would like to thank Małgorzata Fidelis, Emily Greble, and Maureen Healy, as well as the editors of this volume and my fellow workshop participants, for their insightful comments on earlier versions of this chapter.

Kathryn Ciancia, *External or Internal Enemies?* In: *Enemies Within.* Edited by: Harris Mylonas and Scott Radnitz, Oxford University Press. © Oxford University Press 2022. DOI: 10.1093/oso/9780197627938.003.0009

period. National and imperial hysteria about internal enemies—whether citizens or subjects—was commonplace across much of Europe during and immediately after World War I, from France to the Russian empire (Caglioti 2020; Lohr 2003). As that war came to a close, the often violent processes of building successor states in eastern and central Europe, where borders could not possibly align with clear blocs of national groups, were accompanied by the ongoing targeting of non-titular minorities, particularly Jews, who were accused of being disloyal by all sides (Eichenberg 2010; Fink 2004; Hanebrink 2018). Even in peacetime, European elites continued to argue that nationality conditioned a person's loyalty—and therefore his or her right to citizenship. In keeping with their French and Czechoslovak counterparts in Alsace-Lorraine and Bohemia respectively (both borderlands that had, until recently, been part of other states), Polish elites worried that people who did not identify with the titular nation could not be fully trusted and therefore deserved fewer rights (Boswell 2000; Carrol 2018; Zahra 2008). In fact, Polish officials discriminated against national minorities in decisions about citizenship as they sought to simultaneously nationalize their populations and enforce loyalty toward the state (Stauter-Halsted 2021; Zielinski 2004).

If scholars have studied Polish fears of enemies operating within the borders of the Polish Second Republic (as the interwar state was known), however, very little work has been done on the state's attempts to monitor the alleged anti-state behavior of Polish citizens who lived beyond Poland's boundaries. On one level, this tendency reflects a more general obsession among historians with a territorially based definition of Poland. But it is also indicative of the hard-and-fast line that is drawn between the external and internal elements of "fifth columns" more generally. Put simply, definitions tend to rest on the idea that fifth columns are made up of people physically living within the borders of a state whose authority they undermine with the aid of outside groups. The *Merriam-Webster Dictionary*, for example, characterizes a fifth column as "a group of secret sympathizers or supporters of an enemy that engage in espionage or sabotage within defense lines or national borders," while the *Oxford English Dictionary* defines it as "a body of one's supporters in an attacked or occupied foreign country, or the enemy's supporters in one's own country." The definition that Radnitz and Mylonas lay out in the Introduction to this volume also makes this critical distinction between the domestic and external realms. Fifth columns, they state, are "domestic actors who work to undermine the national interest, in cooperation with external rivals of the state" (Radnitz and Mylonas, Introduction, 3). But what does it mean to be a "domestic actor"? To what does the "within" in the phrase "enemies within" refer? And should interpretations of the external and internal components of this phenomenon always be about immediate physical location?

With the aim of exploring these questions, this chapter focuses on a group of people who were simultaneously external and internal to the state: interwar Polish citizens who lived abroad. Despite their place of residence, these people were insiders because they often retained social, political, economic, and emotional connections with the physical territory of Poland and they possessed the rights (and obligations) of citizens, including the right to return home. But they were simultaneously outsiders whose loyalty toward the Polish state—"the first duty of the citizen," according to Article 89 of Poland's 1921 constitution—was considered to be under threat precisely because they operated largely beyond the state's purview. Indeed, for Polish state officials, a multitude of factors, including physical and emotional distance from the motherland; exposure to nationally, ideologically, and morally degenerate foreign environments; and the institutional weakness of the Polish consulates that were supposed to monitor their behavior made these people more likely to act in a subversive manner. Studying how the Polish state viewed them thus allows us to reflect upon what we mean—and what our historical actors meant—by "domestic" and "foreign" threats.

Although Polish citizens could be found in locations across the interwar world, from Germany to Liberia and from Mandatory Palestine to Brazil, I focus here on the industrial and coalmining departments of Pas-de-Calais and Nord in northeastern France where more than two hundred thousand Polish citizens lived. Tracing policies toward these insider-outsider citizens, particularly those implemented by Polish consular officials in the city of Lille, reveals how the state both imagined enemies and mobilized various strategies to prevent those "enemies" from undermining state power over time—from schemes to reduce exposure to communism, to the monitoring of suspected communists who tried to return to Poland, and, eventually, to denaturalization. Since the majority of Polish citizens in northeastern France were considered to be ethnically Polish, looking at this group also suggests that anxieties about citizens becoming internal enemies frequently concerned their alleged susceptibility to communism and left-wing politics more broadly, rather than simply their non-Polish ethnic identity. Yet even when Polish state personnel emphasized political disloyalty, discussions about protecting the state from subversive elements remained stubbornly entangled with assumptions about the disloyalty of non-ethnically Polish citizens, particularly Jews.

## Citizenship and Emigration

Beginning in 1918, representatives of the Polish state attempted to answer two interrelated, but by no means identical, questions: who was entitled to Polish citizenship and who should physically live within the state's borders. Drawing on the various legacies of the Russian, German, and Habsburg empires from which the Second Republic had emerged, as well as on multiple postwar treaties, including the Treaty of Versailles (June 1919) and the Treaty of Riga (March 1921), legislation about citizenship was messy and piecemeal. Nevertheless, in January 1920, the provisional Polish parliament (Sejm) passed a citizenship law that combined elements of *jus sanguinis* and *jus soli*, meaning that people could gain Polish citizenship through, among other criteria, their place of birth, ancestry, marriage to a Polish citizen (if they were a woman), or prewar residency. In some areas, a person could, within a certain time frame, choose between two potential citizenships (e.g., between German and Polish or Polish and Soviet), an arrangement that relied on the liberal idea that citizenship was a bond with the state into which a person entered voluntarily (Lohr 2012: 140). There was no option for dual citizenship; taking the citizenship of another state automatically meant sacrificing Polish citizenship.

If questions about granting citizenship naturally became entangled with the issue of who should be allowed to settle within the state's borders, Polish elites also debated an equally fraught question that had its roots in the pre-1918 empires: which citizens should be allowed to leave (Green 2005; Zahra 2016)? As the sociologist Rogers Brubaker has argued, nation-states generally do not want their citizens to live outside of the state's borders because it means that groups of people are simultaneously subjected to internal and external policies (Brubaker 2010). Yet in light of the problem of domestic overpopulation, Polish elites in the early 1920s deemed it demographically and economically expedient to encourage a relatively large number of Polish citizens to live and work elsewhere. These migrants included "pioneers" in colonial or semi-colonial areas that supposedly required European civilizational influence, such as Brazil or Liberia (Puchalski 2022), as well as seasonal agricultural laborers in nearby states (e.g., in Germany or Latvia) and longer-term industrial and agricultural workers (e.g., in France).

In each case, the precise relationship between citizenship and the act of emigration remained unclear. The January 1920 law, which stipulated how citizenship could be both gained and lost, did not directly mention emigration, although the Polish lawyer and constitutional scholar Stanisław Starzyński argued that the law implicitly dealt with issues of emigration in several of its clauses (Starzyński 1921: 35). For instance, someone could lose Polish citizenship by serving in a foreign army or, in the case of women, by marrying a foreign citizen, both events that were more likely to occur when Polish citizens came into contact with non-Poles outside of the country's physical borders. Moreover, consular personnel and Polish organizations abroad did not always agree upon how ideas of citizenship and national identity should line up. In some cases, people who were not formally Polish citizens were considered Polish, based on ethnic and religious (Roman Catholic) criteria, while those who were Polish citizens but held a non-Polish national identity, such as Jews and Ukrainians, were viewed in less favorable terms (Michalik-Russek 1992; Słyszewska-Gibasiewicz 2011). In 1927, for example, official instructions stated that while the Lille consulate should look after the passport affairs of Polish citizens, in certain cases it could also concern itself with people who were not Polish citizens but were "of the Polish nationality."[1]

## Policing Poles in Northeastern France

Against this backdrop, newly minted Polish citizens moved in droves to northeastern France. Because pre-1914 French anxieties about population decline had been exacerbated by the loss of young men during the recent world war, many French officials, together with leaders of agriculture and industry, supported Polish (as well as Czech, Spanish, and Italian) immigration (Camiscioli 2009). Polish-French labor agreements that were signed in 1919 and 1920 stipulated that Polish citizens could work in France for the same wages as their French counterparts and were entitled to social care from the French state, including health, unemployment, and pension benefits, as well as schooling for their children (Gargas 1926). For Polish workers, the deal was tempting. From under 3,400 in 1921, the number of Polish citizens in Pas-de-Calais reached 91,000, or 8 percent of the region's total population, by 1926 (Slaby 2015: 232).

The French state, acting in concert with a private organization of mine owners and representatives of agriculture, the General Immigration Society (Société générale d'immigration, or SGI), took charge of the process of recruitment, with only 10 percent of Polish migrants between 1919 and 1931 moving to France outside the organizational parameters of the SGI and its precursors (Cross 1983: 60). In the industrial cities of Mysłowice and Wejherowo, which became part of the new Polish state, French authorities set up recruitment centers where the bodies of potential workers were examined. If declared fit, a person would sign a work contract in the recruitment office, which would enable them to obtain an emigration passport from the local Polish state authorities, in turn allowing the French consulate to issue a visa for the journey to France. The contract was usually valid for one year, after which point the worker could either continue working for the same employer or look for work elsewhere (Rozwadowski 1927: 100). Large numbers of highly skilled miners were also recruited directly from Westphalia. Once granted Polish citizenship, these workers could legally live in France and receive the carte d'identité that was required by the French authorities.

In spite of the enthusiasm for such an arrangement, however, French elites also feared the politically and socially subversive effects that might result from so many Polish immigrants living in France. Although French anxieties were much more pronounced in relation to North Africans and even to other (non-Polish) Europeans, the authorities subjected Polish citizens to state surveillance (Lawrence 2000; Lewis 2007; Rosenberg 2006). According to a French law from 1893, immigrant workers had to register with the state authorities (About 2013;

Torpey 2000), a policy that was explained in a Polish-language booklet for new immigrants (Klimowicz 1923: 43–44).

Industrialists in northeastern France certainly sought to prevent immigrants from being infected with what they saw as radical left-wing politics by constructing self-contained Polish "colonies" that were described by one British journalist writing in the London *Times* in 1925 as "villages which are entirely Polish, with Polish shops, Polish banks and Polish churches" (17). Since they associated political radicalization, as well as socially degenerate behavior such as the solicitation of prostitutes, with single young men (Martial 1926: 400), French commentators also encouraged the inhabitants of such colonies to adhere to traditional gender roles. It was almost always the case, wrote one French observer in 1924, that "the married worker whose wife accompanies him to France is more orderly, more stable, [and] more sober than the bachelor." "In particular," he went on, "the Polish wife is hardworking. She works in the garden, the barnyard, [and] on the dairy farm, in occupations that are rarer and rarer in France, where female education, even for the working classes, is a factor of rural depopulation" (Guériot 1924: 433). To further encourage this model of domestic stability, France and Poland signed a protocol in 1928, stipulating that employers cover 60 percent of the costs that would allow the families of immigrants to move to France (Cross 1983: 176).

## Fears of Communist Infiltration

Like their French counterparts, Polish state officials also worried about the political repercussions of immigrant (or, in their case, emigrant) behavior. They too wished to prevent Poles from being exposed to left-wing political agitation in the mines and factories of northeastern France, particularly because they imagined that these people would return to Poland once they could be absorbed into the domestic economy (Albin 1999). Most notably, they feared the effects of exposure to communism, an ideology that Polish elites across the political spectrum, including those on the patriotic socialist left, believed was fundamentally anti-Polish. In Poland itself, the Communist Party was banned and therefore existed only through front organizations, while the so-called Sanacja regime, which was established after Józef Piłsudski's coup d'état in 1926, created an anti-Soviet project that attempted to appeal to Ukrainians living in the neighboring Soviet Union (Snyder 2005).

Although French Communists actually saw Polish workers as rather disappointing—in 1925, the Communist trade union in France was able to organize only 8 percent of Polish miners (Cross 1983: 172)—Polish officials worried that Poles in northeastern France were more susceptible to communism than their fellow citizens in Poland. Like the French industrialists, Polish consular personnel in Lille thus asserted that workers would maintain their Polishness, which was ideally embodied in an adherence to the Polish language, Roman Catholicism, and Polish cultural traditions, if they lived in isolated colonies, resisted secularization, had Polish wives, and did not engage in casual sexual relations. The signs were not good. Reporting back to the Ministry of the Interior in 1926, one doctor stated that venereal diseases and alcoholism were endemic among Polish emigrants in France.[2] That same year, Polish consulates in France organized clinics in an attempt to combat the spread of venereal diseases, which they also believed had been on the rise as a consequence of French licentiousness (Rozwadowski 1927: 191–192). Although they struggled to carry out comprehensive monitoring of social behavior because of a lack of manpower—in 1927, the thirty officials based at the Lille outpost were charged with serving 220,000 Polish citizens (Rozwadowski 1927: 216)—officials did manage to tour mines and interview Polish workers (Slaby 2015: 236).

Despite these efforts, by the early 1930s, consular personnel in Lille were becoming increasingly alarmed by the rise in Communist agitation among Polish workers. They seized Polish-language leaflets published by the Polish Anti-Fascist Committee of Northern France (Polski Komitet Antyfaszystowski Północnej Francji), which claimed that the Polish consul in Lille, Stanisław Kara, was on the side of the fascists.[3] Worryingly, however, much of their work felt like grasping in the dark because the information they received about individuals who allegedly agitated against the Polish state was, by the consulate's own admission, rather "haphazard."[4] Indeed, when Polish citizens went to the consulate in Lille with information about the proliferation of communism among Polish workers, consular officials had to write to the central authorities in Warsaw in an attempt to find out whether the informants were trustworthy. In November 1932, for instance, a member of the local Communist party went to the Lille consulate to offer "comprehensive information about the activities of this [i.e., the Communist] party," prompting the consul to ask officials in Warsaw if the man should be trusted, based on his political history.[5] The following year, another man turned up at the consulate, claiming to have information about communists in the Belgian city of Antwerp who had supposedly smuggled fourteen boxes of Communist leaflets to an address in the Jewish area of Warsaw inside copies of the Yiddish-language newspaper *Haynt*.[6]

As the 1930s progressed, the Polish state's concerns about the disloyalty of some emigrant Poles living in France were compounded by the increasing pressure on Poles to physically return to Poland. Labor agreements had stipulated that foreign laborers could, under certain conditions, be sent back to the country of their citizenship. Between 1931 and 1936, as the Great Depression put pressure on the French government to prioritize domestic workers over their foreign counterparts, around 140,000 Polish citizens had to leave France (Śladkowski

2001: 182). Police officials in Warsaw stated that many returnees brought the subversive ideology of communism back home.[7] Yet, because consular officials had
been unable to track people in France—a failure that underscores the broader
difficulties that states face when they attempt to control citizens who reside beyond their borders (Fahrmeir 2007)—they lacked intelligence about all but the
most high-profile Communist activists and admitted that their knowledge was
both limited and possibly inaccurate.[8] When a specially organized train from
Lille made its way toward the Polish border town of Zbąszyń in August 1934,
the Lille consulate warned local authorities that, although there were certainly
Communist ringleaders on board, it could not identify any of them by name.[9]

## Denaturalization as a Tool of Control

By the midpoint of the 1930s, the Polish government began to promote a more
active approach to controlling supposedly disloyal citizens. As the geopolitical
situation in Europe became more precarious and as the Polish army assumed
a greater degree of control over demographic policies after Piłsudski's death in
1935, the Polish authorities increasingly looked to revoke people's citizenship.
The idea that citizenship could be taken away was itself nothing new. In fact,
denaturalization had long been touted as a legitimate process, if a state could
prove that the citizen has failed to uphold his or her side of the contract (Caglioti
2020; Weil 2013; Weil and Handler 2018). Poland's citizenship law of January
1920 clearly laid out not only the conditions under which citizenship could be
granted, but also those under which it could be revoked. If the initial years after
1918 in Poland were thus informed by the sorting of people into those who were
eligible to become citizens and those who were not, the years just prior to World
War II were dominated by questions about who had proved that they deserved
the rights of the citizen—above all, by showing their loyalty toward the state—
and who had not. Revoking citizenship from people who lived beyond the state's
borders meant removing the last administrative vestige of "insider" status.

Polish officials argued that people would lose their citizenship if and when
their actions proved that they were disloyal, rather than because of an innate
national identity. This approach was in marked contrast to German (and, after
1938, Hungarian) laws that explicitly targeted Jewish citizens. The notorious
Reich Citizenship Laws of 1935, which stripped Jews of their German citizenship
and demoted them to the status of subjects, were based on the racist idea that
Jews constituted an infection within the nation-state and should not therefore
enjoy the same rights as those who possessed "German or kindred blood." In
different ways—and fueled in part by Stalin's own paranoia about internal enemies—large groups of people in the Soviet Union, usually (although not always)

defined by their nationality, were either stripped of their citizenship or were *de facto* deprived of their rights during the late 1930s (Alexopoulos 2006; Martin 1998). In contrast, the Polish government removed citizenship from nominally non-ascriptive groups.

The case of Polish citizens who journeyed from northeastern France to fight on the side of the Spanish Republic during the Spanish Civil War from 1936 onward offers insights into what these dynamics looked like in practice. The consulate in Lille reported that some fighters had been directly recruited from the ranks of left-wing Polish workers in northeastern France, with volunteers allegedly receiving 50 francs for the trip to Paris from whence they had journeyed to Spain via the southern French port of Marseille.[10] Such developments caused consternation in both the local French and the Polish émigré press.[11] In Warsaw, Polish authorities similarly worried about the harmful influence of a "subversive element" that had received both military and political training during the stay in Spain and was now attempting to re-enter Poland.[12] Denaturalization was seen as a viable option. After all, according to Poland's citizenship law of January 1920, entering into the military service of another country without the permission of the Polish government could lead to a loss of citizenship.

Like their counterparts in other modern states, Polish consulates acted as key institutions that policed citizenship by choosing or refusing to issue passports or the stamps that marked a passport as valid (Green 2012; Stein 2016). Concerned that Polish consulates might issue passports to people who had lost their citizenship as a consequence of fighting alongside left-wing forces in Spain, the Polish Ministry of the Interior instructed consular officials in France to carry out interviews with passport applicants. According to the ministry's instructions, they were to ascertain, among other things, the period of a person's service, their position in the army, and the role and character of their military formation. They were also urged to make it clear that decisions about issuing passports that would allow a return to Poland were "dependent, above all, on the sincerity and reliability" of an applicant's testimony.[13] Yet consular officials often gave applicants the benefit of the doubt. In Paris, for instance, Stanisław Kara, who had been the general consul there since his move from Lille in 1935, stated in 1938 that he had recommended issuing a passport to someone who had served in Spain because he had it on good authority that the man was not a Communist, but was rather of low intelligence and had simply been seduced by money and agitation during a long period of unemployment and poverty.[14] Similar decisions to issue passports to former volunteers in Spain who had already been deprived of their citizenship were heavily criticized by officials at the Ministry of the Interior.[15]

Poles in France who lost their citizenship pushed back, not least because the process created two sets of problems: not only were they stripped of their Polish citizenship, but, as a consequence, they could not renew the *carte d'identité* that allowed them to live and work legally in France (Ponty 1988: 342). Importantly, they based their claims for retaining citizenship on national arguments, namely on the idea that fighting against fascism was an expression of deeper Polish patriotic traditions. In one handwritten letter that was sent to the president of Poland in March 1939, a group of citizens, including those from "the Polish emigration in France and Belgium," claimed that they had followed in the footsteps of the "great Polish patriots" who had fought for independence during the period of the partitions when Poland had been under foreign rule. Moreover, they argued that in Spain they had battled against the enemies of both Spanish and Polish independence under the inclusive slogan "For Your Freedom and Ours," which originated in a Polish rebellion against the Russian empire in 1830–1831.[16]

## The Ethnic Politics of Denaturalization

Just as domestic state police files in Poland listed Polish Catholics among those who had been "expelled from France for Communist activity," so many Polish Catholics had their citizenship removed as a consequence of fighting on the side of the Spanish Republic.[17] At the same time, however, officials were more likely to be suspicious of Jews and Ukrainians, thus creating hierarchies of disloyalty among citizens. Discussions of recruitment efforts in relation to the Spanish Civil War certainly highlighted the ways in which Jews were allegedly corrupting ethnic Polish emigrants. In 1937, for instance, personnel at the Polish Ministry of the Interior informed their counterparts at the Ministry of Foreign Affairs that Estera Golde-Stróżecka, a Polish citizen of the Jewish nationality living in France, "takes part in organizing the workers' movement among the Polish emigrants" and was proving instrumental "in organizing care for the Polish militiamen fighting on the side of the government army in Spain."[18]

Such claims were not new, but instead fit into a broader pattern of antisemitic anxieties about Polish citizens of the Jewish nationality. During the early part of the 1930s, consular personnel had already begun to argue that Jews directly corrupted their ethnically Polish fellow citizens who were temporarily severed from the motherland. In a May 1932 letter to Warsaw, the consul in Lille had claimed that a Belgium-based, Jewish-led organization, which ostensibly recruited Jews to farming colonies in Soviet Crimea, had a major influence on the communist centers of northeastern France. Other organizations, he alleged, were made up of Jewish and Ukrainian Polish citizens who tried to incite Polish workers to fight against "our" (i.e., ethnically Polish) organizations, such as the pro-Piłsudski sharpshooters' association, and fostered hostility toward patriotic events organized by the consulate, including Piłsudski's name day and Constitution Day (May 3).[19] The assertion that Jewish Polish citizens, rather than

French Communists, sought to turn ethnically Polish citizens against Poland constituted a new twist on the established theme of Jewish disloyalty that was at the core of right-wing Polish nationalism. In fact, the idea that some ethnic Poles had been corrupted by Jews and were therefore little more than "half-Poles" or even "quarter-Poles" (Dabrowski 2014: 366) was widespread in Poland during the interwar years, as Poles on the right sought to attack their left-wing adversaries with the charge of not being truly Polish (Brykczynski 2016).

In the case of the Spanish Civil War, instructions about which members of a family could have their citizenship revoked indicated how ideas about Jewish disloyalty were codified. Particularly important were rules about how the citizenship status of families was affected by the nationality of the husband and/or father. From the very beginning, Polish citizenship laws had been drawn up based on familial relationships with men. According to the January 1920 law, a woman would lose her Polish citizenship if she married a noncitizen (although male citizens would not lose their citizenship if they married a noncitizen woman), while men could pass on their citizenship to legitimate children (children born out of wedlock, however, would take that of their mother). But while this law applied to all citizens—ethnic Pole or not—a confidential circular issued by the consular department at Poland's Ministry of Foreign Affairs in June 1939, itself based on an edict from May of the previous year, urged differentiation between the treatment of families, according to categories of nationality. The wife, as well as any children under the age of eighteen, of an ethnically Polish man who had fought as a volunteer in Spain would not have their citizenship revoked (the sole exception was if the wife herself had actively and consciously participated in Communist activity, in which case she would also be included in the application for citizenship to be revoked, although the couple's children could retain theirs). However, the wife and children of "Polish citizens of the Jewish nationality" would be included in the application for loss of citizenship, while those of "other non-Polish nationalities" (for instance, Ukrainians) could remain citizens, but only if they physically stayed in Poland.[20] Put simply, men who were deemed to be both politically subversive (a choice) and Jewish (i.e., members of an ascriptive group) would find that their families were punished more severely through denaturalization than those of ethnic Poles.

If this dynamic reflected a long-term ambivalence among Polish officials concerning whether Jewish citizens of Poland were truly members of the Polish nation, despite their legally enshrined rights, it was also indicative of the emergence of antisemitic demographic policies that imagined an international "solution" to the "Jewish question" (Snyder 2015). Indeed, the Polish government made it clear that Polish citizens of the Jewish nationality who lived beyond Poland's physical borders were vulnerable to denaturalization in ways that those who lived within the borders of Poland were not. The clearest sign of this approach came in the wake of Germany's annexation of Austria in March 1938, when Poland passed a new citizenship law that sought to prevent Polish citizens of the Jewish nationality living in Austria from moving to Poland.

On one level, the March 1938 law confirmed that citizenship could be revoked based on a person's actions. According to the law, those who "acted abroad to the detriment of the Polish state" would lose their Polish citizenship. But without alluding to Jews by name, the law was also an attempt to foster a more active definition of loyalty in ways that disproportionately targeted Jews who resided beyond Poland's borders. Most important, it referred to passive reasons for denaturalization, which included living outside of Poland for at least five years and thereby allegedly losing connections with the Polish state. In Lille, consular officials revoked the citizenship of Jews for a number of reasons that pertained to their supposed weak links to Poland—for instance, their lack of fluency in the Polish language—rather than to any proof that they constituted an active fifth column against the Polish state.[21] On the eve of World War II, these people became stateless mainly because they were Jews.

## Conclusion

The story of fifth columns and denaturalization does not, of course, stop here. Given the intensification of fifth-column accusations during periods of warfare, it is not surprising that Poles once again embraced such language during World War II. Indeed, the invasion of Poland in September 1939 and the erasure of the territorial state created new opportunities to accuse non-Polish minorities of betrayal. A Polish propaganda pamphlet that was issued by the Polish Ministry of Information in 1940 argued that Polish citizens of the German nationality had taken advantage of the benefits that they were given as a minority in the Second Republic and had laid the groundwork for the German army's invasion. In the Soviet-occupied area that had constituted interwar Poland's eastern borderlands, Poles accused Jews and Ukrainians of welcoming the Red Army and attacking both the physical symbols and the human representatives of the Polish state (Gross 2000).

But if assumptions about political loyalty continued to be linked to national ascription, as had been the case during World War I, discussions about citizenship in Polish consulates across the world suggested that the story was more complicated. As the territorial Polish state collapsed, the Polish government-in-exile, which was based first in France and then in London, continued to police the borders of Polish citizenship, not least by denaturalizing citizens living abroad, including those who had allegedly refused to fight on behalf of the

Polish state.[22] Although more systematic research needs to be completed on the incidences of Polish citizenship being granted, denied, revoked, and reinstated at those Polish consulates that continued to operate during World War II, a cursory investigation suggests that interwar trends continued: citizenship was revoked both from ethnic Poles who were seen as politically disloyal and from non-ethnically Polish groups whose supposed disloyalty was linked to their ascriptive national identities. In the absence of a territorial Polish state, questions of who was internal and who was external—and what those labels meant—became ever more complicated.

By considering the Polish state's approach to citizens who lived abroad, this chapter has opened up two important areas of exploration in relation to fifth columns more broadly. First, I have argued that, while discussions of fifth columns are generally centered on accusations against disloyal people who physically reside within the geographical confines of a state and are assisted by external elements, we might also consider the role of people who were internal in terms of their citizenship status, even if they physically resided beyond state borders. Tracing the stories of Polish citizens in northeastern France prompts us to more precisely historicize the external and internal components that are fundamental to the definition of fifth columns. Does "internal" always have to equate to permanent physical presence within a state? How do we classify transnational actors who, as citizens, possess internal privileges and are expected to be loyal to their home state, but who, as emigrants, are exposed to external influences that they might bring home? And how is the idea of "enemies within" constructed as people move across physical spaces or lose vestiges of insider status, including their citizenship? As the case of Polish citizens in northeastern France suggests, emigrant citizens, who are neither fully external nor internal, can provoke security anxieties for modern state officials, especially at moments when they cross or attempt to cross borders.

Second, this particular case asks us to contemplate whether interwar Polish anxieties about "enemies within" were based solely on an individual's political actions, regardless of their ethnicity, or on exclusively ascriptive (ethnically based) criteria. The answer seems to be neither. Certainly, the Polish state's liberal constitution codified citizenship as a contract, meaning that denaturalization was officially based on behavior, rather than on one's nationality. Polish citizens whose actions were deemed to be politically subversive, particularly in relation to Communist activity or fighting in the ranks of a foreign army, were seen to have broken the terms of the contract and could thus lose their citizenship. Yet as consular officials sought to police the borders of citizenship abroad, liberal ideas remained in tension with ethnically based criteria. By the mid- to late 1930s, officials increasingly saw Polish citizens who were not of the Polish nationality—Jews, most especially—as more suspicious and less redeemable than their ethnically Polish counterparts.

# Notes

1. "Instrukcja dla urzędnika objazdowego Konsulatu R.P. w Lille," 6. Hoover Institution Archives, Stanford, CA (hereafter HIA), Poland. Konsulat Generalny (Lille, France) Records (hereafter KGL), Box 3, Folder 6. Note that all archival citations are provided in footnotes.

2. "Sprawozdanie Dra. J. Babeckiego o chorobach wenerycznych i warunkach sanit. wśród emigrantów polskich we Francji," submitted to the Ministry of the Interior (December 21, 1926), Archiwum Akt Nowych, Warsaw, Poland (hereafter, AAN) 2/15/0/-/102/10-13.

3. "Do Wszystkich Robotników i Robotnic Nord, Pas-de-Calais, i Anzin," HIA KGL Box 10, Folder 19.

4. Letter from the Lille consulate to the Ministry of Foreign Affairs in Warsaw (April 16, 1934), HIA KGL Box 10, Folder 13.

5. Letter from the Lille consulate to the Ministry of Foreign Affairs in Warsaw (November 11, 1932), HIA KGL Box 10, Folder 11.

6. Letter from Stanisław Kara to the headquarters of the state police in Warsaw (March 10, 1933), HIA KGL Box 10, Folder 12.

7. Letter from the headquarters of the state police to the Ministry of Foreign Affairs (March 9, 1932), HIA KGL Box 10, Folder 11.

8. Letter from the general consulate in Paris to the Ministry of the Interior in Warsaw (September 30, 1937), AAN 2/322/0/-/12345/13.

9. Letter from the Lille consulate to the state police station in Zbąszyń (August 2, 1934), HIA KGL Box 10, Folder 13.

10. Letter from the Polish consul in Lille to the Polish embassy in Paris (October 23, 1936), HIA KGL Box 10, Folder 15.

11. Letter to the Polish embassy in Paris (October 31, 1936), HIA KGL Box 10, Folder 15.

12. Letter from Wacław Żyborski at the Ministry of the Interior to all provincial offices and the government commissariat for the city of Warsaw (April 7, 1939), HIA KGL Box 10, Folder 4.

13. Letter from Wacław Żyborski to the Ministry of Foreign Affairs (August 30, 1938), AAN 2/322/0/-/12345/61.

14. Letter from Stanisław Kara at the Paris general consulate to the headquarters of the state police in Warsaw (June 24, 1938), AAN 2/322/0/-/12345/78.

15. Letter from Wacław Żyborski at the Ministry of the Interior to the Ministry of Foreign Affairs (November 18, 1938), AAN 2/322/0/-/12345/90-1.

16. Letter to the Polish president (March 2, 1939), AAN 2/322/0/-/12333/3.

17. Letter from the provincial police department in Poznań (February 9, 1939), AAN 2/349/1/4.6/231/6.

18. "Golde-Stróżecka-informacja," enclosure with letter from the Ministry of the Interior to the Ministry of Foreign Affairs (October 28, 1937), AAN 2/322/0/-/12345/22-23.
19. Letter from the Lille consul general to the Ministry of Foreign Affairs in Warsaw (May 8, 1932), HIA KGL Box 10, Folder 11.
20. "Okólnik nr. 47 w sprawie b. ochotników, obywateli polskich, w armii 'czerwonej' Hiszpanii" (June 20, 1939), HIA KGL Box 10, Folder 4.
21. See the denaturalization applications put forward by the Lille consulate to the Ministry of the Interior in Warsaw in HIA KGL Box 10, Folder 14.
22. Letter from the British Secretary of State to the British Foreign Office (November 9, 1940), Polish Institute and Sikorski Museum Archives (London), A.42 144/58-61.

# References

About, Ilsen. 2013. "A Paper Trap: Exiles versus the Identification Police in France during the Interwar Period." In *Identification and Registration Practices in Transnational Perspective: People, Papers and Practices*, edited by I. About, J. Brown, and G. Lonergan, 203–223. London: Palgrave Macmillan.

Albin, Janusz. 1999. "La communauté polonaise en France et L'État polonais entre 1920 et 1939." In *La Protection des Polonaise en France: problèmes d'intégration et d'assimilation*, edited by Edmond Gogolewski, 41–56. Villeneuve-d'Ascq: Université Charles-de-Gaulle.

Alexopoulos, Golfo. 2006. "Soviet Citizenship, More or Less: Rights, Emotions, and States of Civic Belonging." *Kritika: Explorations in Russian and Eurasian History* 7, no. 3 (Summer): 487–528.

Boswell, Laird. 2000. "From Liberation to Purge Trials in the 'Mythic Provinces': Recasting French Identities in Alsace and Lorraine, 1918–1920." *French Historical Studies* 23, no. 1 (Winter): 129–162.

Brubaker, Rogers. 2010. "Migration, Membership, and the Modern Nation-State: Internal and External Dimensions of the Politics of Belonging." *Journal of Interdisciplinary History* 41, no. 1 (Summer): 61–78.

Brykczynski, Paul. 2016. *Primed for Violence: Murder, Antisemitism, and Democratic Politics in Interwar Poland*. Madison: University of Wisconsin Press.

Caglioti, Daniela L. 2020. *War and Citizenship: Enemy Aliens and National Belonging from the French Revolution to the First World War*. Cambridge: Cambridge University Press.

Camiscioli, Elisa. 2009. *Reproducing the French Race: Immigration, Intimacy, and Embodiment in the Early Twentieth Century*. Durham, NC: Duke University Press.

Carrol, Alison. 2018. *The Return of Alsace to France, 1918–1939*. Oxford: Oxford University Press.

Chu, Winson. 2012. *The German Minority in Interwar Poland*. New York: Cambridge University Press.

Cross, Gary S. 1983. *Immigrant Workers in Industrial France: The Making of a New Laboring Class*. Philadelphia: Temple University Press.

Dabrowski, Patrice M. 2014. *Poland: The First Thousand Years*. DeKalb: Northern Illinois University Press.

Eichenberg, Julia. 2010. "The Dark Side of Independence: Paramilitary Violence in Ireland and Poland after the First World War." *Contemporary European History* 19, no. 3 (August): 231–248.

Fahrmeir, Andreas. 2007. "From Economics to Ethnicity and Back: Reflections on Emigration Control in Germany, 1800–2000." In *Citizenship and Those Who Leave: The Politics of Emigration and Expatriation*, edited by Nancy L. Green and Françcois Weil, 176–191. Urbana: University of Illinois Press.

Fink, Carole. 2004. *Defending the Rights of Others: The Great Powers, the Jews, and International Minority Protection, 1878–1938*. New York: Cambridge University Press.

"Foreigners in France (By Our Paris Correspondent)." 1925. *Times* (London), December 10, 17–18.

Gargas, S. 1926. "The Polish Emigrants in France." *The Slavonic Review* 5, no. 14 (December): 347–351.

Green, Nancy L. 2005. "The Politics of Exit: Reversing the Immigration Paradigm." *Journal of Modern History* 77, no. 2 (June): 263–289.

Green, Nancy L. 2012. "Americans Abroad and the Uses of Citizenship: Paris, 1914–1940." *Journal of American Ethnic History* 31, no. 3 (Spring): 5–32.

Gross, Jan T. 2000. "A Tangled Web: Confronting Stereotypes Concerning Relations between Poles, Germans, Jews, and Communists." In *The Politics of Retribution in Europe: World War II and Its Aftermath*, edited by István Deák, Jan T. Gross, and Tony Judt, 74–129. Princeton, NJ: Princeton University Press.

Guériot, Paul. 1924. "Politique d'immigration." *Revue politique et parlementaire* 20 (June 10): 419–435.

Hanebrink, Paul. 2018. *A Specter Haunting Europe: The Myth of Judeo-Bolshevism*. Cambridge, MA: Harvard University Press.

Klimowicz, Stanisław. 1923. *Poradnik dla wychodźcy: o czem emigrant zarobkowy we Francji wiedzieć powinien*. Paris: Drukarnia Polska.

Lawrence, Paul. 2000. "'Un flot d'agitateurs politiques, de fauteurs de désordre et de criminels': Adverse Perceptions of Immigrants in France between the Wars." *French History* 14, no. 2 (June): 201–221.

Lewis, Mary Dewhurst. 2007. *The Boundaries of the Republic: Migrant Rights and the Limits of Universalism in France, 1918–1940*. Stanford, CA: Stanford University Press.

Lohr, Eric. 2003. *Nationalizing the Russian Empire: The Campaign against Enemy Aliens during World War I*. Cambridge, MA: Harvard University Press.

Lohr, Eric. 2012. *Russian Citizenship: From Empire to Soviet Union*. Cambridge, MA: Harvard University Press.

Martial, René. 1926. "Le problème de l'immigration." *Revue politique et parlementaire* (December 10): 391–402.

Martin, Terry. 1998. "The Origins of Soviet Ethnic Cleansing." *Journal of Modern History* 70, no. 4 (December): 813–861.

*Merriam-Webster Dictionary* online. https://www.merriam-webster.com/dictionary/fifth%20column.

Michalik-Russek, Barbara. 1992. "Stowarzyszenie 'Opieka Polska nad Rodakami na Obczyźnie,' 1926–1949." *Studia Polonijne* 14: 163–171.

*Oxford English Dictionary* online. www.oed.com/view/Entry/70006.

Polish Ministry of Information. 1940. *The German Fifth Column in Poland*. London: Hutchinson & Co. Ltd.

Ponty, Janine. 1988. *Polonais méconnus: histoire des travailleurs immigrés en France dans l'entre-deux-guerres*. Paris: Publications de la Sorbonne.

Puchalski, Piotr. 2022. *Poland in a Colonial World Order: Adjustments and Aspirations, 1918–1939*. London: Routledge.

The Reich Citizenship Law (September 15, 1935) and the First Regulation to the Reich Citizenship Law (November 14, 1935). 1935. German History in Documents and Images website. http://ghdi.ghi-dc.org/sub_document.cfm?document_id=1523.

Rosenberg, Clifford. 2006. *Policing Paris: The Origins of Modern Immigration Control between the Wars*. Ithaca, NY: Cornell University Press.

Rozwadowski, Jan. 1927. *Emigracja polska we Francji: europejski ruch wychodźczy*. Lille: Wydawnictwo Polskiego Uniwersytetu Robotniczego we Francji.

Schenke, Cornelia. 2004. *Nationalstaat und nationale Frage: Polen und die Ukrainer, 1921–1939*. Hamburg: Dölling und Galitz.

Slaby, Philip H. 2015. "Dissimilarity Breeds Contempt: Ethnic Paternalism, Foreigners, and the State in Pas-de-Calais Coalmining, France, 1920s." *International Review of Social History* 60 (December): 227–251.

Śladkowski, Wiesław. 2001. "Polska diaspora we Francji 1871–1999." In *Polska Diaspora*, edited by Adam Walaszek and Danuta Bartkowiak, 178–195. Kraków: Wydawnictwo Literackie.

Słyszewska-Gibasiewicz, Joanna. 2011. "Opieka konsulatów RP nad emigrantami polskimi w latach 1918–1926." *Studia Ełckie* 13: 229–243.

Snyder, Timothy. 2005. *Sketches from a Secret War: A Polish Artist's Mission to Liberate Soviet Ukraine*. New Haven, CT: Yale University Press.

Snyder, Timothy. 2015. *Black Earth: The Holocaust as History and Warning*. New York: Tim Duggan Books.

Starzyński, Stanisław. 1921. *Obywatelstwo państwa polskiego*. Kraków: Krakowska Spółka Wydawnicza.

Stauter-Halsted, Keely. 2021. "Violence by Other Means: Denunciation and Belonging in Post-Imperial Poland, 1918–1923." *Contemporary European History* 30, no. 1 (February): 32–45.

Stein, Sarah Abrevaya. 2016. *Extraterritorial Dreams: European Citizenship, Sephardi Jews, and the Ottoman Twentieth Century*. Chicago: University of Chicago Press.

Tomaszewski, Jerzy. 1985. *Rzeczpospolita wielu narodów*. Warsaw: Czytelnik.

Torpey, John. 2000. *The Invention of the Passport: Surveillance, Citizenship, and the State*. New York: Cambridge University Press.

"Ustawa z dnia 17 marca 1921 roku—Konstytucja Rzeczypospolitej Polskiej." 1921. *Dziennik Ustaw nr. 44 poz. 267*. http://prawo.sejm.gov.pl/isap.nsf/download.xsp/WDU19210440267/O/D19210267.pdf

"Ustawa z dnia 20 stycznia 1920 roku o obywatelstwie Państwa Polskiego." 1920. *Dziennik Ustaw nr. 7 poz. 44*. http://isap.sejm.gov.pl/isap.nsf/download.xsp/WDU19200070044/O/D19200044.pdf.

"Ustawa z dnia 31 marca 1938 r. o pozbawianiu obywatelstwa." 1938. *Dziennik Ustaw nr. 22 poz. 191*. http://isap.sejm.gov.pl/isap.nsf/download.xsp/WDU19380220191/O/D19380191.pdf.

Weil, Patrick. 2013. *The Sovereign Citizen: Denaturalization and the Origins of the American Republic*. Philadelphia: University of Pennsylvania Press.

Weil, Patrick, and Nicholas Handler. 2018. "Revocation of Citizenship and Rule of Law: How Judicial Review Defeated Britain's First Denaturalization Regime." *Law and History Review* 36, no. 2 (May): 295–354.

Weitz, Eric D. 2008. "From the Vienna to the Paris System: International Politics and the Entangled Histories of Human Rights, Forced Deportations, and Civilizing Missions." *American Historical Review* 113, no. 5 (December): 1313–1343.

Zahra, Tara. 2008. "The 'Minority Problem' and National Classification in the French and Czechoslovak Borderlands." *Contemporary European History* 17, no. 2 (May): 137–165.

Zahra, Tara. 2016. *The Great Departure: Mass Migration from Eastern Europe and the Making of the Free World*. New York: W. W. Norton.

Zielinski, Konrad. 2004. "Population Displacement and Citizenship in Poland, 1918–24." In *Homelands: War, Population and Statehood in Eastern Europe and Russia, 1918–1924*, edited by Nick Baron and Peter Gatrell, 98–118. London: Anthem Press.

# 9

# Imagined Treason and Post–Lavender Scare Politics

## How LGBTQ Communities Respond to Fifth-Column Accusations

*Samer M. Anabtawi*

I wouldn't trust [homosexuals] with a $5 loan, let alone the nation's secrets.

—Rep. Bob Dornan (R-CA)

The [John] Vassal case tended to reinforce the popular notion that Communism has special appeal for homosexuals. And it bore out the commonly held belief that homosexuals are particularly vulnerable to blackmail by Communist agents and therefore are poor security risks. It has been asserted that homosexuals are emotionally immature, that they talk too much, and that they have less resistance than heterosexuals to flattery. (Worsnop 1963)

LGBTQ people all over the world have endured being described as pedophiles, debauched, sodomites, perverts, morally weak, and psychologically disturbed. But perhaps the most dangerous is the accusation that they serve as "agents of foreign enemies," a *political* stigma that has become ubiquitous since the mid-twentieth century. Efforts by state and political actors to produce "a modular LGBT boogeyman" that stands as a "menace to national sovereignty" are on the rise globally.[1] Recent examples of this abound. Take Uganda's president Yoweri Museveni, who famously described homosexuality as "Western perversion" and the result of "random breeding in Western countries" (Fisher 2014). And in Poland, where roughly a third of the country has been declared an "LGBTQ-free zone," President Andrzej Duda said during a campaign speech that the promotion of gay rights is an ideology that is "more destructive" than communism (Ash 2020). Meanwhile in Egypt, the waving of a rainbow flag at a concert led to a swift anti-gay crackdown and the arrest of fifty-seven people over their perceived sexual orientation. Shortly after, Egypt's public prosecutor ordered the State Security Prosecution—a body that

Samer M. Anabtawi, *Imagined Treason and Post–Lavender Scare Politics* In: *Enemies Within*. Edited by: Harris Mylonas and Scott Radnitz, Oxford University Press. © Oxford University Press 2022. DOI: 10.1093/oso/9780197627938.003.0010

specializes in investigating terrorist and national security threats—to investigate the rainbow incident.

Political accusations against LGBTQ people have ranged from being associated with foreign and Western imposition, disloyalty to the nation and its traditional values, and collusion with the nation's enemies. In recent years, scholars have devoted much attention to political homophobia as a strategic tool of nation- and state-building. This chapter builds on this growing body of scholarship to explain how the targets of these homophobic political campaigns respond to their designation as collaborators with their state's own rivals. What strategies do target groups pursue? How do fifth-column accusations constrain the ways in which they may respond? And what explains how target groups frame their collective identity in response to these accusations?

To answer these questions, the chapter focuses on the response of LGBTQ people to fifth-column narratives in two different historical and geographic settings—the United States during the Cold War and Palestine following the 1987 uprising. Both periods saw the ascendancy of conspiracy theories that gays and lesbians were undermining national security and collaborating with the state's rivals. Republicans in Congress spread dodgy claims that hundreds of homosexuals and Communists had infiltrated the federal government— triggering a nationwide "Lavender Scare" and a government-wide purge of homosexuals (Johnson 2004). A few decades later, the nationalist movement in Palestine began warning the public that homosexuality was a gateway to treason and collaboration with Israeli intelligence agencies. Rumors spread widely in each case, placing gays and lesbians[2] in a precarious position and creating a fringe group in society.

The targets of these accusations began to mobilize and strategize about how they could respond to their marginalization. Drawing on social movement research on issue framing, I argue that in both situations, a nascent LGBTQ movement strategically relied on "corrective discourse" as a tool to dislodge the stigma of betrayal and subversion. Corrective discourse encompasses counter-narratives, stories, and literature that aim to "alter the dominant group's perception of the subgroup." (Lindemann 2001). The mobilization that resulted from being political targets and the strategic response that came from early movement leaders ultimately constituted gays and lesbians as a *political community* (D'Emilio 2012). It was through the methodical production of nationalist counter-frames that both of these communities were able to legitimate their presence on the national stage and wade through the waves of political homophobia.

The chapter demonstrates that gay rights leaders in the United States and queer activists in Palestine sought to alter public perceptions of their group from subversives to model citizens in two distinct ways. The first mechanism is framing the movement's claims in ways that align LGBTQ identities with the political mainstream and creating an image of the gay citizen as a bearer of the national flag.[3] The second mechanism entails distancing the movement from the rival nation by producing a nationalist discourse that condemns the latter and links it to the systematic oppression of gays and lesbians.

While both movements pursued similar strategies in responding to politically damaging narratives, the discourses that emerged are vastly different. The homophile movement in the United States quickly pivoted toward an assimilationist discourse and the rejection of radical and confrontational narratives. Queer Palestinian activists, on the other hand, couched their claims in a radical discourse that seeks to dismantle entire power systems.[4] The reader may wonder why we observe this variation in discourse when the two movements were confronting similar stigma. I attribute this variation to the political contexts surrounding the emergence of the two movements, which constrained the menu of viable discursive options. The discourses that emerged, therefore, were a reflection of mainstream political culture in the United States and Palestine.

This chapter proceeds in three sections. First, it discusses the construction of LGBTQ citizens as fifth columns and situates this phenomenon within the broader category of political homophobias. Second, it theorizes targets' responses to marginalization based on national security claims and draws on the social movement literature to explain the strategic framing approaches at their disposal. Third, it leverages case studies from the United States and Palestine— tracing the emergence of the link between fifth-column accusations and homosexuality—and unpacks the way target groups developed a strategic response to alter their image. The cases highlight the agency of the targets vis-à-vis "national security risk" narratives and their ability to pry open a closed political opportunity structure with new counter-narratives that help them gravitate toward the mainstream.

## Situating Anti-LGBTQ Fifth-Column Claims within the Framework of Political Homophobia

Homophobia, as a broad construct, denotes a set of attitudes and discourses that are hostile to homosexuality irrespective of the underlying premise for such attitudes. Recent scholarship has homed in on a subset of these discourses—ones that are utilized and promoted tactically by political actors. This chapter joins this growing literature in stressing that the stigmatization of sexual minorities[5] is not exclusively a sociocultural phenomenon based in religion, tradition, or "scientific" claims. In many cases—like the ones I explore here—the construction of sexual minorities as a collective identity was inextricably linked to complex *political* processes and nationalist narratives that painted homosexuals as a "morally subnormal" group (Lindemann 2001; Korycki and Nasirzadeh 2013; Bosia and Weiss 2013).

All three elements of fifth-column politics identified earlier in this book—discourse, policy, and mobilization—are present when it comes to the othering and exclusion of sexual minorities in the twentieth century. During the Cold War, for example, there was an abundance of inflammatory statements made by politicians and mainstream media accusing sexual minorities of undermining the state from within. In the United States and the UK, fifth-column discourses were multifaceted. Some fifth-column claims were based on identity traits. US officials argued that "the underlying reason why a homosexual is classified as an unsuitability risk is, of course because homosexuality involves potential blackmail by the enemy leading to disclosures that betray the security of the United States" (Taylor 1969). Moral backlash against homosexuality coincided with fears of spying and entrapment, and those two fears became increasingly intertwined. This was a direct result of the public officials and mainstream media contending that "effeminate" characteristics and "sexual indiscretion" associated with homosexuals made them identifiable targets for entrapment by external enemies. This presumption alone—the prospect of disloyalty—was sufficient to justify the wholesale exclusion of all homosexuals and banning gays and lesbians from serving in sensitive government positions. Other fifth-column claims against gays and lesbians were based on ideology—as many Republican representatives in Congress assumed an ideological alignment between homosexuals and Communists, rendering them a nefarious subversive group. The spread of fifth-column claims resulted in exclusionary policies such as President Eisenhower's executive order banning gays and lesbians from getting security clearances and opening the door for the FBI to carry out sweeping purges of suspected homosexuals from the State Department.

Several stories along these lines in the 1950s and 1960s became fodder for fifth-column politics. In the UK, there were the infamous "Cambridge Spies" Guy Burgess and Donald Maclean, both British diplomats who spied for the Soviet Union and later fled to Moscow. When it was discovered that they were living in Moscow in 1956 as Communists, their alleged homosexuality became the focal point of public commentary, which gave fodder to ideology-based fifth-column accusations against those identified as homosexuals.

In 1962, another espionage scandal exacerbated the coupling of homosexuality and treason. John Vassall, a British diplomat, was arrested for spying on behalf of the Soviets. It was later revealed that Vassall had participated in a gay orgy in Moscow before Christmas of 1954 (Worsnop 1963). Vassall detailed in a newspaper interview how he was blackmailed by the KGB, who photographed him in compromising positions with multiple men. Unlike the Cambridge Spies, Vassall was not a Communist. His story fueled a nationwide media frenzy and moral panic, which contributed to the framing of homosexuality as a national security risk. Also indicative of such panic was the publication of a newspaper article in the *Sunday Mirror* in 1962 titled "How To Spot a Possible Homo," featuring a photograph of Vassall. The article claimed to give the public and the British government advice on how to identify homosexual men, both the "obvious" type and the "concealed." Reading the list of stereotypical descriptions offered in the piece reveals how fifth-column claims rely on a foundation of familiar cultural constructs (see Radnitz and Mylonas, Introduction) to construct an identity category that the nation is called upon to expose and ostracize.

But the exclusion of sexual minorities and their entanglement with fifth-column politics is not an exclusively Cold War phenomenon. The rise of queer visibility and the diffusion of gay and lesbian identities around the world in recent years have presented political leaders with an opportunity to scapegoat and target LGBTQ citizens and declare their lifestyles as foreign impositions. And it is not only state actors who promote fifth-column narratives. Take, for example, the secretary general of the Lebanese armed militia, Hezbollah, who stated in a televised speech that "there are societies abroad that were ruined by homosexuality, and now they are exporting it to Lebanon and to the Arab and Islamic world."

The framing of gay rights as a by-product of Western hegemony or a "foreign menace to national sovereignty" (Bosia 2013: 40) is now a common phenomenon across many Middle Eastern, African, and East European states. The alignment of sexual minorities and "Western" interests can be directly linked to the global convergence toward institutionalizing LGBTQ rights and their promotion by supranational bodies like the European Union and the United Nations. While this type of fifth-column discourse does not invoke accusations of treason or nefarious behavior, it does rest on claims of disloyalty and ideological Western infiltration that tears at the fabric of the nation and its core values. Bosia and Weiss's definition of "political homophobia" characterizes this phenomenon:

> as purposeful, especially as practiced by state actors; as embedded in the scapegoating of an "other" that derives processes of state building and retrenchment; as the product of transnational influence peddling and alliances; and as integrated into questions of collective identity and the complicated legacies of colonialism.... [It is] a remarkably similar and increasingly modular phenomenon across a wide range of countries. (Bosia and Weiss 2013: 2)

This definition entails a strategic decision on the part of those who promote this rhetoric, which is perhaps most evident in authoritarian settings where incumbents scapegoat queer people in order to divert attention from their own failures or signal the repressive capacity of the state's security apparatus. Focusing on Iran, Korycki and Nasirzadeh contend that state-sponsored homophobia "as a tool of conservative backlash is proving useful yet again as a means of reasserting the coercive power of the state, in the wake of liberalization" (2013: 188). The targeting of gay youth has become a tool to "signal revolutionary resolve, create national narratives, and extend the state's reach into private morality" (Korycki and Nasirzadeh 2013: 188). Egypt's military regime under Sisi has arguably followed a similar approach in scapegoating sexual minorities to foster an image of the regime as the moral guardians of the people against Western deviance (Ayoub and Page 2019).

Political homophobia, however, is a multifarious phenomenon. This catch-all concept obscures major distinctions that merit closer attention. For example, political leaders may frame homosexuality as a foreign influence or decry the spread of gay rights norms globally as a pretext for meddling in their countries' internal affairs and destroying the moral fabric of their societies. We have seen this play out in Egypt, Lebanon, Uganda, and Zimbabwe, among other countries. And in some instances, particularly when states are experiencing international conflict, leaders go beyond moral claims and work deliberately to construct queer citizens as collaborators, traitors, informants, and internal agents of their rival states. It is this particular subset of anti-LGBTQ narratives that I focus on in the rest of the chapter because it has distinct implications for the target group and the way it chooses to respond to its own marginalization. The historical cases in this chapter illustrate how fifth-column narratives thrust challengers into unique positions that limit the menu of strategic options and counternarratives from which they may choose to respond. My research on state repression of LGBTQ people in other countries shows that in the absence of widespread fifth-column accusations, LGBTQ movements have a broader set of strategic choices to make when it comes to framing their claims and demands.

## How Do Target Groups Respond?

What determines how LGBTQ social movements respond to stigmatizing narratives that paint them as subversives and security threats and how do we understand their responses? Extant research on social movements helps us understand the way marginalized groups confront widely circulated narratives discrediting their collective selfhood. In their fight against systematic exclusion from society, stigmatized groups resort to a variety of framing tactics in dismantling damaging narratives. What they say and do is inextricably linked to the discursive context in which they are embedded. A relational approach to understanding social action suggests that the relationship between a stigmatized group and the broader society conditions the discursive options of identity movements.

Put simply, when stigmatized communities advocate for themselves, they do so by crafting a "counterstory" to dominant societal narratives—one that ultimately helps them restore their moral status in society and facilitates their inclusion.

Frames, according to Snow et al., denote "'schemata of interpretation' that enable individuals 'to locate, perceive, identify, and label' occurrences within their life space and the world at large" (Snow et al. 1986: 464). Framing itself is "the process through which movement actors engage in interpretive work to produce and maintain meaning for movement participants and potential supporters, as well as antagonists" (Snow and Benford 1988; Hewitt and Mccammon 2005:152). Understood as such, the production of discourse and meaning-making is both deliberate and strategic, a process activists pursue in order "to garner bystander support, and to demobilize antagonists" (Snow and Benford 1988).

When it comes to damaging narratives dominant among "inimical publics," as is the case with homosexuals becoming a symbol for national security risk, scholars contend that movements should counter outsiders' claims about their "deviant" qualities by pursuing public activities that alter the group's public identity" altogether. The process through which they do so is referred to as "counterframing," which Benford defines as the attempt to "rebut, undermine, or neutralize a person's or group's myths, versions of reality, or interpretive framework" (cited in Benford and Hunt 2003). It follows that one of the crucial framing tasks for these movements is "a call to arms or rationale for engaging in ameliorative or corrective action" (Benford and Snow 2000: 617).

Two frame-alignment mechanisms are particularly important for LGBTQ social movements. The first is "frame extension" (Snow et al. 1986: 472). This discursive tactic is deployed when the claims of the movement are not rooted in a common societal experience and may have "little if any bearing on the life situations and interests of potential adherents" (472). In other words, a non-core group can make its claims resonant by couching them in broader terms that are congruent with the values of the dominant group. For example, gay rights activists can frame their group identity as an *oppressed minority* struggling to achieve its *civil rights*. This renders their cause more resonant among those who lack an understanding or interest in the lives of gays and lesbians but find themselves sympathetic to civil rights causes. The other frame-alignment tactic is "frame transformation." Activists resort to this tactic when the values they advocate for are "antithetical to conventional lifestyles or rituals and new values may have to be planted or nurtured, old meanings or understandings need to be jettisoned, and erroneous beliefs reframed" (473). Goffman referred to this process as "*keying*," meaning the systematic alteration or definition of existing understandings about certain activities or groups and it involves the radical reconstitution of social reality (Goffman 1974: 43).

These framing theories have influenced the research of other scholars who examine the rise of LGBTQ activism and movements advocating on behalf of stigmatized groups. Philip Ayoub and Agnes Chetaille most recently argued that "framing processes are shaped across time by movement/countermovement interactions, as well as discursive opportunities derived from multi-level contexts" (Ayoub and Chetaille 2017). Similarly, Hilde Lindemann claims that:

> because identities are narratively damaged, they can be narratively repaired. The morally pernicious stories that construct the identity according to the requirements of an abusive power system can be at least partially dislodged and replaced by identity-constituting counterstories that portray group members as fully developed moral agents. . . . Counterstories, then, are tools designed to repair damage inflicted on identities by abusive power systems. They are purposive acts of moral definition, developed on one's own behalf or on behalf of others. They set out to resist, to varying degrees, the stories that identify certain groups of people as targets for ill treatment. Their aim is to reidentify such people as competent members of the moral community and in doing so to enable their moral agency. (Lindemann 2001: xii–xiii)

Analyzing the macro-historical context in which gays and lesbians emerged as an identity group can help us understand how gay rights activists go about repairing their damaged identities across various societies. Understanding the response of LGBTQ movements therefore requires a careful historical tracing of the conditions early movers in the movement's history encountered, the political climate in which they existed, and the internal discussions that took place within their ranks on how to best frame and construct their identities to the broader public. This framework demonstrates that movement rhetoric and actions are best understood if examined contextually with regard to the geopolitical, cultural, and social environment in which activists operate (Ray 1998; Swidler 2000; Polletta and Jasper 2001; Hunt et al 1994; Rucht 1996; Moussawi 2015). The relational theoretical framework I build on in my process tracing of LGBTQ movement evolution in the United States and Palestine requires us to bring activists back in, and to examine their rational calculus in framing their processes.

## From the Mattachines to Al-Qaws: Queer Liberation and Gravitating toward the Mainstream

For the purposes of this chapter, I compare how gay rights advocates in the United States and Palestine responded to fifth-column accusations in the early phase of their movement's life cycle. While it is rare to find comparative studies between

Palestine and the United States, these cases lend themselves to the purposes of this analysis. In both cases, homosexuality was directly linked to accusations of subversion and undermining national security. Both countries were experiencing international conflict when fifth-column claims spread among the public, and in both cases, the propagation of fifth-column accusations preceded the mobilization of an LGBTQ movement and queer visibility. Both cases diverge on the outcome of interest—how targets responded to those accusations. In the United States, the movement pivoted rapidly toward an assimilationist discourse, while queer activists in Palestine held onto a discourse of radical liberation. Although this difference is crucial, I draw a parallel in these cases between how the movements responded. They adopted the same strategic approach to framing by gravitating toward the political mainstream.

The contextual constructionist framework (Best 2003; Gamson and Modigliani 1989; Van Gorp 2007) laid out in the previous section leads us to examine how the political fields in the United States and Palestine molded LGBTQ activists' discursive repertoires. More specifically, we have to interrogate how heightened nationalist sentiment during the Cold War and the first Palestinian uprising mattered for the evolution of the gay rights movement and the framing of its claims in each case.

As stated earlier, when LGBTQ people are portrayed as a fifth column, they often rely on the strategic production of new discourses to resist the systems of power oppressing them. Against the backdrop of rising state repression in the United States during the Lavender Scare period (1947 onward), homophile activists founded the first sustained gay rights organization, the Mattachine Society, a secret foundation that was modeled after the cells of the Communist Party in the 1930s. The spread of conspiracy theories that homosexuals were conspiring with the Reds against their own country and being tapped to spy on their own government sparked intra-movement disputes within Mattachine over strategy and framing. The founders—who had championed a more radical approach—resigned and gave way to a new leadership that sought assimilation into the mainstream. The homophile movement began the wide circulation of magazines and periodicals that aided homosexuals in producing new narratives surrounding their sexuality and citizenship and in reasserting their national loyalty and denouncing Soviet subversion.

In the case of Palestine, queer activists in the early 2000s packaged their claims in language that they deemed resonant locally, drawing on a host of familiar political and cultural repertoires to illicit solidarity and allyship from mainstream rights and advocacy organizations. When LGBTQ Palestinians were accused of being tapped as informants by the Israeli intelligence system, Palestinian LGBTQ activists used the "frame expansion" tactic. By highlighting the notion of queer intersectional politics as the cornerstone of their ideological foundation, they were able to convince a broad segment of civil society that gay rights claims (framed as a struggle for queer liberation) are part and parcel of a broader project of national liberation. Their application of the concept of intersectionality was based on framing their organizations' work as anti-Zionist, anti-imperialist, and anti-occupation—systems of oppression they argued interlocked with patriarchy and queerphobia.

Significant in both of these cases is activists' emphasis on discourse and identity construction. The primary tool through which they built collective consciousness was the circulation of periodicals and hundreds of newspaper and magazine articles. Such tools helped citizens whose sexual orientation or gender identity were deemed non-normative make sense of their own queerness. These periodicals also allowed gays and lesbians to imagine themselves as members of an oppressed cultural minority and members in a political community with a common cause. As the cases illustrate, spreading new narratives through print and media played a prominent role in the efforts of both movements to forge new discourses and transform their group identity.

## The Construction of the LGBTQ Community as a Fifth Column in the United States

The post-Stonewall era of the 1970s is often credited for the rise of a gay liberation movement. The Stonewall riot of 1969—the famous rebellion against police raids on bars where gays congregated—became a commemorated event nationally through annual pride parades. Homosexuality became decriminalized in many states, gays and lesbians had their first national march on Washington, and LGBTQ organizations proliferated across the United States, protesting alongside the civil rights movement, anti–Vietnam War activists, and the activists of the New Left. But the 1970s were hardly the origin of the modern LGBTQ community and movement in the United States. Much of the groundwork had been laid as far as back as the late 1940s in the critical years following World War II.

The second half of the twentieth century saw the evolution of homosexuality as an identity rather than a mere vice, and later a category of self-identification. Large numbers of men discovered their homosexuality during military service and chose to reside in urban centers like Los Angeles and San Francisco upon their return. A *community* of homosexuals began to take form with both rising visibility and repression (Stulberg 2018: 5). In the first half of the 1950s, the federal government started labeling homosexuals a "national security risk," and the first two self-sustained gay and lesbian organizations—the Mattachine Society and the Daughters of Bilitis—were founded in part as a response to rampant police and state repression of homosexuals.

It is in those early years that homosexuals engaged in deliberate community-building, assembling collective identity, and contesting their national subjectivity. And there were serious disputes on strategy, discourse, and approach. The theoretical framework laid out in the previous section leads to the conclusion that the gay and lesbian identities that would emerge in the 1950s are both the product of internal contestation and *intra*-movement disputes as well as a reflexive process that was punctuated by the political climate of the 1950s, dominant group fears, and national anxieties.

## The Lavender Scare: How Queers Became a Synonym for Security Risk

During the first half of the twentieth century, the military depicted "homosexual tendencies" as markers of "psychopathic personalities" and stigmatized homosexuals as "lawbreakers and loonies" (Faderman 2015). The psychiatric community treated homosexuality as a psychological pathology, while officials in the military argued that homosexual acts were a result of mental instability and that they threaten order, morale, and discipline within the ranks of the armed forces (Canaday 2009: 183). This medical view of homosexuality led to a great deal of state repression, particularly in the armed forces where it was grounds for dismissal.

Other political actors during the Cold War began to redefine the risk they believed homosexuals posed to the state and to society. In 1950, Senator Joseph McCarthy made the famed claim that hundreds of "card-carrying Communists" had infiltrated the State Department while highlighting that homosexuals were a prominent group among them (Johnson 2004: 16). As historian David Johnson highlights in his book *The Lavender Scare*, McCarthy asserted that homosexuality was "the psychological maladjustment that led people toward Communism" (16). Following pressure from McCarthy and his Senate Republican colleagues in February 1950, Assistant Secretary of State John Peurifoy announced during testimony to the US Senate that the State Department had ousted ninety-one "security risks" on the grounds of their homosexuality. Johnson goes on to note:

> Like McCarthy's charges, the revelation that ninety-one homosexuals had been dismissed from the State Department unleashed a flurry of newspaper columns, constituent mail, public debate, and congressional investigations throughout 1950 about the presence of homosexuals in government and their connections to Communists. The revelation set in motion a chain of events that would have widespread repercussions for government security policies and the

millions of people affected by them for the next twenty years. . . . Headlines warning of "Perverts Fleeing State Dept." peppered newspapers throughout the country. While members of Congress held hearings to determine how to "eradicate this menace," jokes circulated about "the lavender lads" in the State Department. (18)

Nationwide, the news of these firings gave credibility to McCarthy's attacks on Truman's State Department and his assertion that Communists and subversives had access to the nation's highest security clearances (18). However, historians prior to Johnson often miss the fact that the campaign to confront this "red infiltration" had already been underway at the State Department since 1947. It was in June of that year when purges of Communists and alleged homosexuals from the department began with legal cover from "McCarran rider," which permitted the Secretary of State to fire any employee at his "absolute discretion" if doing so was deemed in the interest of national security (21). By 1953, the federal government and particularly the State Department appeared convinced that they had become haven for sexual misfits and deviants. The State Department reported it had fired 425 employees on grounds of homosexuality.[6]

It is crucial to note here that the State Department and Senate Republicans were not the only ones trying to paint homosexuals as security risks. In fact, the FBI had played a major role in sponsoring a special program to influence public opinion on the threat of Soviet infiltration since early 1946 (Charles 2015: 70). And the framing of who constitutes "national security risk" according to the FBI and congressional conservatives was fairly broad to encompass all homosexuals. Republican Senators argued that "a man doesn't have to be a spy or a Communist to be a bad security risk. He can be a drunkard or a criminal or a homosexual" (Johnson 2004: 23). Homosexuals were believed to be vulnerable to blackmail by the Reds because their "deviant" behaviors could make them targets for recruitment. Gay employees were seen as lacking moral standards, leading a perverted lifestyle, vain, greedy, and lacking emotional stability, rendering them incapable of holding the nation's secrets.

As a consequence of the ideological polarization of the Cold War and the emergence of an imagined binary between a "Judeo-Christian West" and an "atheistic communism," society came to regard any immoral behavior as "part of the Communist plot to hasten moral degeneracy of America" (Johnson 2004: 37). Tabloids and newspapers circulated these claims as well and drilled them into the minds of Americans. Homosexuals were painted as " 'pink pansies' who 'shriek, scream, cry and break down into hysterical states of psychoses when they are called upon to carry arms to defend our shores from the enemy.' " Homosexuality thus became known as "Stalin's Atom Bomb" (37).

The logic of this linkage is perhaps most clear in a news article by former US Ambassador to Switzerland Henry Taylor:

The homosexual is the enemy's marked man. Every nation's espionage appa-ratus, including our Central Intelligence Agency, concentrates on homosexuals if its agents can find such a target. Thus the homosexual receives the brunt of the enemy's concentrated (and patient) attention and surveillance. Attracting such concentration is, of itself, too dangerous . . . homosexuals have a famous tendency to support and shield other homosexuals. The evidence is endless which shows tragic cover-ups of security violations and entanglements buried within what practically amounts to nests and cells of homosexuals, sometimes even when they are strangers to one another. Any idea that the homosexuals' lives are "their own business," that dismissing them is "discrimination," etc., is known throughout the world to be utterly ruinous to the security of any nation.

A 1950 Senate report titled "Employment of Homosexuals and Other Sex Perverts in Government" concluded that there were over 5,000 alleged homosexuals who "polluted" the rank of the federal bureaucracy and who needed to be removed. This influential report gave way to President Eisenhower signing Executive Order 10450, banning gay employees from the federal government. Johnson puts the number of gay employees who were affected by the subsequent purge to be between 5,000 and 10,000. This executive order remained in effect until 1995, when it was rescinded by Bill Clinton and replaced with the infamous "Don't Ask Don't Tell" policy for the military. The legacy of the Lavender Scare is that this linkage between gays and external enemies of the state became the dom-inant political rhetoric for decades to come. For instance, when asked at a local event in 1973 about whether gays should be allowed to hold a security clearance, then-US Senator Joe Biden hastily stated: "My gut reaction is that they are secu-rity risks, but I must admit I have not given this much thought" (Cutler 1973).

## The Evolution of the Mattachine Society: Mutiny or Strategic Adaptation?

The Lavender Scare undoubtedly left a devastating impact on the lives of the cit-izens it targeted. But it also contributed to the rise of solidarity among them and the growth of the modern gay rights movement. The most prominent figure in the rise of the movement was a Los Angeles–based American Communist and labor activist named Harry Hay. Hay—a member of the Communist Party—had long been aware of his same-sex desires. But by 1948, Hay's reading of Alfred Kinsey's famous study *Sexual Behavior in the Human Male* inspired him to envision the growth of bonds among what would be millions of homosexuals in America.[7] In that same year, Hay became aware of the homosexual purges taking places in the State Department and argued, "these are the next scapegoats, to replace Negroes and Jews" (Faderman 2015: 54). He believed both of those groups at that time had allies and defenders, but that homosexuals were a more convenient target because they were a marginalized people "who did not even know they were a group" (54).

Hay believed that if homosexuals conceived of themselves as a *minority group*, they could perhaps mobilize and resist the government's persecution campaigns. Hay envisioned an organization that would have social, cultural, and political programing. But his ideas met with little enthusiasm given public paranoia and sodomy laws that criminalized homosexuality.

By 1950, Hay identified four other collaborators: Dale Jennings, Bob Hull, Chuck Rowland, and Rudi Gernreich. Four of the five were Communists. As the five began deliberating how to organize, they were in search of a rallying claim that would allow them to mobilize membership and translate grievances into collective action. Hay famously articulated an effective framing: "We are an oppressed cultural minority," he proclaimed (57). As their group began to grow with new members joining, they agreed to form an organization with the name Mattachine Society, based on a French secret fraternity from the early Renaissance era that held "clandestine dances and rituals" (59). Hay realized the magnitude of the risk he was taking in mobilizing a movement advocating for homosexuals and understood that it would be a costly undertaking. On the one hand, it entailed a departure from the Communist Party given its hostility toward homosexuality. On the other hand, he believed that leading such a movement would put his family at risk given the political climate. Nonetheless, Hay's determination to mobilize against the state and its dangerous accusations led him to leave both his family and the Communist Party. Hay's decisions in this period embody the desire to resist and confront political homophobia which dominated the early stage of gay organizing in America.

In order to initiate a social movement, Hay and his peers had to draw on their prior organizational expertise to expand their membership. Like any identity-based social movement, their ability to expand would require an effective way to allow the targets of discriminatory policies to conceive of themselves as a group and develop a collective rights consciousness—a process that is central to the study of subject formation and group boundary construction. The founders of Mattachine learned important lessons from the Communist Party—that secrecy and convolution were key to the survival of their group. They insisted on the secrecy of the members and the society's top leadership—what they called the Fifth Order. Mattachine would approach men in areas where they cruised other men—on the beach and in parks—and invited them to public discussion groups across Los Angeles to discuss the Kinsey study, and used those meetings to recruit members. Mattachine's Communist founders extrapolated from the Marxist notion of "false consciousness" to "foment among homosexuals a critical awareness of their oppression, as a group, at the hands of the majority society" (Epstein 1998: 35).

Mattachine's first public fight against state homophobia took place when one of its founders, Dale Jennings, fell victim to police entrapment. The organization formed a "Citizens Committee to Outlaw Entrapment" and began soliciting funds to raise the legal fees to hire a lawyer in defense of Jennings. The group began printing leaflets and leaving them in bars, public restrooms, and venues where homosexuals were known to congregate (Charles 2010: 266). These leaflets were asking for help with the legal fees but also served to "make clear to the world that homosexuals were not ipso facto lewd and dissolute" (Faderman 2015: 64). This was the first time that homosexuals in the United States used the framing of "civil rights," a language their leaflets had emphasized. Their fighting words inspired homosexuals whose lives had been filled with fears of police entrapment and arrest. The money began pouring in. Soon enough, Jennings would claim before a jury that he indeed was a homosexual, but that he did not commit any acts he was accused of. The city attorney dismissed the charges against Jennings. The latter would go on to articulate the significance of his case for the rise of homosexual unity, and indeed, what later became known as a *homophile* movement: "A Bond of brotherhood is not mere blind generosity. It is unification for self-protection. Were all homosexuals and bisexuals to unite militantly, unjust laws would crumble in short order" (Faderman 2015: 66).

This first legal victory led to hundreds of new incomers to Mattachine discussion groups, and scores of new members in the secret society. Meetings began to take place in other parts of California and quickly spread throughout the United States, numbering twenty-seven national discussion groups. With the rise of anti-homosexual paranoia, the Mattachine Society could no longer eschew public attention.

The simultaneous growth of this group and the spread of the Lavender Scare and fear of Communists also forced change within the organization fairly rapidly—a move away from a strategy of resistance and confrontation toward a desire for assimilation within the national body. A dramatic shift occurred in 1953, when a major divide took place between the cadre of leaders and the organization's rank-and-file members (Meeker 2001). Scholars note that "the new Mattachine Society abandoned the radical critique of the original group, and instead embraced conservative politics advocating that homosexuals adjust to life in a homophobic society" by adopting an *accommodationist* strategy in lieu of the radical confrontational approach of the original founders (Meeker 2001: 79). After forcing its Communist founders to resign, secrecy was disavowed, and the post-1953 Mattachine arguably adopted *conservative* politics.

After 1953, the homophile movement consisted of three related entities: Mattachine, the Daughters of Bilitis (a lesbian organization founded with the help of Mattachine leaders), and *One* magazine. The core of the gay and lesbian movement for the remainder of the decade focused much less on sexual difference or queerness, and more on how gays and lesbians are similar to their fellow heterosexual citizens and share their allegiance and loyalty to the nation and its laws. Their periodicals and publications became a powerful tool for the development of gay and lesbian political identities—ones that navigated with great caution the tumultuous political grounds of the 1950s and attempted to reposition gay and lesbian subjects vis-à-vis the nation.

Homophile periodicals in that latter period often took on a nationalist and anti-Soviet tone not only as a way to reassert the movement's national loyalty but to distance itself from the legacy of its Communist founders (Serykh 2017). Periodicals at the time constructed a mainstreamed image of the "homonational"[8] subject—one that is progressive, liberal, and "eager to stand with his or her nation against the common enemy" but who was part of a minority group that demanded legal reforms and concessions from a corrupt US government whose behavior mirrored Soviet homophobia (913). This was done with the purpose of aligning homosexual identities and national identity and presenting a discourse that shows the two as compatible. Discourse analysis of the depiction of the Soviet region in homophile periodicals shows that authors used "negative constructions of the region both to align themselves with the U.S. nation against the Soviet Union and communism and to criticize the U.S. government's Soviet-like treatment of its citizens" (911). This discursive legacy can later be traced to nationalist frames and rhetoric in movement periodicals in the 1960s and 1970s, long after the accommodationist phase had ended.

Historical accounts converge on the idea that the major shift from Mattachine's radical first three years toward a more conservative approach was a product of a growing and more diverse membership that had come to encompass moderates and businessmen who were suspicious of Soviet influence. But more recent accounts argue the shift is far more complex. Martin Meeker, for example, argues that

> rather than being a cowardly retreat, the Mattachine Society's presentation of a respectable public face was a deliberate and ultimately successful strategy to deflect the antagonisms of its many detractors. . . . This strategy allowed the organization to speak simultaneously to homosexuals and homophobic heterosexuals and to communicate very different ideas to each population,

during a time when the latter exerted considerable power over the former. This practice of dissimulation disarmed some of the antigay sentiment in American society while it also enabled the homophiles to defend and nurture the gay world. . . . In the case of the Mattachine Society, daring and successful politics emerged from an apparently conservative ideology. (Meeker 2001: 81)

Meeker's account cuts against the grain of historical conventional wisdom as recounted by D'Emilio and Faderman. It posits that what many saw as a shift that rendered the Mattachine Society obsolete and led to the dissolution of the national organization a decade after its founding was merely a strategic adaptation of *public* discourse. It was an effective movement tactic that accomplished success and secured the movement in its early years from detrimental backlash. It gave room for gays and lesbians to construct an identity and mobilize in a way that paved the road for the more confrontational phase of gay liberation activism in the 1960s. Whereas historians may offer conflicting accounts of the determinants of discursive shifts in the movement's early years, there is no doubt that the 1950s saw major victories for gay rights and a remarkable ability of a small emergent movement to survive and endure nationalist homophobia that discredited gays and lesbians and put their existence in danger.

After all, homophile periodicals such as *One* magazine, the *Mattachine Review*, and the *Ladder* had flourished and managed to shape collective identities and influence the public's perception on the place of homosexuals in the nation. Homophile periodicals in the 1950s and 1960s "opened the option of anonymous participation in a community" at a time when direct non-anonymous participation would have come at a high price (Milo 2017).

## Reframing LGBTQ Identity in Palestine

Palestinian gay rights organizing shares many similarities with the birth of the homophiles in the United States. It too emerged in a contested political period in which fears of Israeli infiltration of the national movement in the 1990s became rampant. Those fears became the basis for a systematic effort to mobilize against moral perversion in society and linking homosexuality to national security risk. But the Palestinian movement's founders were careful to position themselves at the core of the national movement as champions of both queer and national liberation. Their strategic response to fifth-column accusations was premised on linking the two in the national imaginary and creating interlocking symbols of queerness and Palestinian desire for freedom and liberation.

## A Palestinian Lavender Scare?: How Queer Palestinians Became a Fifth Column

Homosexuality—a social and cultural taboo—was hardly the subject of controversy in Palestinian society prior to the 1980s. But that would quickly change during periods of heightened political tensions and militarized conflict. Two time periods are particularly relevant for the evolution of the Palestinian LGBTQ community and movement: the first Palestinian national uprising or Intifada (1987–1993) and the second national uprising (2000–2006). Palestinian LGBTQ activists repeatedly note that in both periods, Israel pursued policies that entangled gay identity in nationalist discourse in a way that further reinforced homophobia and attached a distinct fifth-columnist dimension to it.

The first Intifada constitutes a turning point in the Israeli-Palestinian conflict. In the preceding period (1967–1987), Israel had primarily relied on its intelligence units to target small militarized cells of the PLO and splinter armed groups that targeted Israelis. During the first Intifada, Palestinian resistance to military occupation took a more "popular" approach and opened the door for broader participation in civil disobedience, stone-throwing, the formation of localized national committees, and rioting. In response, Israeli security agencies realized that in order to counter this wave of popular violence, it needed to gain access to local information about the leaders of this emerging movement and who stood behind orchestrating these actions at the local level. Security forces leaned heavily on recruiting larger numbers of local informants and spies who would collaborate with the Israeli security agencies and offer them information.

In 2014, a story appeared in the *New York Times* and the *Guardian* reporting that forty-three veterans from an elite, secretive Israeli military intelligence unit had penned a letter to their commanders and Israel's prime minister declaring their refusal to participate in reserve duty. The veterans' letter detailed how their unit blackmailed gay Palestinians and threatened to expose them to their communities unless they collaborated secretively with Israeli military officers (Rudoren 2014). This entrapment practice became known widely among Palestinians as *Isqat* (to make one fall) (Ritchie 2015).

Though this policy was not well documented until recently, stories of entrapment and collaboration penetrated Palestinian national narratives during the first Intifada. As a result, an equivalence between the homosexual and the occupier emerged among Palestinians, even though the policy targeted other vulnerable Palestinians such as those who required permits to enter Israel for medical purposes, and women or men who had sexual affairs outside of marriage.

In the early 2000s, newly formed queer Palestinian groups such as Al-Qaws were keenly aware of how entrenched those negative associations had become; and for more than a decade, they had to grapple with ways to disrupt fifth-column narratives. As international media sources began highlighting Israel's entrapment policies, queer Palestinian activists grew more concerned that such coverage would only further perpetuate the link between collaboration and homosexuality. In 2014, Al-Qaws issued a statement in response to the growing coverage on the topic, warning that focusing exclusively on the entrapment of LGBTQ people has dangerous consequences. The statement argued:

> Indeed, this pervasive linking of non-normative sexuality and Palestinian collaboration has become a term and identity of its own in the Palestinian imaginary and reality: isqatat. While it is sometimes true that Israel succeeds in using sexuality as a lever to coerce some Palestinians into becoming collaborators, this is not the primary way in which collaboration is enlisted, nor is it the only option for living a viable queer life in Palestine. In our work, we struggle constantly to resist and dismantle this oppressive stereotype that links queer people in Palestine with Israel, and instead to promote the complex understanding of "homophobia" that emphasizes its relationship with colonialism. (Al-Qaws 2015)

In an ethnography on queer Palestinian activism, Sa'ed Astshan, a queer Palestinian anthropologist, further documents the impact of entrapment on queer Palestinians and their portrayal in society. He notes:

> the fact that Israel also utilizes Palestinian homosexual bodies for Israeli state projects puts queer Palestinians in an even more precarious position. . . . The Israeli intelligence history of entrapment of LGBTQ Palestinians from the West Bank and Gaza Strip has contributed to the further stigmatization of queerness in Palestinian society because of the subsequent association of homosexuality with betrayal and collaboration with Israel. (Atshan 2020: 44–45)

In a historical publication on the status of queer Palestinians during the Intifada, Al-Qaws activists detailed the response of the Palestinian national movement to the policy of entrapment at the time. The national response entailed an "awareness" campaign that took the motto of "moral corruption leads to national corruption," feeding into existing narratives and sexual taboos, and promoting intolerance of homosexuality and other practices as "gateways" to national betrayal (Al-Qaws-Jadal). In reality, there was no documented decline in the numbers of those who collaborated with Israel as a result of this campaign; to the contrary, gay Palestinians became even more vulnerable to blackmailing in a less tolerant society that consciously associated "gay" with treason and foreign occupation.

During the second Intifada, the process of further linking gay Palestinians to an occupying state was boosted through a new set of PR projects adopted by the Israeli government and pro-Israel campaigns. This PR project emerged with a 2005 marketing campaign led by the Israeli government known as "Brand Israel" that sought to show an image of Israel as "liberal," "cosmopolitan," and "modern" through what Israel describes as a welcoming attitude toward LGBTQ people. The campaign was targeting LGBTQ communities internationally, gaining Israel a new supportive constituency among progressives, and boosting Israeli tourism (Schulman 2011).

Queer activists and theorists refer to these efforts as "pinkwashing," an attempt by the state to diffuse criticism of its human rights abuses by touting its record on LGBTQ rights. In the Israeli-Palestinian context, pinkwashing takes the form of corporate or government-sponsored activities seeking to "draw attention to Israel's purportedly progressive record on LGBT rights to detract attention from Israel's gross violations of Palestinian human rights" (Atshan and Moore 2014). Examples of pinkwashing include campaigns by the Israeli Ministry of Tourism marketing Tel Aviv as the world's top gay destination, Brand Israel's funding of pride festivals and gay films that depict Israel as the only safe haven for gays in a homophobic neighborhood, and the production of Israeli gay porn in villages that Palestinians deem to be ethnically cleansed in 1948. What promoted further fifth-column narratives was a growing number of films and stories about gay Palestinians who had been outed and forced to flee to Israel for protection. And although Israel does not grant LGBTQ Palestinians asylum or legal protection, accounts of queer Palestinians hiding in Israel have reinforced the coupling of homosexuality and collaboration with the Israeli state. All in all, the active promotion of Israel as "gay-friendly" nation in contrast to "backward" and "repressive" Arabs served to reinforce in the Palestinian imaginary, once again, a link between the homosexual and the colonizer, a "Western-other" who has no place in the Palestinian nation.

Understanding the effect of these two policies—Isqat and pinkwashing—on public perceptions of LGBTQ Palestinians make it easier to understand why LGBTQ Palestinians seek to disavow any discourse or strategy that would reinforce the link between "gay" and "outsider." Additionally, the fact that several Western gay rights groups bought into the narrative that Palestinian queers are persecuted by their "gay-bashing" communities and whose rights are only protected by Israel rendered these groups unlikely allies to LGBTQ Palestinians.

## Crafting a Queer Response to Pernicious Claims

Palestinian gay rights activism took root in the early 2000s with the establishment of two queer organizations, Al-Qaws (in 2007) and Aswat (in 2002). Both organizations, headed by feminist Palestinian women, sought to advance the rights of LGBTQ communities and were among the first in the Middle East to bring these issues into the public sphere. Haneen Maikey, one of the pioneers behind Al-Qaws, had been working with LGBTQ Palestinians prior to Al-Qaws' founding as a staff member in the Jerusalem Open House—an Israeli gay rights organization (Atshan 2020: 47). Since their establishment, both Al-Qaws and Aswat have been publishing periodicals, newspaper and online articles, and educational content targeted at both queer Palestinians but also the broader society.

Al-Qaws places significant emphasis on the importance of discourse and narrative. The organization has repeatedly articulated that its philosophy is grounded in queer politics rather than in identity politics. Five prominent themes appear regularly in their public discourse, which has become increasingly visible in recent years. First, that the Palestinian struggle against the occupation is a queer issue. Second, that the Palestinian struggle for freedom is intersectional. Third, that the Palestinian queer movement focuses on the rights of all Palestinians who are subject to the violence of a settler colonial state and not just one subset of Palestinians. Fourth, that the movement considers resisting the occupation through exposing pinkwashing narratives a top priority. Fifth, that the movement considers the Palestinian Authority as an agent of the colonial Israeli state apparatus.

Until recently, the discourse of Al-Qaws placed more emphasis on national liberation and confronting Israel's military control and pinkwashing than on any other topic. The organization's emphasis on the word "queer," as opposed to gay or LGBTQ, bears tremendous significance. Historically, the term has been used by activists in the United States since the Stonewall riots of 1969 as a political identifier, signaling opposition to mainstream societal expectations and radical opposition to existing power structures. In an interview explaining the mission behind Al-Qaws, Maikey emphasizes the importance of the organization's development of a *political* collective identity that is both Palestinian and queer and one that prioritizes the promotion of national principles as its primary purpose:

> I co-founded Al-Qaws to promote BDS [campaign for Boycotts, Divestment and Sanctions against Israel] to more Palestinian queers and to be visible in every campaign related to Palestinian struggle, both nationally and internationally, physically and virtually. We are everywhere. We are challenging the current image of what "Palestinian gay" means—images of "victimhood" or

"the exotic"—challenging this binary and bringing in more of the activist spirit. (Al-Qaws 2012)

The term "queer," which Maikey and other prominent figures prefer, continues to bear a revolutionary meaning and is often deployed to signal a nonconformist identity and vision of politics. Even when writing in Arabic, publications of Al-Qaws transliterate the word "queer" and use it in Arabic. As noted earlier, such wording draws on a particular discursive "toolkit" or "repertoire" utilized by sexual minority activists elsewhere. In explaining their affinity to the expression queer, Palestinian activists often highlight that the term signals solidarity with a revolutionary community that also fought for decades against capitalism and state abuse in the United States.

Besides Al-Qaws, the queer feminist organization Aswat espouses similar principles. Both organizations have become more visible in recent years and relied on social media to craft a consistent narrative that aligns the queer movement with Palestinian resistance against the occupation. Following the first queer rally in July 2020, Aswat published a video of queer Palestinian activists marching with a Palestinian flag and chanting about queer liberation and the Palestinian right of return. This assemblage of national and queer causes is complemented with a powerful graphic design framing the video they circulated. The square frame consists of the Palestinian Kaffiya—a national symbol for all Palestinians that is often associated with PLO revolutionaries like Yasser Arafat—with the drawing of *Handhala* (a famous child caricature by the assassinated Palestinian leftist artist Naji Al Ali)—an emblem that signifies Palestinian defiance. The bottom part of the video frame has the phrase "resist" in Arabic calligraphy, in which the central letters are drawn in the shape of a rifle. Both the literal and figurative *framing* of the first Palestinian queer march in Haifa are an obvious testament to the queer movement's strategy of challenging any association between LGBTQ identity and collaboration with the occupation. The movement's social media and public outreach create a powerful visual connection between queer rights and the Palestinian national movement.

The movement's framing of LGBTQ issues in Palestine as a "queer anti-colonial anti-racist struggle," and putting emphasis on words like queer, neoliberal, anti-colonial, national, and Israeli-apartheid are not just a mere reflection of its ideology. The discourse it advances is also a strategic tool used to emphasize that LGBTQ Palestinians are not agents of an external "other," but rather an integral part of the Palestinian people and their lived experiences.

The radical queer discourses described earlier have been deployed strategically to accomplish three *keying* goals, all of which center on transforming identities and perceptions of the self. They serve to (1) distance the movement from Israeli state attempts at co-opting the struggles of queer Palestinians against homophobia; (2) shift the movement toward the mainstream of Palestinian civil society and emerging Palestinian youth movements in which the national liberation frame is highly salient; and (3) counter existing understandings in the Palestinian imaginary that link "gay" to "Western, colonial, and foreign" and replace these links with new narratives that link "homophobia" to "occupation" and "colonialism."

Atshan's book on the Palestinian queer movement concludes that the movement's approach allowed queer activists "to claim their share of successes, particularly in forming alliances with progressive movements around the world and undermining the standing of the Israeli state among leftist queer activist communities" (Atshan 2020: 6). This assessment of the movement's success in aligning queer identity with the Palestinian national movement finds support in major developments that took place in the summer of 2019 when the Palestinian Authority police issued an order banning Al-Qaws from holding events in the West Bank and incited citizens to remain "vigilant" and report any such subversive or homosexual activity to the police. Within a short period, a collective of civil society organizations in Palestine issued a statement together condemning the Authority's crackdown on Al-Qaws and LGBTQ rights and some publicly declared their support for Al-Qaws and LGBTQ Palestinians. Al-Qaws itself issued a response to the police statement in which it condemned its portrayal of the organization as a suspicious entity and called on the police spokesperson to read the organization's mission statement and articulated goals. It described itself as an organization of the masses and encouraged the police force to focus its efforts on resisting the Israeli occupation. Following pushback from Al-Qaws and other NGOs, the police statement against Al-Qaws was withdrawn. These recent events are indicative of the queer movement's growing ability to forge ties with powerful societal forces and to survive in the face of rising political homophobia. The role Al-Qaws and Aswat have played in articulating an authentic Palestinian voice and repositioning queer Palestinians vis-à-vis Palestinian society writ large shows a remarkable ability to disarm fifth-column politics.

## Conclusion

Sexual minorities confront widely circulated narratives that damage their identities. Framing theories suggest that the political and social context, or the stigma that marginalized groups experience, compels them to engage in counterframing in order to reestablish their legitimacy. I argue that this is often the case, and that contextual differences across movements explain much of how activists act as agents of social change. The differences between the queer liberationist narratives of the Palestinian LGBTQ NGO Al-Qaws and the accommodationist discourse of the American homophile movement in the 1950s might appear as radically different responses to a common form of oppression—the fifth-column designation. However, the historical tracing of the political climates in which LGBTQ communities emerged reflects how political and contextual constraints in each case shaped early movement narratives and brought about a rapid "counterframing" response to transform the identity of queer people in the national imaginary—from subversive groups to loyal bearers of the national flag.

The strategic response in both of those cases was gravitation toward the national mainstream. This was done in two ways: "othering" the nation's rivals in the movement's discourse, periodicals, and publications, and constructing an image of a patriotic homonational citizen whose loyalty to the nation is demonstrably clear. Both movements, thus, ultimately resorted to producing a discourse that aligned their community's identity with the political values of the dominant group in both respective periods. The Mattachine Society, the first homophile organization in the United States, expelled its founders with Communist ties and amended its constitution to express opposition to "subversive elements" and affirm loyalty to the United States. LGBTQ periodicals became a method of narrative repair and a means to articulate a homonational identity. The periodicals analyzed from that period took on a tone that was anti-Soviet and heavy on reasserting the national loyalty of the rising homophile movement. This assimilationist tactic allowed the rising group to secure major victories including a favorable Supreme Court decision during a pivotal and precarious moment in the movement's early history. In Palestine, the LGBTQ movement came into being with fifth-column discourses already lurking in the background. The first queer organizations Al-Qaws and Aswat perhaps unknowingly took a page out of the US homophile movement's playbook. Their strategic approach focused on avoiding the pursuit of narrow identity politics. Instead, it highlighted the imperative of resisting military occupation domestically and in international forums and tying queer emancipation to national liberation. This approach allowed them to become embedded in a broader national resistance network, and to utilize their position in this network to gain legitimacy and gradually enter the political mainstream. This discursive approach made the movement more resilient to homophobic backlash and shielded it from retaliation as it became increasingly more visible.

## Notes

1. See Bosia 2013. In this seminal book, the authors argue that homophobia, as practiced by the state, captures "the totality of strategies and tools, both in policy and in

mobilizations, through which holders of and contenders over state authority invoke sexual minorities as objects of opprobrium and targets of persecution" (31).

2. A note on terminology: prior to the adoption of the identity labels "gay" and "lesbian," the word "homosexual" was used to describe people who had relations with someone of the same sex. Activists in the 1950s preferred to call themselves *homophiles* to de-emphasize sex. The term "homophile" fell out of favor in the 1970s with some members of the community choosing to identify as gays and lesbians, while others preferred to reclaim the word "queer," which had been used as an insult. Originally, the term "queer" signified opposition to normative institutions and a radical political identity, but due to conceptual slippage, many use the term as an umbrella term that encompasses all the letters of the acronym LGBTQ. I reserve the use of the term "queer" to its original meaning since many gays and lesbians do not identify with radical politics.

3. The gay rights movement in the United States was initially focused on people with same-sex desires. The inclusion of Trans and gender nonbinary people occurred much later. In Palestine, queer activism has always encompassed in its discourse LGBTQ people.

4. The strategic and ideological divide over assimilation versus liberation continues to be a source of debate for many LGBTQ movements outside the United States today.

5. LGBTQ people were not always perceived as a minority group. In the first half of the twentieth century, homosexuality was considered an act rather than an identity. Many LGBTQ activists around the world do not prefer the designation of LGBTQ people as a "minority."

6. A similar process of designating homosexuals as security risks was taking place in Canada in the 1950s and 1960s, where hundreds of gays and lesbians lost their civil service jobs and the RCMP collected the names of over 9,000 suspects of homosexuality. "Were You Affected by the Anti-Homosexual Security Campaign?," *Bi-line*, February 1995, 10. *Archives of Sexuality and Gender*. https://link-gale-com.proxygw.wrlc.org/apps/doc/GAWOLG607691424/AHSI?.

7. Kinsey concluded in his study at the time that at least 10 percent of men in America had engaged in homosexual acts throughout their lives.

8. Puar (2007) coined the term "homonationalism" in reference to the assimilation of gays and lesbians into national-normative projects such as the "War on Terror." The term has become a catchword that academics and activists use to describe how the United States and European states—among others—deploy the inclusion of gays and lesbians as model citizens for racist, anti-immigrant, and Islamophobic ends.

# References

Al-Qaws. 2012. "Queer Politics and Haneen Maikey." Al-Qaws, January 22. http://www.alqaws.org/articles/Queer-Politics-Haneen-Maikey?category_id=0.

Al-Qaws. 2014. "Al-Qaws Statement re: Media Response to Israel's Blackmailing of Gay Palestinians." Al-Qaws, September 19. http://www.alqaws.org/articles/alQaws-Statement-re-media-response-to-Israels-blackmailing-of-gay-Palestinians.

Al-Qaws. 2015. "Individuals, Bodies, and Sexuality: through a futuristic critique from the past of Palestinian leftist movements." *Jadal Journal* (24), October. https://mada-research.org/wp-content/uploads/2015/11/JDL24-Full.pdf.

Ash, Lucy. 2020. "Inside Poland's LGBT-free Zones." *BBC News*, September 20.

Atshan, Sa'ed. 2020. *Queer Palestine and the Empire of Critique*. Stanford, CA: Stanford University Press.

Atshan, Sa'ed, and Darnell L. Moore. 2014. "Reciprocal Solidarity: Where the Black and Palestinian Queer Struggles Meet." *Biography* 37 (2): 680–705.

Ayoub, Phillip M., and Agnès Chetaille. 2017. "Movement/Countermovement Interaction and Instrumental Framing in a Multi-Level World: Rooting Polish Lesbian and Gay Activism." *Social Movement Studies* 19 (1): 21–37.

Ayoub, Phillip M., and Douglas Page. 2019. "When Do Opponents of Gay Rights Mobilize?: Explaining Political Participation in Times of Backlash against Liberalism." *Political Research Quarterly* 73 (3):696–713.

Benford, Robert D., and Scott A. Hunt. 2003. "Interactional Dynamics in Public Problems Marketplaces: Movements and the Counterframing and Reframing of Public Problems." In *Challenges and Choices: Constructionist Perspectives on Social Problems*, edited by James A. Holstein and Gale Miller, 153–186. New York: Walter de Gruyter.

Benford, Robert D., and David A. Snow. 2000. "Framing Processes and Social Movements: An Overview and Assessment." *Annual Review of Sociology* 26 (1): 611–639.

Best, Joel. 2003. "But Seriously Folks: The Limitations of the Strict Constructionist Interpretation of Social Problems." In *Challenges and Choices: Constructionist Perspectives on Social Problems*, edited by James A. Holstein and Gale Miller, 51–69. New York: Walter de Gruyter

Bosia, Michael J. 2013. "Why States Act: Homophobia and Crisis." In *Global Homophobia: States, Movements, and the Politics of Oppression*, edited by Meredith L. Weiss and Michael J. Bosia, 30–54. Urbana: University of Illinois Press.

Bosia, Michael J., and Meredith L. Weiss. 2013. "Political Homophobia in Comparative Perspective." In *Global Homophobia: States, Movements, and the Politics of Oppression*, edited by Meredith L. Weiss and Michael J. Bosia, 1–29. Urbana: University of Illinois Press.

Canaday, Margot. 2009. *The Straight State: Sexuality and Citizenship in Twentieth-Century America*. Princeton, NJ: Princeton University Press.

Charles, Douglas. 2010. "From Subversion to Obscenity: The FBI's Investigations of the Early Homophile Movement in the United States, 1953–1958." *Journal of the History of Sexuality* 19 (2): 262–287.

Charles, Douglas. 2015. *Hoover's War on Gays: Exposing the FBI's" Sex Deviates" Program*. Lawrence: University Press of Kansas.

Cutler, Hugh. 1973. "Gay Activist Gets Biden's Gut Reaction." *Morning News*, September 25. https://freebeacon.com/wp-content/uploads/2019/04/The_Morning_News_Tue__Sep_25__1973_.pdf.

D'Emilio, John. 2012. *Sexual Politics, Sexual Communities: The Making of a Homosexual Minority in the United States 1940–1970*. Chicago: University of Chicago Press.

Epstein, Steven. 1998. "Gay and Lesbian Movements in the United States: Dilemmas of Identity, Diversity, and Political Strategy." In *The Global Emergence of Gay and Lesbian Politics*, edited by Barry Adam, Jan W. Duyvendak, and Andrew Krouwel, 30–90. Philadelphia: Temple University Press.

Faderman, Lillian. 2015. *The Gay Revolution: The Story of The Struggle*. New York: Simon & Schuster.

Fisher, Max. 2014. "Report: President of Uganda Says Homosexuality Is Caused by 'Random Breeding' in Western Countries." *Washington Post*, January 17. https://www.washingtonpost.com/news/worldviews/wp/2014/01/17/report-president-of-uganda-says-homosexuality-is-caused-by-random-breeding-in-western-countries/.

Gamson, William A., and Andre Modigliani. 1989. "Media Discourse and Public Opinion on Nuclear Power: A Constructionist Approach." *American Journal of Sociology* 95 (1): 1–37.

Goffman, Erving. 1974. *Frame Analysis: An Essay on the Organization of Experience*. Cambridge, MA: Harvard University Press.

Hewitt, Lyndi, and Holly J. Mccammon. 2005. "Explaining Suffrage Mobilization: Balance, Neutralization, and Range in Collective Action Frames." In *Frames of Protest: Social Movements and the Framing Perspective*, edited by Hank Johnston and John A. Noakes, 33–52. Lanham, MD: Rowman & Littlefield.

Hunt, Scott A., Robert D. Benford, and David A. Snow. 1994. "Identity Fields: Framing Processes and the Social Construction of Movement Identities." In *New Social Movements: From Ideology to Identity*, edited by Enrique Laraña, Hank Johnston, and Joseph R Gusfield, 185–208. Philadelphia: Temple University Press.

Johnson, David K. 2004. *The Lavender Scare: The Cold War Persecution of Gays and Lesbians in the Federal Government*. Chicago: University of Chicago Press.

Korycki, Katarzyna, and Abouzar Nasirzadeh. 2013. "Homophobia as a Tool of Statecraft: Iran and Its Queers." In *Global Homophobia: States, Movements, and the Politics of Oppression*, edited by Meredith L. Weiss and Michael J. Bosia, 174–195. Urbana: University of Illinois Press.

Lindemann, Hilde. 2001. *Damaged Identities, Narrative Repair*. Ithaca, NY: Cornell University Press.

Meeker, Martin. 2001. "Behind the Mask of Respectability: Reconsidering the Mattachine Society and Male Homophile Practice, 1950s and 1960s." *Journal of the History of Sexuality* 10 (1): 78–116.

Milo, Sage. 2017. "'But Oh! What Tales': Portraying the Middle East in US Homophile Periodicals of the 1950s and 1960s." *Journal of Homosexuality* 64 (7): 889–907.

Moussawi, Ghassan. 2015. "(Un)Critically Queer Organizing: Towards a More Complex Analysis of LGBTQ Organizing in Lebanon." *Sexualities* 18 (5/6): 593–617.

Polletta, Francesca, and James Jasper. 2001. "Collective Identity and Social Movements." *Annual Review of Sociology* 27: 283–305.

Puar, Jasbir. 2007. *Terrorist Assemblages: Homonationalism in Queer Times*. Durham, NC: Duke University Press.

Ray, Raka. 1998. "Women's Movements and Political Fields: A Comparison of Two Indian Cities." *Social Problems* 45 (1): 21–36.

Ritchie, Jason. 2015. "Pinkwashing, Homonationalism, and Israel-Palestine: The Conceits of Queer Theory and the Politics of the Ordinary." *Antipode* 47 (3): 616–634.

Rucht, Dieter. 1996. "The Impact of National Contexts on Social Movement Structures: A Cross-Movement and Cross-National Comparison." In *Comparative Perspectives on Social Movements*, edited by Doug McAdam, John D. McCarthy, and Mayer N. Zald, 185–204. Cambridge, MA: Cambridge University Press.

Rudoren, Jodi. 2014. "Veterans of Elite Israeli Unit Refuse Reserve Duty, Citing Treatment of Palestinians." *New York Times*, September 12. https://www.nytimes.com/2014/09/13/world/middleeast/elite-israeli-officers-decry-treatment-of-palestinians.html.

Schulman, Sarah. 2011. "Israel and 'Pinkwashing.'" *New York Times*, November 22. https://www.nytimes.com/2011/11/23/opinion/pinkwashing-and-israels-use-of-gays-as-a-messaging-tool.html.

Serykh, Dasha. 2017. "Homonationalism before Homonationalism: Representations of Russia, Eastern Europe, and the Soviet Union in the US Homophile Press, 1953–1964." *Journal of Homosexuality* 64 (7): 908–927.

Snow, David A., and Robert D. Benford. 1988. "Ideology, Frame Resonance, and Participant Mobilization." In *From Structure to Action: Comparing Social Movement Research Across Cultures*, edited by Bert Klandermans, Hanspeter Kriesi, and Sidney Tarrow, 197–217. Greenwich, CT: Jai Press.

Snow, David A., E. Burke Rochford Jr., Steven K. Worden, and Robert D. Benford. 1986. "Frame Alignment Processes, Micromobilization, and Movement Participation." *American Sociological Review* 51 (4): 464–481.

Stulberg, Lisa. 2018. *LGBTQ Social Movements*. New York: John Wiley & Sons.

Swidler, Ann. 2000. "Cultural Power and Social Movements." In *Culture and Politics: A Reader*, edited by Lane Crothers and Charles Lockhart, 269–283. New York: Palgrave Macmillan.

Taylor, Henry. 1969. "Henry J. Taylor Says Homosexual, Security Risk." *Kingston Daily Freeman*, May 24.

Van Gorp, Baldwin. 2007. "The Constructionist Approach to Framing: Bringing Culture Back In." *Journal of Communication* 57: 60–78.

Worsnop, Richard. 1963. "Homosexuals Vs. Security Becomes Issue in Britain: Can Morals Be Official Concern?" *The Courier* (Waterloo, Iowa), July 15.

# 10

# When Fifth Columns Fall

## Religious Groups and Loyalty-Signaling in Erdoğan's Turkey

*Kristin E. Fabbe and Efe Murat Balıkçıoğlu*

On July 15, 2016, a coup attempt on Recep Tayyip Erdoğan's Justice and Development Party (AKP) government left hundreds dead and the Turkish public stunned. After a chaotic and bloody night, the putschists had been defeated. Erdoğan was visibly shaken but still in power, as evidenced by a bizarre instance of political theater in which he phoned a news anchor on *CNN Turk*, live, using FaceTime, to assert his authority and urge people to resist the coup plotters. On air, he alleged that the night's violent events had been instigated by a dangerous "parallel state" operating within Turkey—an oblique reference to an influential religious group that followed the teachings of Fethullah Gülen, a Turkish imam self-exiled and living in Pennsylvania since 1999.

Known to its followers as Hizmet (Service), and often referred to simply as the Gülen movement or Gülenists, the group had been gaining influence and prominence in Turkish politics for decades. Ironically, the movement's supporters had originally helped bring Erdoğan to office. Furthermore, prosecutors with ties to the group had aided Erdoğan's consolidation of power through a series of political show trials starting in 2008 designed to silence AKP opponents. As the events of July 15 made abundantly clear, however, relations between the AKP and Gülenists eventually soured to the point of violence. According to recent Turkish government discourse, the Gülenists had come to represent a dangerous subversive fifth column in Turkish politics. Erdoğan lambasted the group as "traitors" and "heretics," alleging that they had nefariously infiltrated the Turkish bureaucracy and created a dangerous "parallel state." In the year before the failed coup, Erdoğan had designated the Gülenists as a terrorist organization, rebranding the group as FETÖ (short for Fethullah Terrorist Organization).

In the wake of the failed coup, Erdoğan's government directly accused Gülen-affiliated army officers, possibly with the backing of the United States, of trying to assassinate him and topple his regime (*The Economist* 2016; *Yeni*

Kristin E. Fabbe and Efe Murat Balıkçıoğlu, *When Fifth Columns Fall* In: *Enemies Within*. Edited by: Harris Mylonas and Scott Radnitz, Oxford University Press. © Oxford University Press 2022. DOI: 10.1093/oso/9780197627938.003.0011

*Şafak* 2016a).[1] He vowed to cleanse Turkey of the Gülen "virus," describing the group as "like a cancer" that needed to be eradicated from the body politic (Nakhoul and Yackley 2016). Above all, Erdoğan moved swiftly to attack the alleged fifth column and eradicate Gülenist influence from Turkish politics. A state of emergency law (Olağanüstü Hal or OHAL) was enacted across the entire country, commencing the largest witch-hunt and purge in Turkish history. Nearly half a million alleged Gülenists (along with other opponents) were removed from the bureaucracy, as well as government and state offices (Kirby 2016; Hansen 2017).

This chapter examines the role that fifth-column claims play in authoritarian politics. Specifically, we document how escalating fifth-column claims against the Gülenists have transformed the relationship between the Turkish state and both official (state-sanctioned) and unofficial Islamic actors in Turkey since July 2016. As fifth-column discourse against the Gülenists reached a fever pitch, Turkey's religious bureaucracy, known as the Presidency for Religious Affairs (Diyanet İşleri Başkanlığı, hereafter the Diyanet), became increasingly politicized. The office and its officials were simultaneously elevated to a new position of authority *and* put under increasing scrutiny from the Turkish state. Furthermore, as Diyanet officials worked to vilify the Gülen movement and demonstrate state loyalty, they subjected an increasing number of religious groups to heightened state scrutiny and, in some instances, state repression.

Documenting this transformation in Turkey's religious politics sheds light on broader theoretical discussions highlighted in this volume about alleged ideological fifth columns in three ways. First, this chapter illustrates an important underlying consequence of fifth-column politics for authoritarian consolidation by documenting how regimes reconfigure *collusive* claims about fifth columns into *subversive* ones as they consolidate power. During the AKP's initial period of alliance with Gülenists, political opponents of the two groups used collusive claims suggesting that an elite cabal of political Islamists with backing from Western governments was engineering a takeover of Turkish politics. After the AKP broke with its Gülenist allies, however, it worked to downplay its own pact with the group and produced a new narrative based on subversive fifth-column claims. These subversive claims accused Gülenists of seeking to overthrow the government, carrying out acts of sabotage, and damaging the national interest from within, with external backing from the West—all while misleading the AKP about its original motives. The shift from *collusive* to *subversive* claims illustrates one way in which former allies are recast and discarded from emerging authoritarian coalitions as power becomes increasingly centralized under an autocratic leader.

Second, we explore how actors proximate to alleged fifth columns respond when authoritarian leaders work to shift the narrative about a vilified group from collusive claims into subversive claims. We find that transforming accusations against ideological fifth columns into subversive claims facilitates authoritarian consolidation as groups and individuals proximate to the target compete to demonstrate loyalty to the state—and escape its wrath. Specifically, Erdoğan's reframing of fifth-column allegations and discourse against Gülenists into subversive claims expanded the scope of *guilt-by-proximity* and gave rise to a *loyalty-signaling dynamic* within the Diyanet and broader community of political Islamists. In an effort to signal loyalty, the Diyanet's leadership and rank and file publicly defended Erdoğan's actions against Gülenists. They also pushed to expand state inquiry to target any and all other religious threats to Erdoğan's rule. Yet, as the Diyanet joined Erdoğan's crusade to eradicate the alleged fifth-column threat posed by Gülenists, it simultaneously exposed itself to heightened scrutiny, perpetuated an increasing number of fractures among Turkish political Islamists, and elicited new forms of state intervention against other Sunni religious brotherhoods (*cemaat*s) and traditional orders (*tarikat*s) operating in Turkey, further consolidating Erdoğan's authority. Ultimately, the Diyanet leadership's efforts to signal loyalty by uncovering other religious threats to the state have contributed to its own subjugation to Erdoğan's authoritarian politics.

Third, and relatedly, a focus on the shift from collusive to subversive fifth-column claims reveals the extreme form of political pragmatism often guiding autocratic rule. Although Erdoğan's regime has long been viewed through the prisms of desecularization, Islamic political revival, and neo-Ottoman ambition, these analytical frameworks obscure the fact that his political project is first and foremost about eliminating competing sources of domestic authority. As referenced in the Introduction to this volume, Erdoğan had been leveling fifth-column accusations against opponents for years. He targeted liberals, journalists, academics, and ethnic minorities—sometimes with the backing of his onetime Gülenist supporters—long before uncovering the alleged subversive Gülenist threat. Erdoğan's recent hostility toward the Gülen movement and other *cemaat*s and *tarikat*s shows that his efforts to root out "the enemy within" has now turned against other political Islamists, many of them once erstwhile regime allies. The upshot has been an increasingly "nativist" phase of Erdoğan's rule, in which many of the AKP's original Islamist supporters have been sidelined in favor of a group of right-wing nationalists set on confronting enemies at home and abroad.

The remainder of the chapter proceeds as follows. First, it briefly introduces *tarikat*s and *cemaat*s as religio-political actors in the Turkish context and documents the rise, transformation, and fall of the alleged Gülenist fifth column. Next, through a detailed investigation of how top Diyanet officials responded to the government's anti-Gülenist campaigns based on subversive claims, it examines what happens once an alleged fifth column falls. The chapter documents guilt-by-proximity and loyalty-signaling dynamics in detail, showing how Diyanet officials' efforts to defend themselves often involved exposing new potential religious threats to the state and inciting government crackdowns against them. The chapter closes by summarizing key insights about fifth-column claims and responses in an authoritarian context.

## *Tarikats* and *Cemaats* in Turkey: Background and Context

The post-2016 period was not the first time that *tarikats* and *cemaats* had come under state scrutiny in Turkey. One of the most radical reforms of modern Turkey's founding father, Mustafa Kemal Atatürk, was to abolish the dervish lodges (sing. *tekke* and *zaviye*) and tombs (sing. *türbe*) on November 30, 1925, under the Law number 677. This law was designed to squelch the power of the various religious organizations that had traditionally gathered around the lodges and tombs scattered across Turkey.[2] The law also made it impermissible to use titles such as *dede*, *şeyh*, and *emir*. As a result, most *tarikats* and *cemaats* effectively went underground or fled Turkey's borders. In 1949, two amendments to Article 1 of the Law on the Abolition of Lodges and Tombs added a mandatory sentence of six months of imprisonment or exile for those who claimed to be shaykhs, *babas*, and caliphs; but it also excluded from the law a number of significant tombs belonging to key historical figures (*T.C. Resmî Gazete* 7234 1949, 16383). This led to the reopening of the tombs of various Ottoman Sultans, princes, and bureaucrats, as well as thirteenth-century legendary folk figures and mystics such as Nasreddin Hoca, Rumi, and Hacı Bektaş Veli.

After the Democratic Party's (Demokrat Parti, DP) victory in the general elections in 1950, some religious orders moved out of the shadows and began to operate more openly. Later, after the 1980 coup, these groups were able to legalize themselves as associations (sing. *dernek*) and foundations (sing. *vakıf*) due to Prime Minister Turgut Özal's neo-liberal policies, which gave more autonomy and power to religious foundations as legal entities (Din İşleri Yüksek Kurulu 2019, 13–14). Starting in the 1990s, *cemaats* entered into a new phase, going corporate (sing. *şirket* or *holding*) and forming new networks of private colleges, Qur'anic schools, media outlets, and business associations, as well as increasingly occupying certain positions in government (Öğreten 2019). According to a recent report, around 2016 there were some 30 *tarikats* and 500 *cemaats* in Turkey with 2.5 million affiliates. Furthermore, by way of education, the *tarikats* exercise influence over more than 1 million students (Balcı 2018).

## The Rise and Fall of Fethullah Gülen's Alleged Fifth Column: From Collusive to Subversive Claims

It was in this context that an offshoot of Said Nursi's Nurcu movement calling itself Hizmet (Service) emerged around the popular preacher Fethullah Gülen and spread throughout Turkey in the 1970s and 1980s. Gülen worked as a Diyanet-appointed *imam* from the late 1950s to the 1980s, during which time he learned how to position himself with other state actors and avoid restrictions on his community of followers (Lord 2018: 221). Gülen's movement grew in popularity as a result of his oratory skills and writings, and recordings of his preaching began circulating widely in the 1980s. To further build its base, the movement typically established individual charitable foundations with no official link to one another, although they were all tacitly and ideologically connected to Gülen's teachings. In this manner, the group developed a formidable network of subsidized private educational institutions, preparatory schools, and dormitories, which filled widening gaps in the state's developmental capabilities (White 2002: 207). Gülenists also used charitable trusts, non-governmental organizations, firms, business associations, media outlets, and certain arms of the state bureaucracy to expand their influence and support (Lord 2018: 219–225).

Despite being avowedly apolitical, Gülen's group supported the AKP's rise to power and its first two terms in government, although rumors circulated that the alliance was opportunistic and purely strategic. Many notable AKP politicians visited Gülen's compound in Pennsylvania and Gülenist media outlets, such as *Zaman*, were overwhelmingly supportive of the AKP's policies.

Notably, many of the AKP's opponents distrusted Gülen's group profoundly, perhaps even more so than they distrusted the AKP itself. Such distrust had historical roots: Gülen had been arrested and released after the military coups in 1971 and 1980, and his relations with certain Kemalist elites were punctuated by tension. In the 1990s Kemalist columnists such as Hikmet Çetinkaya and Uğur Mumcu, who was assassinated in 1993,[3] wrote about Gülenist infiltration into the army and police forces in the daily *Cumhuriyet* (Alkaç and Ünlü 2017; Mumcu 2006). Distrust of the movement grew further, in no small part, because Gülen began permanently residing in the United States in 1999, thereafter managing his growing network of followers from abroad.[4] Gülen cites that he traveled to the United States to receive medical treatment for a heart condition. While abroad, Turkish prosecutors opened a case against him in absentia and eventually issued a warrant for his arrest for conspiring to overthrow the country's secular government—a crime punishable by death—thereby compelling Gülen to remain in exile in the United States indefinitely (Hendrick 2013; Frantz 2000).

By the early 2000s, expressions of distrust evolved to allegations of a collusive fifth- column alliance linking Gülen, the AKP, and foreign intelligence services.

For example, on the eve of his assassination in 2002, Necip Hablemitoğlu, a professor of the early Turkish Republic, was preparing a book entitled *Köstebek* ("Spy"), which details Gülen's infiltration of the Turkish state, as well as the group's ties to Mossad and the CIA. The book was published posthumously in 2003. Hikmet Çetinkaya also went on to publish several books between 2000 and 2009 detailing ties between Gülen, the CIA, and the AKP—leveling accusations of a collusive fifth-column alliance between the groups and facing a slew of lawsuits as a result (2000, 2007, 2009). Other journalists writing on the Gülenist's collusive ties in the 2010s, including Ahmet Şık and Nedim Şener, were imprisoned in 2011 at the initiative of powerful Gülenist supporters in the state bureaucracy (Lord 2018), who had helped launch a series of sham trials designed to silence the AKP's opposition that were largely applauded by the West.

By 2013, uneasiness and distrust had given way to outright hostility, and the Gülenists came to be seen as a fifth column by most elements in Turkish society, including the AKP, which now actively worked to downplay collusive claims made by the opposition by recasting the Gülenists as a subversive fifth-column threat. Whereas AKP opponents resented the group's influence in education and the bureaucracy, and were concerned about criminality and threats to secularism, AKP-supporters loyal to Erdoğan worried that the group had grown too powerful. The ultimate reasons behind the split between the Gülenists and the AKP are murky, but the division eventually ended in catastrophe and bloodshed. What began in 2013 as a seemingly innocuous disagreement over the AKP's moves to shut down a lucrative network of Gülenist-affiliated preparatory schools (*dershaneler*) quickly transformed into a fierce struggle among various branches of the Turkish state that threatened the country's stability. According to the Gülenists themselves, the movement broke with Erdoğan over his increasingly anti-democratic tendencies and particularly the violent suppression of the anti-government Gezi Park protests that took place in 2013. According to Ahmet Şık, however, the open antipathy between the AKP and the Gülenists had started when the Gülenists attempted to place their cadres in the National Intelligence Service. Realizing that the Gülenists had gained an uncontrollable degree of power within the state, Erdoğan broke his covenant with the movement and turned on it (Şık 2013). The Gülenists are said to have retaliated first by disclosing secret negotiations between the PKK and the Turkish government to the media, next by revealing corruption involving a gas-for-gold deal between Iran and Turkey (which breached US sanctions against the Iranian regime), and finally by launching the failed coup of 2016.

In an ironic and bizarre twist of fate, after the AKP-Gülenist split, writers that had previously exposed collusive claims linking the Gülenists, the AKP, and foreign actors were now targeted as Gülenist sympathizers by Erdoğan. For example, the aforementioned Hikmet Çetinkaya was taken to court by Erdoğan's prosecutors in 2017 for supporting FETÖ, presumably in an attempt to silence the Kemalist *Cumhuriyet* and to convert collusive fifth-column claims into a new subversive claim about Gülenist activities. Çetinkaya's defense testimony captured how he had become ensnared in Erdoğan's recasting of the Gülenists as a subversive threat simply by virtue of reporting on the collusive activities of the group for decades. He stated:

> Based on the views that I had expressed, Fethullah Gülen was prosecuted for establishing and running a criminal organization. Now through an indictment prepared by prosecutors who have forgotten about the past, I am being prosecuted on charges of aiding and abetting the FETÖ terrorist organization. (*Expression Interrupted*)

Similarly, Ahmet Şık was imprisoned again in 2016, this time over allegations of being a propogandist *for* the Gülen movement, despite having written for years about the dangers posed by the group at its collusive activities.

### When Fifth Columns Fall: The Diyanet, Mehmet Görmez, and a New Era of Loyalty-Signaling

In the immediate aftermath of the failed coup in the summer of 2016, as Erdoğan worked to reframe the Gülen movement as a subversive threat, official Islamic actors at the Diyanet became increasingly visible players in domestic politics. Both the head of the Diyanet, Mehmet Görmez, and government-related media agencies claimed that the Diyanet had played a key role in defeating the coup attempt. The Diyanet urged state preachers to broadcast *sela,* a prayer normally called from the minarets to commemorate the deceased or a significant event, all night during the coup attempt in an effort to bolster religious solidarity and encourage people to resist the putschists. According to an op-ed published in the state-run *Anadolu Ajansı* just a few months after the incident, the Diyanet, possibly for the first time in its history, took a stance against "coup ideology" (Okumuş 2016). Initially, then, the Diyanet and its leadership seemed to be vindicated by its activities during the coup attempt, at least in the eyes of the government.

Nonetheless, many of Mehmet Görmez's actions in the immediate aftermath of the failed coup can be also interpreted as defensive efforts to signal his loyalty to the regime and distance himself from any current and potential religious "enemies of the state" as the dragnet against the Gülenists widened to include proximate actors. As this section demonstrates, although Görmez endeavored to bolster his anti-Gülenist credentials through both action and rhetoric, his offices at the Diyanet became embroiled in the government purges and he himself became targeted as a potential Gülenist sympathizer. What is more, as Görmez and other official religious actors at the Diyanet worked to vilify the Gülen movement and demonstrate state loyalty, they exposed an increasing number of religious groups to heightened state scrutiny and, in some instances, state repression, thereby tightening Erdoğan's grip on power.

In the months following the coup attempt, Görmez's Diyanet engaged in acts of loyalty-signaling to support Erdoğan's newfangled claims of subversive fifth-column activity by the Gülenists. First, it organized a summit for *cemaat*s to inform them about the Gülenist "conspiracy" and the government's policies on the unfolding Gülenist purge. This summit marked the first time that the Diyanet had held an official meeting with various *tarikat*s and *cemaat*s. Representatives from thirty-two religious groups—including all substantial Nurcu groups other than the Gülenists, most branches of the powerful religious order Naqshbandiyya, such as the Süleymancılar, and the politically powerful İskenderpaşa which has had historical ties to former prime minister Erbakan's Millî Görüş movement—were in attendance. Many of these groups had previously been deemed to be unofficial (and maybe even illegal) from the state's perspective. The mere act of convening these groups caused controversy. Erdoğan's political opponents and a number of religious groups objected to the summit because the Diyanet refrained from inviting *tarikat*s and *cemaat*s known to be overtly critical of Erdoğan and the Diyanet's centralizing religious discourse (*Akit* 2016; *CNNTürk* 2017; *Sol Haber* 2017). The meeting demonstrated the fact that the Diyanet was signaling loyalty, in part, by taking on an enhanced role in convening state supporters and by monitoring any threats to the state from across the religious landscape of *cemaat*s and *tarikat*s, even beyond Gülenists (*T24* 2017).

Next, Görmez sought to bolster his anti-Gülenist credentials by commissioning a report about the Gülenist terror threat at the Diyanet. This report on the Gülenists, which was eventually published openly in July 2017, gave a scathing critique of the group and its efforts to undermine the Erdoğan regime (Din İşleri Yüksek Kurulu 2017). The report spans a period from the early 1970s into the early 2010s, and comprises recordings from different parts of Turkey, such as İzmir, Yozgat, Erzurum, İstanbul, and Ankara. Besides Gülenist political tactics, Görmez's report concerns problematic aspects of the group's ideology from the Diyanet's perspective. The religious accusations mentioned in the report included:

- Fethullah Gülen is seeking a synthesis between Islam and Christianity, which is described as a heinous scheme cloaked in the guise of interfaith dialogue. The report claims that the scheme was supported by the intelligence services of various Christian countries in an effort to destroy Islam in Turkey. The report also suggests that Gülen's interest in non-Muslim traditions was not restricted to Christianity and the second coming of Jesus, but also extended to Judaism, Ancient Greek mythology, antinomian forms of Islam such as Hurufism, and parapsychological trends in the West (56–57, 73–84, 82–83, 125–126).

- Fethullah Gülen and his followers are hiding their real religious motivations and agenda, that is, performing *takiyye*, according to which they might adopt behavior forbidden in Islam—such as drinking alcohol—so as not to be spotted when infiltrating a particular state agency or department (112, 88). In this regard, the report suggests that Gülenists engaged in systematic deception and the manipulation of religion to infiltrate state offices.

- Fethullah Gülen is exhibiting messianism and presenting himself as above the Prophet, by claiming the ability to have a direct relationship to God through divine inspiration and intuition. The report also writes that, in his sermons and recordings, Gülen intentionally uses pseudo-Sufi terms encrypted with secret messages for his disciples, which undermine God's divinity and Muhammad's prophethood (102, 119, 126, 134).

By the time Görmez's report on the Gülenists was published, the group had been silenced in Turkey as result of purges and crackdowns. Perhaps more surprisingly, however, Görmez and many of his employees also came under increasing scrutiny and pressure from the state around the time when the report was issued given their past proximity to group. For one, the Diyanet found itself caught up in the Gülenist purges. While the scope of the purge in the Diyanet was small compared to that in other state organs, thousands of religious civil servants working at the Diyanet were dismissed (*Anadolu Ajansı* 2016; *Yeni Şafak* 2016b; *T.C. Resmî Gazete* 30472 2018). Second, Görmez's assistant Mehmet Emin Özafşar was removed from his post with the signatures of President Erdoğan and Prime Minister Binali Yıldırım, which many interpreted as evidence of government pressure on Görmez to resign. Some media sources even targeted Görmez himself as a Gülen supporter, using Görmez's visit to Gülen's Pennsylvania home as proof of his ties to the group (*Karar* 2018a). This was made all the more confusing by the fact that Mehmet Görmez had previously been a target of Gülenist media outlets. For example, a prominent smear campaign launched by pro-Gülenist media some six months before the failed coup attempt accused Görmez of using religion to legitimize incest (*Sabah* 2016).

Despite going to great lengths to vilify the Gülenist movement and demonstrate his commitment to the Turkish state, Görmez increasingly fell out of the government's favor. After Erdoğan took an openly critical stance on his tenure and the Diyanet's incapacity to eliminate the Gülenist FETÖ threat (*Milliyet* 2017a), Görmez resigned from his office, despite having another three years to serve. Former president Abdullah Gül (*Hürriyet* 2017a) and former prime minister Ahmet Davutoğlu (*Hürriyet* 2017b) tweeted supportive messages, claiming that he was a man of merit with no ties to the FETÖ whatsoever. In his last sermon (*Diyanet.tv* 2017) and later speeches, Görmez continued to underline his constant fight against the Gülenists during his term.

Domestically, Görmez's fall from grace was highly publicized. Some observers surmised that he had probably become a target of the regime because he was particularly disliked by the populist, conservative Naqshbandis of the İsmailağa group that had long backed the AKP. Görmez was rumored to have acted against the İsmailağa's infiltration of the Diyanet cadres and did not favor their religious interpretations (*CNNTürk* 2018a; *Karar* 2017a; *Dini Haber* 2020). Perhaps this group wanted to take him out to make way for their own imams in state ranks? Others speculated that Görmez had become a target because of his openness toward Turkey's Kurds and his support of the AKP's ultimately failed Kurdish initiative (Oğur 2017). Perhaps Görmez was only now drawing criticism by virtue of Erdoğan's renewed alliance with right-wing nationalist groups like the Nationalist Movement Party (MHP), which objected to such reconciliation with Kurdish factions? Regardless, it was Görmez's past proximity to Gülenists that rendered him vulnerable.

Notably, in his farewell sermon, Görmez made a historically significant comment targeted at Erdoğan's government. He implied that the Turkish state must decide whether the Diyanet is merely a bureaucratic apparatus or whether it is a genuinely independent religious entity with merit above all political pressures (Çakır 2017). Clearly, Görmez had not been able to avoid such political pressures himself during his tenure in office while the alleged Gülenist fifth column was being felled. After functioning for nearly a century as a relatively unremarkable bureaucratic arm of the state, the Diyanet had been thrust into the political spotlight with the failed coup, the cascade of anti-Gülenist fifth-column claims, the purges, and then with Görmez's resignation. These schisms and accusations of disloyalty within the AKP's former support base of Islamists riled domestic politics, for they illustrated that even compliant regime loyalists, like Görmez, could be sidelined using the authoritarian logic of guilt by association. These schisms also showed how the evolution from collusive to subversive fifth-column accusations against Gülenists coincided with the Erdoğan regime's shifting toward a more nationalist phase of rule, one that was marked by reduced reliance on a number of informal Islamic organizations and groups that had once been its core supporters.

## From Behind the Shadow of an Alleged Fifth Column:
## New Threats, New Alliances

In September 2017, Ali Erbaş became the new head of the Diyanet. His first Friday sermon used fifth-column rhetoric, accentuating the perils of those who betrayed their own country by exploiting religion, specifically the concept of *Hegira* (The Migration, which refers to the journey of the Islamic prophet Muhammad and his followers from Mecca to Medina) (*Karar* 2017b). This appeared to be a direct jab at Gülen, who sometimes used the concept in his sermons to describe his own plight and that of his followers as exiles who had been persecuted by the Turkish state.[5]

This opening move proved indicative of what was to follow, as subversive fifth-column claims against Gülenists and loyalty-signaling further fueled authoritarian consolidation. The Diyanet worked to fully extinguish the fifth-column threat by confiscating various Gülenist dorms and turning them into Diyanet facilities (*Milliyet* 2017b). The bureaucracy organized mandatory workshops, seminars, and conferences to raise awareness about the religious threat at various universities and regional gatherings (*Diyanet Haber* 2018; *Karar* 2017c; *Karar* 2018b). It also published a legal opinion (*fetva*) and produced sermons against the Gülenists (*Karar* 2018c). Finally, the Diyanet oversaw campaigns through Turkey's Maarif Foundation (a state-run international education initiative), to wage a legal battle against hundreds of Gülenist schools abroad (Zorlu 2019; Güvendik 2019).

As the targeting of Gülenists and other religious opponents deepened and expanded with Erbaş at the Diyanet, the authoritarian logic of loyalty-signaling led men from informal religious orders to become more vocal about their political allegiances. One illustrative example is Professor Cevat Akşit, the head of the Erenköy group of the greater Naqshbandi-Khalidi branch, known for his close ties to the Millî Görüş movement, as well as the AKP. In 2018, Professor Akşit turned up at an official event to commemorate the martyrs of the 2016 coup attempt alongside President Erdoğan, the Diyanet head Ali Erbaş, the minister of defense, and the head of the Turkish armed forces. Professor Akşit, who was the only attendee at the event representing religious orders, was photographed praying alongside Erdoğan and Erbaş—notably squatting next to them in an inferior-looking position. The sociologist Tayfun Atay, writing for the daily *Cumhuriyet*, interpreted the photo as symbolic of the trend of unofficial religious groups succumbing to state authority under Erdoğan and the Diyanet (2018).

Soon after the public appearance in support of the government, Akşit began casting allegations of guilt by proximity against other religious groups in a widely publicized and controversial interview. He argued that some religious orders, such as the Süleymancılar and the Menzilciler, had the potential to become the new FETÖ since they both had managed to infiltrate various state institutions and could exercise great power over youth through their respective networks of religious schools, organizations, and dormitories. These groups lacked the Gülenists' ties to the United States, but they were targeted nonetheless given their perceived similarities with and proximity to the group. Interestingly, Akşit also made some backhanded comments about the Diyanet itself, stating that he had personally declined the opportunity to become the head of the Diyanet twice because of the office's politically controversial history (*İnternet Haber* 2018). Within a year of making these comments about the Diyanet, a leading Turkish discount retail chain with known ties to Professor Akşit's Erenköy group was facing pressure from the regime for exercising a price monopoly—pressure that was interpreted by the keen observer of Turkish politics Ruşen Çakır as evidence that even the AKP's religious allies were becoming targets of Erdoğan's increasingly authoritarian government (Çakır 2019). No amount of loyalty-signaling, it seemed, could shield religious actors from the pragmatism of authoritarian rule.

While the indirect pressure on the Akşit's Erenköy group did not escalate much further, Erdoğan and the Diyanet went after other prominent religious figures. Somehow, these figures had not been questioned previously by the AKP. But now, after the fall of the Gülenist fifth column, they became more vulnerable and exposed because their charisma, popularity, and ability to wield alternative interpretations of religious authority resembled that of Gülen. One example in this regard was Nurettin Yıldız, a controversial "YouTube imam" with a sizable following and founder of the Sosyal Doku Foundation. Yıldız gained prominence for his controversial religious prerogatives, which covered a wide variety of topics including the possibility of marrying seven-year-old girls, the "lustful" practice of consuming ketchup, and the use of elevators by unmarried people. In March 2018, Erdoğan criticized him openly, requesting that the Diyanet cease Yıldız's operations and annul his legal opinions (*CNNTürk* 2018b). After Erdoğan's speech, an MP from the main opposition party CHP (Republican People's Party), on the contrary, claimed that Yıldız had long-been protected by the AKP, calling him the AKP's "new Gülen" (*Cumhuriyet* 2018).

A second example involves the bizarre Muslim televangelist Adnan Oktar. Known for his messianic claims and televised shenanigans involving an audience of female followers he called "kittens," Oktar had long garnered media attention and criticism for his lifestyle, his self-proclaimed cult of personality, and his provocative remarks. He also appeared to enjoy AKP protection for many years. In January 2018, however, Erbaş made a speech condemning Oktar as immoral and accusing him of exploiting religion (*Karar* 2018d). In the following months, Oktar and his cult became an open target of the state. Erdoğan, the Diyanet, and the Diyanet syndicate *Diyanet-Sen* all took aim at Oktar (*Karar* 2018e; *Hürriyet* 2018). It is worth noting that Oktar's antics were hardly new: he had been operating his cult for some thirty-five years with considerable political immunity due to his early financial support for Islamist parties. Oktar's downfall suggested that Erdoğan no longer needed his support, as the regime was increasingly seeking legitimacy and power in coalition with right-wing, nationalist elements in the Turkish government. Instead, Oktar would be branded as a charlatan and potential threat. Like other religious groups that failed to partake in obsequious loyalty-signaling to the regime, his fate would be turned into a cautionary tale used to intimidate other pseudo-religious cults and figures that could potentially threaten Erdoğan's claim to ultimate authority through religion (Çakır 2018).

## Naming Names: A Secret Report Surfaces . . .

It was in this context of regime crackdowns on other religious actors that a report on "Religio-Social Organizations, Traditional Religio-Cultural Formations, and New Tendencies in Religion" landed with a splash in March 2019, mysteriously released online and circulated through various Internet channels in Turkey. The body of the report described all prevalent Sunni religious brotherhoods (*cemaat*s) and traditional orders (*tarikat*s) operating in Turkey, and listed the names of influential preachers with a considerable following. The report's cover stated that it was the work of Turkey's High Council for Religious Affairs (Din İşleri Yüksek Kurulu), acting under the Diyanet.

A brief preface to the report purported to explain the document's origins: By the end of 2016, the drafters of the aforementioned Görmez report on the Gülenists decided to extend the scope of their research to include all other organized religious groups operating in Turkey. The drafters believed it was important to effectively monitor the religious, political, and economic activities of informal religious communities and deliver intelligence on them. Such monitoring, the report argued, was needed to prevent such groups from manipulating religion for their own political purposes, and could ultimately help prevent future coup attempts like the one attempted by the Gülenists (Din İşleri Yüksek Kurulu 2019: 7–8).

The "Religio-Social Organizations" report seems to have been deliberately leaked, since it was downloadable from a number of prominent media outlets. Notably, other Diyanet reports with similar content—such as the Diyanet's aforementioned Gülen report and its ISIS report—also had been announced in the press and made publicly available.[6] Furthermore, the mysterious report was designed to appear as a supplement to the Gülen report, using the same formatting, font, chapter headings, and evaluation criteria. The leak of the report therefore signaled that, going forward, no unofficial religious organizations in Turkey—even AKP allies—would escape the Diyanet's and the state's gaze.

As for its contents, the report "named names." It listed a total of forty-nine religio-social groups operating in Turkey, dividing them into eight categories: "Islam of the Qur'an," "The Salafis," "Messianism," "Traditionalists," "Religio-Political Organizations," "Risale-i Nur Groups" (i.e., the Nurcu groups besides the Gülenists), "Traditional Religio-Cultural Formations (*Tarikat*s)," and "Others." In addition to the categorization, the report included patchy and uneven information about each religious group's history, members' biographies, prominent views and activities, as well as an evaluation section with a summary of each group's religio-political tendencies and deviant views (if any).

The report concluded that all political, social, and economic activities of religio-social formations and organizations—whether religious orders (*tarikat*s), groups (*cemaat*s), or personality cults—should be regularly monitored by the Turkish state to prevent new Gülenists from emerging. Crucially, then, the report used the regime's allegations of subversive fifth-column activities by Gülenists as a justification for suppressing other groups preemptively. The report referred to a number of groups with foreign connections, such as Hizbu't-tahrir with ties to Jordan, Turkish Hizbullah/Mustazaflar Hareketi with ties to Iran, Davet ve Kardeşlik Vakfı with links to Egypt's Muslim Brotherhood, and the pro-Barzani Nurcu group Med-Zehra. It also drew attention to certain Salafi preachers, such as Ebu Hanzala and Abdurrahman Kuytul, who were accused of having links to al-Qaeda, the Islamic State, and other Salafi groups outside Turkey—ties that they both deny (Din İşleri Yüksek Kurulu 2019: 74, 79, 117–118, 122, 133; Eroğlu 2017, 2018; Çakır 2017).

Notably, the report also revealed a strategy of "fifth-columning by association"—singling out a handful of other groups as obviously dangerous state threats by implying links to or similarities with the alleged Gülenist fifth column. For example, the Süleymancılar,[7] a Naqshbandi *tarikat* with historic ties to Turkish Islamist networks both in Turkey and Germany, were openly compared to FETÖ (Din İşleri Yüksek Kurulu 2019: 194). The section on Süleymancılar was the longest in the report and included numerous details about the group's controversial history, political entanglements with right-wing politics and Turkey's National Intelligence Organization (Millî İstihbarat Teşkilatı, MİT), religious views and activities, as well as its pervasive networks of Qur'anic schools and dormitories, which indirectly likened them to Gülenists (Din İşleri Yüksek Kurulu 2019: 188–195). The Yeni Asya group, a Nurcu offshoot, was also described as potentially having ties to FETÖ. And the Salafist Alparslan Kuytul and his Furkan group were alleged to have a pro-coup and pro-FETÖ mentality. The report also made ambiguous remarks to suggest that the Menzilciler, another Naqshbandi *tarikat*, had the capacity to become a threat because they enjoyed significant popular support through their network of religious schools, educational institutions, and social organizations.

The ultimate message of the report was that no group would escape the new wave of heightened scrutiny by the state. Indeed, based on our analysis and that of others, at least one organization in every category listed in the report faced arrests or significant media slander around the time the report was released, deeming them all potentially guilty by association (Ağcakulu 2019). It was therefore unsurprising that the report recommended that government departments, such as Turkey's Ministry of Education (Milli Eğitim Bakanlığı), the Council for Higher Education (Yükseköğretim Kurulu, YÖK), and the Diyanet, should be utilized to check religio-social formations' activities. It also advised monitoring the content of their rhetoric and pronouncements in order to prevent dangerous innovations (*bidat*) and misbeliefs (*hurafe*) in religion that conflicted with the Diyanet's discourse (Din İşleri Yüksek Kurulu 2019: 224–225).

Although the report was supposedly drafted on Mehmet Görmez's watch as head of the Diyanet, he long remained silent on the report's origins and eventually disavowed any involvement in its drafting. Ali Erbaş, Görmez's successor at the Diyanet, also declined to comment on the report—neither rejecting nor accepting it as the work of the Diyanet—though he repeatedly capitalized on the report's contents to solidify regime authority (Şahin 2019). In the months following the release of the report, the Diyanet for the first time in its history signed a protocol concerning Qur'an reciters with the Ministry of Education, thereby extending its influence and education facilities into Qur'anic teaching, a domain that had previously been dominated by informal religious groups (*Diyanet Haber* 2019). The Diyanet also banned Süleymancı activity that lacked Diyanet permission, organizing a gendarmerie raid on a Süleymancı fair and prohibiting the sacrifice of animals and the sale their skins for charity during the religious holiday "Festival of Sacrifice" (*Elbistan Pusula* 2019; *Veryansın TV* 2019a). In December 2019, the Diyanet went a step further, publishing an official directive prohibiting all religious groups from opening Qur'anic schools and dormitories without Diyanet authorization and approval in order to prevent religious exploitation, brainwashing, and unearned income. The move was widely interpreted as a blow to the Süleymancı and as evidence of state efforts to silence potential religious competitors to state authority on the basis of the report's contents (*Anadolu Ajansı* 2019; *Kronos* 2019; *El-Aziz* 2019; *Independent Türkçe* 2019).

Perhaps more important for the purposes of this chapter, the intentional ambiguity surrounding the report's origins and authors suggested that the regime was using informal leaks as a strategy for raising fifth-column claims. Many of the groups listed in the report had tacitly supported the AKP or at least been shielded by it during its first decade in power. Informal leaks thus absolved the government of having to answer difficult questions about past cooperation with these newly identified subversive forces. Leaking such accusations without clear attribution, as opposed to making them directly, thus served the underlying authoritarian logic for centralizing power: accrue and then discard coalition partners—all while maintaining a façade of plausible deniability.

Political wrangling about who penned the report eventually found its way to the Turkish parliament. When Sezgin Tanrıkulu, an MP from the main opposition party (CHP), demanded answers, Turkish vice president Fuat Oktay officially denied that the report was ever prepared or released by the Diyanet (*Gazete Duvar* 2019). The report also caused significant controversy in the media, and the myriad explanations for its origins illustrated the climate of confusion, suspicion, and conspiracy that had come to grip the Turkish state as a result of fifth-column politics. Some writers saw the report as representative of a long-standing state tradition of repressing religion, something that harkened back to the Kemalist ideology of the early Turkish Republic (Alpman 2019).[8] Although the Kemalist old guard no longer dominated Turkish politics, was this their attempt at revenge? Other columnists argued that the report's use of various labels to describe religious groups, such as "pro-state" (*devlet yanlısı*) and "national" (*millî*), belied the fact that it was a product of Erdoğan's new presidential regime, under which there was now only one acceptable *tarikat*: Erdoğanism (Çelik 2019). The timing of the report was deemed especially important: it was released only a few months before the contentious 2019 municipal elections. Perhaps it represented an attempt to intimidate those *cemaat*s and *tarikat*s opposed to the AKP?

The media outlet *Yeni Asya*, which is owned by one of the religious groups listed in the report, had a slightly different take. A writer at the outlet speculated that the report was an "inside job" linked to a shadowy but influential advisor to Erdoğan named Doğu Perinçek. Perinçek, a neo-Nationalist with decades-long legal and illegal ties to the Turkish state, had long been hostile to religion (Güleçyüz 2019a, 2019b, 2019c).[9] Such speculation seemed to be corroborated by the fact that Perinçek's news organ *Aydınlık* enthusiastically prepared lengthy commentary about the report on its website, and later published the report as a book.[10] The Diyanet applied to Turkish courts to have the book removed from circulation, but failed (*Veryansın TV* 2019b). Perhaps Perinçek had a nefarious agenda of his own. Was he seeking to disempower both informal religious organizations and the Diyanet in an effort to strengthen the new political "*yerli ve millî*" (local and national) coalition between Erdoğan's AKP and Bahçeli's Nationalist Movement Party (MHP) that had helped secure Erdoğan's presidential election in 2018? The increased scrutiny on informal religious organizations indeed signaled the dominance of the "*yerli ve millî*" coalition—and thus a nativist and nationalist turn in Turkish politics—though the Diyanet continued to toe the regime line. Perhaps more important, however, these debates revealed that after years of fifth-column accusations being leveled by the regime, few political actors—even compliant loyalists at the Diyanet and Erdoğan's key advisors—could avoid the paranoia, suspicion, and allegations of ulterior motives that had come to dominate the Turkish scene.

## Conclusion

We may never know the true origins of the report on "Religio-Social Organizations, Traditional Religio-Cultural Formations, and New Tendencies in Religion" or its intended purpose. Nonetheless, the report's very existence indicates that the fall of one alleged religious fifth column—in this case the Gülenists—created a domino effect spilling across Turkey's religio-social spaces. Interestingly, the pressures on Islamic groups documented in this chapter apply to many of the AKP's traditional allies and tacit supporters, thereby illustrating how nativist and nationalist elements within Erdoğan's ruling coalition triumphed at the expense of other ideological currents—whether democratic, liberal, or Islamist. In part, this move was achieved by recasting the collusive fifth-column claims made against the AKP during its alliance with the Gülenists as subversive claims against the Gülenists themselves and those groups proximate to them.

The leak of the report, its convoluted origins, the competing theories surrounding its release, and the fractures that it caused in Turkey reveal a great deal about how authoritarian politics operate in practice. First, authoritarians use fifth-column claims, that is, the threat of an enemy within with external backing, as an excuse to crackdown more broadly on all forms of possible dissent that are proximate and similar. Second, elites that are "close" to the alleged fifth column often facilitate authoritarian consolidation by engaging in defensive loyalty-signaling. That is to say, as elites seek to distance themselves from the accused and proclaim their loyalty to the state, they not only bolster the regime directly but also force others to take sides or risk being accused themselves. Third, over the long term, autocrats govern using extreme forms of political pragmatism, characterized by cycles of coalition building and coalition busting that reconfigure collusive fifth-column claims into subversive ones. That is why, in the topsy-turvy world of autocratic governance—whether in Turkey or elsewhere—today's regime ally is likely to be tomorrow's enemy within.

## Notes

1. It should be mentioned that allegations about possible Gülen-CIA ties had long circulated in the in Turkish news media, starting in the 1980s. Furthermore, these

allegations were also made by members of the opposition. For example, a recent brochure published by the CHP focuses on the Erdoğan-Gülen alliance and includes the first page of the National Intelligence Services' 1991 report on Gülen's possible ties to the CIA (see *21 Soruda FETÖ'nün Siyasi Ayağı* (July 2020), 17. https://chp.azureedge. net/f1f1653063fd405cb411cf94c7f91edc.pdf. The report was banned in October 2020 by the current Turkish government. We elaborate below.

2. See the original document in the official website of the Turkish state: https://www. mevzuat.gov.tr/MevzuatMetin/1.3.677.pdf.

3. A recent confessional video released by Sedat Peker, a controversial mafia leader turned whistleblower who fled Turkey and now broadcasts about government scandals and crimes from YouTube, contains allegations that Mumcu's murder was instigated by the Turkish deep state and the minister of the interior. See https://www.youtube. com/watch?v=7ivcvcWmOPI&ab_channel=RE%C4%B0SSEDATPEKER.

4. Sources, both Gülen-affiliated and otherwise, have reported different timelines regarding his arrival and permanent residence in the United States. Others have cited the group's intentional use of "strategic-ambiguity." See Hendrick, Joshua, *Gülen: the ambiguous politics of market Islam in Turkey and the world* (New York University Press, 2013).

5. For examples of Gülenist sermons on this topic, see http://www.herkul.org/tag/cebri-hicret/.

6. *Anadolu Ajansı* and other news cites removed the PDF from their website but the full content of the report is still available at https://medyascope.tv/2019/05/31/turkiyed eki-dini-sosyal-tesekkuller-geleneksel-dini-kulturel-olusumlar-ve-yeni-dini-akim lar-baslikli-raporun-tam-metni/. The other Diyanet reports, such as the Gülen and the IS reports, are also still available. For the Diyanet's Gülen report see https://www. aa.com.tr/uploads/TempUserFiles/haber/2017/07/KENDI-DILINDEN-FETO-2017 0725son.pdf. For the IS report, see https://www.aa.com.tr/tr/turkiye/diyanetten-deas-raporu/665270.

7. For more on the group see: Ruşen Çakır, "Süleymancılık nedir? Ne değildir?," *Medyascope* (November 30, 2016), https://medyascope.tv/2016/11/30/suleym ancilik-nedir-ne-degildir/; and İsmail Saymaz, "Kurslar ve yurtlar imparatorluğu Süleymancılar," *Hürriyet* (December 11, 2016). https://www.hurriyet.com.tr/gundem/ kurslar-ve-yurtlar-imparatorlugu-suleymancilar-40303192.

8. For a similar view from the opposite political spectrum, see the Salafi preacher Ebu Hanzala's angry response video published on his YouTube channel *Tevhid Dersleri* under the title "Diyanetin cemaatler hakkında hazırladıkları raporu nasıl değerlendiriyorsunuz?" *Tevhid Dersleri* (September 5, 2019). https://www.youtube. com/watch?v=DEZuIhagtRg.

9. On Perinçek's nebulous and winding career see Rüşen Çakır, "Doğu Perinçek 60'lu yıllardan bu yana doğru zamanlarda hep yanlış tercih yapmış bir siyasetçidir," *Medyascope* (October 19, 2017). https://medyascope.tv/2017/10/19/rusen-cakir-dogu-perincek-60li-yillardan-bu-yana-dogru-zamanlarda-hep-yanlis-tercih-yapmis-bir-siyasetcidir/.

10. See the statist-nationalist *Aydınlık*'s August 2019 article series on "Diyanet İşleri Başkanlığı'nın gizli tarikat raporu 1–4" that supports the state's move to monitor them:    https://www.aydinlik.com.tr/diyanet-isleri-baskanligi-nin-gizli-tarikat-rap oru-2-liderleri-ve-one-cikan-faaliyetleri-ozgurluk-meydani-agustos-2019-2.

# References

Ağcakulu, Ali. 2019. "Diyanet'in gizli raporu ve cemaatleri bitirme planı." *Ahval News*, May 23. https://ahvalnews.com/tr/diyanet/diyanetin-gizli-raporu-ve-cemaatleri-biti rme-plani.

*Akit*. 2016. "Diyanet'in 'Cemaat Zirvesi'ne katılacak temsilciler belli oldu." *Akit*, September 25. https://www.yeniakit.com.tr/haber/diyanetin-cemaat-zirvesine-katila cak-temsilciler-belli-oldu-215547.html.

Alkaç, Fırat, and Bahar Ünlü. 2017. "FETÖ'yü yazdım inandıramadım." *Hürriyet*, July 27. https://www.hurriyet.com.tr/gundem/fetoyu-yazdim-inandiramadim-40533126.

Alpman, Polat S. 2019. "Diyanet'in Raporu: Dinime Dahleden Bari Müslüman Olsa." *Birikim*, June 6. https://www.birikimdergisi.com/haftalik/9533/diyanetin-raporu-din ime-dahleden-bari-muselman-olsa.

*Anadolu Ajansı*. 2016. "Yeni KHK'lar Yayınlandı." *Anadolu Ajansı*, October 29. https:// www.aa.com.tr/tr/turkiye/yeni-khklar-yayinlandi/674800.

*Anadolu Ajansı*. 2019. "Diyanet İşleri Başkanlığından Kur'an kursları ve yurtlara ilişkin yönerge." *Anadolu Ajansı*, December 20. https://www.aa.com.tr/tr/turkiye/diyanet-isl eri-baskanligindan-kuran-kurslari-ve-yurtlara-iliskin-yonerge/1679742.

Atay, Tayfun. 2018. "Meşihat Makamı." *Cumhuriyet*, July 18. https://www.cumhuriyet. com.tr/yazarlar/tayfun-atay/mesihat-makami-1029910.

Balcı, Esengül. 2018. "Eğitim'de Tarikat ve Medrese Gerçeği: 1 Milyon Öğrenci Tarikatlerin Elinde." Unpublished report. https://yadi.sk/i/bEVcDjdTbSCBzg.

Çakır, Elif. 2017. "Diyanet İşleri bağımsız ve tarafsız bir yapıya kavuşturulmalı!" *Karar*, August 8. https://www2.karar.com/yazarlar/elif-cakir/diyanet-isleri-bagimsiz-ve-taraf siz-bir-yapiya-kavusturulmali-4642.

Çakır, Ruşen. 2017. "Furkan Vakfı Olayı." *Medyascope*, April 24. http://rusencakir.com/ Furkan-Vakfi-olayi/6448.

Çakır, Ruşen. 2018. "Benim tanıdığım Adnan Oktar." *Medyascope.tv*, July 11. http://rus encakir.com/Tanidigim-Adnan-Oktar/6780.

Çakır, Ruşen. 2019. "BİM Olayı." *Medyascope*, March 11. https://medyascope.tv/2019/03/ 11/bim-olayi/.

Çelik, Özlem Akarsu. 2019. "'Dini Oluşumlar' Raporu'nu Diyanet kime yazdı?" *Gazete Duvar*, May 31. https://www.gazeteduvar.com.tr/yazarlar/2019/05/31/dini-olusumlar-raporunu-diyanet-kime-yazdi/.

Çetinkaya, Hikmet. 2000. *Fethullah Gülen'in 40 Yıllık Serüveni*. Istanbul: Günizi.

Çetinkaya, Hikmet. 2007. *Fethullah Gülen, ABD ve AKP*. Istanbul: Günizi.

Çetinkaya, Hikmet. 2009. *Amerikan Mızıkacıları*. Istanbul: Cumhuriyet.

*CNNTürk*. 2017. "Diyanet 30 tarikatlarla görüştü: 'Cemaatler ve Tarikatlar Buluşması düzenleyecek." *CNNTürk*, January 5. https://www.cnnturk.com/turkiye/diyanet-30-tarikatlarla-gorustu-cemaatler-ve-tarikatlar-bulusmasi-duzenleyecek.

*CNNTürk*. 2018a. "Mehmet Görmez ne 'Ulu Hakan'dı ne 'Kızıl Sultan.'" *CNNTürk*, November 12. https://www.cnnturk.com/turkiye/mehmet-gormez-ne-ulu-hakandi-ne-kizil-sultan.

*CNNTürk*. 2018b. "Cumhurbaşkanı Erdoğan'dan Nurettin Yıldız'a tepki." *CNNTürk*, December 11. https://www.cnnturk.com/turkiye/cumhurbaskani-Erdoğan dan-nurettin-yildiza-tepki.

*Cumhuriyet*. 2018. "Ankara Milletvekili Aylin Nazlıaka: Nurettin Yıldız yeni dönemin Fetullah Gülen'idir." *Cumhuriyet*, March 8. https://www.cumhuriyet.com.tr/haber/ankara-milletvekili-aylin-nazliaka-nurettin-yildiz-yeni-donemin-fetullah-gulenidir-938953.

Din İşleri Yüksek Kurulu, ed. 2017. *Kendi dilinden FETÖ: Örgütlü bir din istismarı*. Ankara: Diyanet İşleri Başkanlığı Yayınları. https://www.aa.com.tr/uploads/TempUs erFiles/haber/2017/07/KENDI-DILINDEN-FETO-20170725son.pdf.

Din İşleri Yüksek Kurulu, ed. 2019. *Dinî-Sosyal Teşekküller, Geleneksel Dinî-Kültürel Oluşumlar ve Yeni Dinî Yönelişler* (March). https://medyascope.tv/2019/05/31/turk iyedeki-dini-sosyal-tesekkuller-geleneksel-dini-kulturel-olusumlar-ve-yeni-dini-akimlar-baslikli-raporun-tam-metni/.

*Dini Haber*. 2020. "Cübbeli Ahmet'in bilmediği. . . ." *Dini Haber*, January 17. https://www. dinihaber.com/diyanet-haber/cubbeli-ahmetin-bilmedigi-diyanetin-gizli-ibareli-cemaatler-ve-tarikatlar-h144258.html.

*Diyanet Haber*. 2018. "Cumhurbaşkanı Yardımcısı Fuat Oktay, 'Diyanet, dini istismar eden FETÖ, DAEŞ gibi örgütlere karşı seferberlik ilan edilmiştir.'" *Diyanet Haber*, October 31. https://www.diyanet.gov.tr/tr-TR/Kurumsal/Detay/12093/cumhurbask ani-yardimcisi-oktay-diyanet-dini-istismar-eden-feto-deas-gibi-orgutlere-karsi-sefe rberlik-ilan-etmistir.

*Diyanet Haber*. 2019. "Diyanet'le Milli Eğitim arasında hafızlık protokolü." *Diyanet Haber*, May 2. https://www.diyanethaber.com.tr/diyanet-haber/diyanetle-mill-egitim-arasinda-hafizlik-protokolu-h5292.html.

*Diyanet.tv*. 2017. "Diyanet İşleri Görmez'den veda." *Diyanet.tv*, August 2. https://www. diyanet.tv/diyanet-isleri-baskani-gormezden-veda.

*The Economist*. 2016. "Don't Lose the Plot." *The Economist*, August 27. https://www. economist.com/leaders/2016/08/27/dont-lose-the-plot.

*El-Aziz*. 2019. "Cemaat ve tarikat kursları kontrol altına alınıyor." *El-Aziz*, December 22. https://el-aziz.com/tr/gundem/detail/2467.

*Elbistan Pusula*. 2019. "Süleymancılara operasyon!" *Elbistan Pusula*, August 2. https:// www.elbistanpusula.com/haber/4838810/suleymancilara-operasyon-jandarma-bas kin-yapti-iste-detaylar.

Eroğlu, Doğu. 2017. "Ebu Hanzala: IŞİD'in 'Türkiye emiri' değil, vaizlerin en etkilisi." *Diken*, November 11. http://www.diken.com.tr/ebu-hanzala-isidin-turkiye-emiri-degil-vaizlerin-en-etkilisi/.

Eroğlu, Doğu. 2018. "Tevhid Dergisi kadrosu anlatıyor (2): IŞİD sempatisi yok." *Diken*, April 25. http://www.diken.com.tr/tevhid-dergisi-kadrosu-anlatiyor-2-isid-sempat isi-yok/.

*Expression Interrupted*. N.d. "Hikmet Çetinkaya." *Expression Interrupted*, https://www. expressioninterrupted.com/hikmet-cetinkaya/.

Frantz, Douglas. 2000. "Turkey Assails a Revered Islamic Moderate." *New York Times*, August 25.

Frost, Lillian. 2022. "Report on Citizenship Law: Jordan." Global Citizenship Observatory (GLOBALCIT) Country Report, 2022/2, European University Institute, February. https://cadmus.eui.eu/handle/1814/74189.

*Gazete Duvar.* 2019. "Cumhurbaşkanlığı 'gizli cemaat raporu'nu reddetti!" *Gazete Duvar*, July 23. https://www.gazeteduvar.com.tr/gundem/2019/07/23/cumhurbaskanligi-gizli-cemaat-raporunu-reddetti/.

Güleçyüz, Kâmil. 2019a. "Diyanet'ten açıklama bekliyoruz." *Yeni Asya*, August 6.

Güleçyüz, Kâmil. 2019b. "Diyanet niye susuyor?" *Yeni Asya*, August 9.

Güleçyüz, Kâmil. 2019c. "Cemaat Raporu ve Diyanet." *Yeni Asya*, August 20.

Güvendik, Evrin. 2019. "Maarif Continues Legal Battle against FETÖ Schools Abroad." *Daily Sabah*, August 17. https://www.dailysabah.com/war-on-terror/2019/08/17/maarif-continues-legal-battle-against-feto-schools-abroad.

Hablemitoğlu, Necip. 2003. *Köstebek*. Istanbul: Birharf.

Hansen, Suzy. 2017. "Inside Turkey's Purge." *New York Times*, April 13. https://www.nytimes.com/2017/04/13/magazine/inside-turkeys-purge.html.

Hendrick, Joshua. 2013. *Gülen: the ambiguous politics of market Islam in Turkey and the world.* New York University Press.

*Hürriyet.* 2017a. "Abdullah Gül'den Mehmet Görmez açıklaması." *Hürriyet*, August 1. https://www.hurriyet.com.tr/gundem/abdullah-gulden-mehmet-gormez-aciklamasi-40537654.

*Hürriyet.* 2017b. "Ahmet Davutoğlu'ndan Mehmet Görmez açıklaması." *Hürriyet*, August 1. https://www.hurriyet.com.tr/gundem/ahmet-davutoglundan-mehmet-gormez-aciklamasi-40537374.

*Hürriyet.* 2018. "Diyanet-Sen'den Oktar açıklaması." *Hürriyet*, December 17. https://www.hurriyet.com.tr/gundem/diyanet-senden-adnan-oktar-aciklamasi-bundan-boyle-istismar-edemeyecek-40894377.

*Independent Türkçe.* 2019. "Diyanet'in Kur'an kursu kararı Süleymancılara yönelik bir tedbir olabilir." *Independent Türkçe*, December 21. https://www.indyturk.com/node/106916/haber/%E2%80%9Cdiyanet%E2%80%99-kuran-kursu-karar%C4%B1-s%C3%BCleymanc%C4%B1lara-y%C3%B6nelik-bir-tedbir-olabilir%E2%80%9D.

*İnternet Haber.* 2018. "Cevat Akşit'ten bombalar! Diyanet, Süleymancılar, Menzil . . . Yeni FETÖ kim?" *İnternet Haber*, July 30. https://www.internethaber.com/cevat-aksitten-bombalar-diyanet-suleymancilar-menzil-yeni-feto-kim-1892521h.htm.

*Karar.* 2017a. "Mehmet Görmez'in görevinden ayrılmasına en çok Cübbeli Hoca sevindi." *Karar*, August 1. https://www2.karar.com/guncel-haberler/mehmet-gormezin-gorevinden-ayrilmasina-en-cok-cubbeli-hoca-sevindi-558817.

*Karar.* 2017b. "Diyanet İşleri Başkanı Erbaş'tan İlk Cuma Namazı ve Hutbe." *Karar*, September 2. https://www2.karar.com/ankara/diyanet-isleri-baskani-erbastan-ilk-cuma-namazi-ve-hutbe-605777.

*Karar.* 2017c. "FETÖ'nün İslam'a verdiği zarar 8 dilde anlatılacak." *Karar*, December 21. https://www2.karar.com/-haberleri/fetonun-islama-verdigi-zarar-8-dilde-anlatilacak-697465.

*Karar.* 2018a. "Eski Diyanet İşleri Başkanı Mehmet Görmez'den dikkat çeken FETÖ açıklaması." *Karar*, July 18. https://www2.karar.com/guncel-haberler/eski-diyanet-isleri-baskani-mehmet-gormezden-dikkat-ceken-feto-aciklamasi-917595.

*Karar.* 2018b. "Erbaş: Diyanet İşleri Başkanlığı olarak din istismarı ile mücadele programı başlattık." *Karar*, May 23. https://www2.karar.com/istanbul/erbas-diyanet-isleri-baskanligi-olarak-din-istismari-ile-mucadele-programi-baslattik-863013.

*Karar.* 2018c. "Diyanet İşleri Başkanı Erbaş'tan Asya Endonezya Sufi Ulema Meclisi Başkanı'na FETÖ fetvası için teşekkür." *Karar,* October 19. https://www2.karar.com/ankara/diyanet-isleri-baskani-erbastan-asya-endonezya-sufi-ulema-meclisi-baskan ina-feto-fetvasi-tesekkuru-1003645.

*Karar.* 2018d. "Diyanet İşleri Başkanı'ndan Adnan Oktar'a sert sözler: Akli dengesi bozulmuş." *Karar,* January 31. https://www2.karar.com/guncel-haberler/diyanet-isl eri-baskanindan-adnan-oktara-sert-sozler-akli-dengesi-bozulmus-741210.

*Karar.* 2018e. "Diyanet'ten FETÖ ve Adnan Oktar açıklaması." *Karar,* August 10. https://www2.karar.com/guncel-haberler/diyanetten-feto-ve-adnan-oktar-aciklamasi-940062.

Kirby, Paul. 2016. "Turkey Coup Attempt: Who's the Target of Erdoğan's Purge?" *BBC,* July 20. http://www.bbc.com/news/world-europe-36835340.

*Kronos.* 2019. "Diyanet dışında açılan Kur'an kursları yönerge ile yasaklandı." *Kronos,* December 22. https://kronos34.news/tr/diyanet-disinda-acilan-kuran-kurslari-yone rge-ile-yasaklandi/.

Lord, Ceren. 2019. *Religious Politics in Turkey: From the Birth of the Republic to the AKP.* Cambridge: Cambridge University Press: 2018.

*Milliyet.* 2017a. "Cumhurbaşkanı Erdoğan'dan Diyanet İşleri Başkanı Görmez ile ilgili açıklama." *Milliyet,* July 24. https://www.milliyet.com.tr/yerel-haberler/ankara/cumhur baskani-Erdoğan dan-diyanet-isleri-baskani-gormez-ile-ilgili-aciklama-12184742.

*Milliyet.* 2017b. "FETÖ'nün ek konulan yurdu Diyanet Vakfı'na tahsis edildi." *Milliyet,* August 28. https://www.milliyet.com.tr/yerel-haberler/kutahya/fetonun-el-konulan-yurdu-diyanet-vakfina-tahsis-edildi-12247954.

Mumcu, Uğur. 2006. *Bütün Yazıları Serisi: Son Yazıları.* Ankara: Uğur Mumcu Araştırma ve Gazetecilik Vakfı.

Nakhoul, Samia, and Ayla Jean Yackley. 2016. "Turkish President Gains Upper Hand in Power Struggle." *Reuters,* July 24. https://www.reuters.com/article/us-turkey-security-gulen-insight-idUSKCN10407W.

Öğreten, Tunca. 2019. "Türkiye'de 'şirketleşen' tarikat ve cemaatler." *Deutsche Welle,* August 4. https://www.dw.com/tr/t%C3%BCrkiyede-%C5%9Firketle%C5%9Fen-tari kat-ve-cemaatler/a-49885320.

Oğur, Yıldıray. 2017. "Ehliyet, liyakat, sadakat, Diyanet." *Karar,* August 2. https://www.karar.com/yazarlar/yildiray-ogur/ehliyet-liyakat-sadakat-diyanet-4594.

Okumuş, Ejder. 2016. "15 Temmuz Darbe Girişimi ve Diyanet'in Tarihi Rolü." *Anadolu Ajansı,* October 7. https://www.aa.com.tr/tr/15-temmuz-darbe-girisimi/15-temmuz-darbe-girisimi-ve-diyanet-in-tarihi-rolu/660288.

*Sabah.* 2016. "FETÖ medyasından Görmez'e iftira kampanyası." *Sabah,* January 8. https://www.sabah.com.tr/gundem/2016/01/08/feto-medyasindan-gormeze-iftira-kam panyasi.

Şahin, Sefa. 2019. "Diyanet İşleri Başkanı Erbaş: Dini Sosyal Teşekküller Şeffaf bir Yapıya Kavuşturulmalı." *Anadolu Ajansı,* November 28. https://www.aa.com.tr/tr/turkiye/diyanet-isleri-baskani-erbas-dini-sosyal-tesekkuller-seffaf-bir-yapiya-kavusturulm ali/1658272.

Şık, Ahmet. 2013. "AKP'nin elinde Gülen cemaatini bitirecek bir arşiv mi var?" *T24,* November 25. http://t24.com.tr/haber/ahmet-sik-akpnin-elinde-gulen-cemaati-bitire bilecek-bir-arsiv-var,244702.

*Sol Haber.* 2017. "Diyanet'ten İsmailağa tarikatına özel toplantı." *Sol Haber,* January 12. https://haber.sol.org.tr/toplum/diyanetten-ismailaga-tarikatina-ozel-toplanti-182124.

*T24.* 2017. "Diyanet İşleri'nden tarikatlar ve cemaatlere 5 ilke." *T24*, January 5. https://t24. com.tr/haber/diyanet-islerinden-tarikatlar-ve-cemaatlere-5-ilke,381231.

*T.C. Resmî Gazete* 7234. 1949. "Kanunlar." *T.C. Resmî Gazete*, June 16. https://www.resm igazete.gov.tr/arsiv/7234.pdf.

*T.C. Resmî Gazete* 30472. 2018. "KHK/701 Kanun Hükmünde Kararname." *T.C. Resmî Gazete*, July 8. http://213.14.3.44/20180708/20180708-1.pdf.

*Veryansın TV.* 2019a. "Süleymancılar devleti tanımadı!" *Veryansın TV*, August 14. https:// veryansintv.com/suleymancilar-devleti-tanimadi/.

*Veryansın TV.* 2019b. "Diyanet 'Tarikatlar Raporu'nu neden toplatmak istedi?" *Veryansın TV*, December 3. https://veryansintv.com/diyanet-tarikatlar-raporunu-neden-toplat mak-istedi/.

White, Jenny. 2002. *Islamist Mobilization in Turkey: A Study in Vernacular Politics.* Seattle: University of Washington Press.

*Yeni Şafak.* 2016. "Darbe girişiminin arkasında ABD var." *Yeni Şafak*, July 17. https://www. yenisafak.com/gundem/darbe-girisiminin-arkasinda-abd-var-2495442.

*Yeni Şafak.* 2016b. "Kamudan ihraç edilenlerin tam listesi." *Yeni Şafak*, September 2.     https://www.yenisafak.com/gundem/kamudan-ihrac-edilenlerin-tam-listesi-2521060.

Zorlu, Faruk. 2019. "Turkish Foundation, Aid Body Ink Protocol to fight FETÖ." *Anadolu Ajansı*, January 1. https://www.aa.com.tr/en/todays-headlines/turkish-foundation-aid-body-ink-protocol-to-fight-feto/1365489.

# Conclusion

## The Politics of Fifth Columns Revisited

### Scott Radnitz and Harris Mylonas

The specter of fifth columns is evergreen. Their lineage goes back at least to the origins of modernity and the nation-state, as they are inextricably bound with notions of sovereignty and national belonging. Fifth columns emerge from the convergence of two sites: the territorially bound spaces where insiders reside, and an extraterritorial world from which outsiders attempt to exercise influence.

*Enemies Within*, though global in scope, applied various local lenses to understand how the politics of fifth columns operates on the ground. The cases in this volume revealed similar patterns of accusations, policy responses, and mobilization across multiple locations and time periods, but also showed how a common form contains important variations in content, duration, and effects.

In the remainder of this chapter, we take stock of what we have learned and look ahead. First, we summarize and synthesize the major insights from the preceding chapters. Then, we highlight counterintuitive patterns and unanswered questions, which act as a jumping-off point to sketch out a research agenda. We conclude by considering how global developments are likely to influence fifth-column politics in the coming years.

## What Have We Learned?

Collectively, the preceding chapters reveal several important insights into the politics of fifth columns. First, fifth-column politics is highly adaptable, appearing in various guises and institutional contexts: in both strong and weak states, in democracies, autocracies, and hybrid regimes, and under various institutional arrangements. Not only does democracy not constrain fifth-column politics, but the electoral motive and the impetus to produce winning coalitions at times create strong incentives for politicians to partake in the construction of such enemies within. Irrespective of regime type, security considerations often motivate state elites to put forth accusations on the basis of ascriptive or ideological differences (Kuo and Mylonas). But civil society associations may drive

Scott Radnitz and Harris Mylonas, *Conclusion* In: *Enemies Within*. Edited by: Harris Mylonas and Scott Radnitz, Oxford University Press. © Oxford University Press 2022. DOI: 10.1093/oso/9780197627938.003.0012

security perceptions as much as governing elites or geopolitics do (Bulutgil and Erkiletian).

Second, whether a group is targeted with fifth-column claims is not a matter of simple demography. Fifth-column politics typically rests on a foundation of familiar ideological constructs that political actors can selectively invoke. Just as different types of states give rise to similar patterns of fifth-column politics, tropes may recur and resonate over time *within the same state*, even as regimes, coalitions, and individuals change. This has been the case in Poland (Charnysh),[1] where claims about Jews endured more or less continuously, as well as in Hungary (Jenne et al.), where anti-globalist and antisemitic tropes were revived after a liberal interregnum. In the geopolitical context of Afghanistan, Crews found that imperial legacies and colonial meddling were superimposed on religious and tribal groupings to produce a lexicon of accusations of treachery that endured across generations. In other cases, fifth-column discourses appeared *de novo*, as with the backlash against new immigrant populations in the United States (Bulutgil and Erkiletian).

Third, the preceding chapters show that there is often no clear-cut distinction between targeting opponents based on their ideological or ethnic differences. The fuzzy conceptual boundary comes across most clearly in cases in which religion ostensibly marks out difference. For example, the targeting of Polish Jews on the basis of their ethnicity, ideology, or Zionist inclinations depending on the instrumental needs of the moment highlights how elites incorporate strategic ambiguity when formulating fifth-column claims (Charnysh; Ciancia). A similar fluidity applied to Uyghurs, whose Muslim-ness was strategically used to reframe anti-regime opposition as part of a transnational jihadist movement (Kuo and Mylonas). In Afghanistan, class, nation, and religion were all invoked in the 1970s (Crews). In another instance, the focus on a single individual, George Soros, was apparently intended to inflame public opinion through the simultaneous invocation of antisemitism, externally imposed democratization, and the specter of international financial interests (Jenne et al.).

In other cases, the narrative of ideological betrayal prevailed when salient political cleavages emerged within an ethnic group. This was the case with collusive fifth-column claims involving Russia's democrats and President Trump, where allegations of nation-selling—to use Crews's term—were leveled based on the premise of self-interest, greed, and toadyism rather than ties stemming from heritable traits or national origins (Radnitz). Yet even in those cases, to the extent that the charges resonated, it was due to widely held animus against the supposed external puppet-masters rooted in nationalist sentiments. One form of fifth-columnization initially appears to slip through the categorial cracks: sexual minorities in the United States and Palestine that are neither ideological nor ethnic (Anabtawi). Again, however, their resonance as a fifth column rather than merely a marginalized group came from claims that they covertly represented the interests of an ideological enemy (the Soviet Union) or a settler colonial state (Israel)—both of which stoked animosities based on nationalism.

Fourth, the absence of fifth-column claims in certain instances helped to identify variables that account for their presence in others. In the case of Japanese Americans in Hawaii and German Americans in Chicago, their purported linkages with America's perceived enemies in active hostilities made these populations vulnerable, but their previous civic integration and ability to mobilize spared them from being targeted (Bulutgil and Erkiletian). Hungary appeared to overcome its previous legacies of fascism and minority exclusion in the 1990s, until a new and retrograde form of legitimation became politically expedient (Jenne et al.). In Russia, collusive fifth-column claims waned when Putin took power, but subversive claims—which have a long lineage dating back to Soviet times but were rarely deployed by the Yeltsin government—reappeared with a vengeance after 2005. Due to Russia's relative weakness internationally and the angst over the Soviet collapse, there was only a brief moment in which neither the opposition nor the regime sought to build coalitions based on fifth-column politics (Radnitz). In Jordan, leaders used a fifth-column frame to restrict the country's nationality law in the mid-2010s, in contrast to 1954, when a more inclusive nationality law aligned with domestic elite interests. In both cases, policymakers could convincingly frame Palestinian groups as fifth columns, but in only one of these cases did they decide to do so.

Fifth, the insider-outsider distinction, which lies at the heart of the definition of a fifth column, is not always so straightforward. This is evident when the allegedly subversive fifth column temporarily resides outside the territorial boundaries of the state where fifth-column claims originate (Ciancia). This ambiguity challenges the very notion of an "enemy within." "Outsider" status can also be endogenous to fifth-column claiming and framing itself, as political entrepreneurs sometimes seek to construct a targeted group's position in ways that may differ from their "objective" circumstances. For instance, Hungarian-born American philanthropist George Soros is recognizable as a *foreign* actor but is also an "insider" (Jenne et al.). Similarly, Frost challenges our understanding of the insider-outsider distinction by focusing on the category of protracted refugee groups, some of whom have received nationality in their host state while others have not. These groups exemplify the ambiguous categories of *noncitizen insiders*, who have lived in a country for generations despite lacking formal citizen status, as well as *citizen outsiders*, who may be citizens in the eyes of the state but informally lack the stature of national insiders. Both types of groups can become the object of fifth-column frames.

Sixth, the authors have detailed several ways in which the targets of subversive fifth-column claims respond to their depiction by powerful actors as enemies within. Anabtawi as well as Fabbe and Balıkçıoğlu describe two strategic responses in the face of otherwise unpalatable options. LGBTQ groups in the United States and Palestine sought to demonstrate their "normalcy" by embracing patriotic and nationalist rhetoric. In Turkey, religious groups, fearing becoming the target of a presidential purge, signaled their loyalty to the regime by portraying other groups as more extreme and potentially threatening. These cases do not demonstrate the full range of possible responses—for example, fighting or fleeing are also available—but they indicate that it may be possible for groups to effectively confront, or at least mitigate, smear campaigns against them.

Seventh, few contributors placed much causal emphasis on psychological factors. Social psychological theories note the tendency for people to make quasi-automatic inferences about ethnicity based on subtle cues, and to assess behavior based on group stereotypes (Brubaker et al. 2004; Hale 2008). Moreover, distrust increases as a function of threat and secrecy (Huddy et al. 2005; Maoz and McCauley 2008). But, while social identity, self-esteem, and group comparisons are presumably activated in instances of fifth-column politics, they are not sufficient to explain our outcomes of interest: fifth-column discourses, policies, and mobilization. It would be misleading to map the micro-mechanisms of prejudice onto the level of state policy, but politicians can deliberately activate latent biases in society, or deepen social and cultural cleavages out of political expediency. The most convincing causal stories therefore lie in the domestic political realm.

Finally, arguments resting on geopolitics are hard-pressed to explain patterns of *public* fifth-column politics. Geopolitical threats are better suited to explain the target selection and timing of *covert* policies against perceived fifth columns. This is because states faced with internal security threats will tend to favor surveillance and coercive measures and have little incentive to publicize their actions, as that could provoke external intervention. Taken together, these insights into the phenomenon of fifth-column politics can facilitate further theory-building.

## A Research Agenda

When the compelling findings of our case studies are examined together, a new research agenda emerges. There is good reason to believe that institutions matter, but how they do so is unclear. One possible source of variation is regime type, yet there is no theoretical reason why fifth-column politics might be more common in some types than others. Autocracies may struggle to achieve legitimacy through conventional practices like elections, leading them to rely more on invoking threats to justify their rule. Yet in democracies, institutionalized competition invites invidious "othering" discourses (e.g., Dickson and Scheve

2006; Wilkinson 2006; Hafner-Burton et al. 2014). As several chapters have shown, the United States is not only vulnerable to fifth-column claims, but it also is well suited for it. As a settler colonial nation and a multicultural polity, the boundaries of citizenship—and the vaguer but loaded question of what it means to be American—have been constantly contested (Smith 1997; Theiss-Morse 2009; Schildkraut 2011; Lepore 2018). Similar debates about national belonging have recently emerged across Europe (Krastev 2017; Mylonas and Tudor 2021). Given the broad potential for fifth-column politics, it may make sense to reframe the question from what institutions give rise to fifth-column politics to how institutions shape its manifestation.

Beneath the arena of elite politics, civic organizations can shape the salience of group differences and their potential to be politicized. Civic organizations may succeed in advancing their exclusionary goals when politicians are sensitive to lobbying and compelled to appeal to voters. The variation that Bulutgil and Erkiletian find in fifth-column policies and mobilization appears to be conditional on the federal structure of the United States, which channels political coalition-building into local arenas with their own power dynamics and demographic (and other) concerns. The politics of fifth columns at the sub-national level therefore represents another fruitful avenue for further research.

Other non-state actors may also be implicated in fifth-column politics, especially when it comes to the promotion of collusive claims. As Radnitz and Crews both demonstrate, the mobilization of counter-hegemonic fifth-column claims can involve coalitions including opposition party leaders, political activists, cultural and religious elites, media personalities, and militant groups. Such "grassroots" mobilization does not require formal democratic institutions to operate. In fact, because it develops outside of institutional channels, it may be an especially potent form of opposition under authoritarian regimes. The role of non-state actors in the politics of fifth columns is worth exploring further.

Another promising direction of future research area is uncovering the ways that the historical construction of "groupness" impacts fifth-column dynamics. Although targets have agency, their actions may be constrained by the degree to which individuals are identified with a group and the corresponding status of the group in the polity. Historical and institutional processes shape how solidified a target's "groupness" is. These observations point to a relationship between fifth-column politics and the formation of group boundaries (Wimmer 2013; Chandra 2012; Brubaker 2018). At both the stages of defining groups and contesting categorization, scholars must be able to trace the interaction of institutions—especially ones that mark out and constrain identities—and the agency of the targets of fifth-column politics.

Last, important questions remain about the effectiveness of fifth-column claims. While our contributors focused mainly on the production of fifth-column politics, at times they suggest what tactics "work" and why, pointing to an important research program. Yet it is difficult to identify the direct effects of policy or rhetoric on relevant outcomes and distinguish them from other plausible explanations. In general, politicians are assumed to be savvy in their efforts to set agendas and establish narratives. Yet they operate in conditions of uncertainty and limited information, and may misread the public opinion landscape. For example, Charnysh notes that efforts to undermine Solidarity in the 1980s using antisemitism failed because the ruling party had already lost its legitimacy. Also, Radnitz suggests that Democrats' emphasis on Trump's ties to Russia may have diverted them from more compelling lines of attack. More research is needed to systematically test how and when fifth-column politics succeeds in advancing the claimants' goals.

## The Future of Fifth-Column Politics

At a time when exclusionary nationalism and illiberalism are ascendant, it is unlikely the politics of fifth columns have played themselves out. Technological developments have made external linkages more accessible than ever while abetting the spread of (mis)information domestically and across borders. Waves of protest against state repression and policing, and fears brought on by the global spread of Covid-19, offer new material for salvos against groups that were already on the political margins (Mylonas and Whalley 2022).

The demise of the so-called liberal international order (Ikenberry 2018) multiplies state insecurity as well as opportunities to meddle in other countries' affairs. A redistribution of the global balance of power brought on by a rising China and an inward-looking United States provides opportunities for revisionist powers like Russia and Iran to challenge institutional arrangements and alliance patterns. Within this context, new assemblages of global winners and losers provide incentives for political entrepreneurs to identify new clients abroad and targets of blame at home.

Global developments that privilege an ethnocultural understanding of the nation threaten to make the politics of fifth columns even more pervasive. Ethnonationalist appeals have gained purchase in democratic states like India and the United States, as well as in authoritarian or hybrid regimes like Russia, China, and Hungary. When an ethnoculturally defined understanding of nationhood replaces class or party ideology as the salient political cleavage, politics comes to revolve more around the axis of core versus non-core groups. By itself, this is not sufficient to bring about fifth-column politics, but it does not take a major leap to connect internal foes to external threats to sovereignty, even if that threat is imagined or merely symbolic. The proliferation of dual-citizenship regimes as well as diaspora engagement policies ensures that extraterritoriality will remain prominent in domestic politics (Gamlen 2019; Mylonas 2020).

Even where ethnicity does not structure political divides, polarization by education or urban-rural divisions can lead to a politics of demonization and scapegoating. Where right-wing populists are in power, the claim to rule on behalf of the "somewheres" can lead naturally to hostility against cosmopolitan "anywheres"(Goodhart 2017). When the latter espouse political ideologies deemed anti- or supra-national, their loyalty can be challenged. The external patrons may be "globalist" financiers, powerful states like the United States, or inter-governmental bodies like the EU or the IMF whose policies may conflict with national-level laws and norms. These accusations are likely to take subversive form when right-wing populists are in power and collusive form when they are in opposition.

An aggravating factor in a world where institutions are under threat from within is the influx of people from the global south to the global north. According to UNHCR, as of 2020, 80 million people were defined as refugees.[2] Increasingly, climate change will lead to political instability and destroy livelihoods, driving people to seek better opportunities elsewhere. As these refugees tend to be culturally distinct from the majority population, this mass movement will further stoke concerns about borders and sovereignty that may animate fifth-column politics. In Europe, politicians in states with large or rapidly growing Muslim populations have "successfully" (in an electoral sense) deployed fifth-column allegations. Following the post-9/11 "war on terror," the failure of the Arab Spring, and the rise and fall of the Islamic State, politicians have benefited from making openly Islamophobic appeals linking the threat of refugees to promises to defend national sovereignty or culture (Hafez 2014). Thus, migration coupled with anxieties over demographic change threaten to hand demagogues a script based on the defense of national borders and the promise to maintain ethnic hierarchies.

The Covid-19 pandemic contributes to this set of dynamics. Historically, epidemics have led to responses like those described in this volume: ostracization, expulsion, and violence (Snowden 2019; Weindling 2000). In addition to concerns about sovereignty, Covid-19 feeds into biopolitics: the notion that outsiders, however defined, can "infect" the body politic (Bieber 2022). Moreover, as Mylonas and Whalley (2022) argue, external actors may use internal cleavages over the politicization of vaccination and lockdowns to undermine the stability of their competitors, while internal competitors continue to politically instrumentalize these issues to their benefit. Early on in the pandemic, Indian officials blamed a Muslim sect for seeding the virus from abroad

(Singh 2022), and the Trump administration used Covid-19 as a pretext to halt claims for asylum from the southern border. Covid-19 has left traumatized and aggrieved populations in its wake, potentially presaging a new surge of fifth-column politics even in a post-pandemic world.

A final trend that will preoccupy policymakers and scholars alike is the emergence and proliferation of "digital fifth columns." With the advent of social media and mass connectivity through the Internet, there are a growing number of global communities that defy internationally recognized borders. At the same time, many governments have surveillance capabilities to monitor these groups (Feldstein 2019; Greitens 2020). Fifth-column claims have already been made about citizens who are members of organizations with a digital component, especially jihadist networks (Kuo and Mylonas; Vacca 2019). Whether the politics of digital fifth columns operates according to the same logics as conventional fifth-column politics is a question researchers are likely to confront.

We undertook this interdisciplinary and comparative project to establish a framework for the study of a phenomenon that spans continents, regimes, and group identities. We put forth *fifth-column politics* as a category of analysis and distinguished it both from the colloquial use of the term *fifth column* as well as other related concepts. The contributions presented in *Enemies Within* have provided insights on this understudied but consequential phenomenon and may ultimately yield implications in the policy realm. We hope this volume encourages others to join in on this venture.

## Notes

1. Throughout this chapter, when we refer to earlier chapters in this volume, we list only the authors' name(s).
2. https://www.unhcr.org/en-us/figures-at-a-glance.html.

## References

Bieber, Florian. 2022. "Global Nationalism in Times of the COVID-19 Pandemic." *Nationalities Papers* 50 (1): 13–25.

Brubaker, Rogers. 2018. *Trans: Gender and Race in an Age of Unsettled Identities.* Princeton, NJ: Princeton University Press, 2018.

Brubaker, Rogers, Mara Loveman, and Peter Stamatov. 2004. "Ethnicity as Cognition." *Theory and Society* 33 (1): 31–64.

Chandra, Kanchan, ed. 2012. *Constructivist Theories of Ethnic Politics*. Oxford: Oxford University Press.

Dickson, Eric S., and Kenneth Scheve. "Social Identity, Political Speech, and Electoral Competition." *Journal of Theoretical Politics* 18, no. 1 (2006): 5–39.

Feldstein, Steven. 2019. *The Global Expansion of AI Surveillance*. Washington, DC: Carnegie Endowment for International Peace.

Gamlen, Alan. 2019. *Human Geopolitics: States, Emigrants, and the Rise of Diaspora Institutions*. Oxford: Oxford University Press.

Goodhart, David. 2017. *The Road to Somewhere: The Populist Revolt and the Future of Politics*. London: Hurst & Co.

Greitens, Sheena Chestnut. 2020. "Surveillance, Security, and Liberal Democracy in the Post-COVID World." *International Organization* 74 (S1): E169–E190.

Hafez, Farid. 2014. "Shifting Borders: Islamophobia as Common Ground for Building Pan-European Right-Wing Unity." *Patterns of Prejudice* 48 (5): 479–499.

Hafner-Burton, Emilie M., Susan D. Hyde, and Ryan S. Jablonski. 2014. "When Do Governments Resort to Election Violence?" *British Journal of Political Science* 44 (1): 149–179.

Hale, Henry E. 2008. *The Foundations of Ethnic Politics: Separatism of States and Nations in Eurasia and the World*. Cambridge: Cambridge University Press.

Huddy, Leonie, Stanley Feldman, Charles Taber, and Gallya Lahav. 2005. "Threat, Anxiety, and Support of Antiterrorism Policies." *American Journal of Political Science* 49 (3): 593–608.

Ikenberry, John. 2018. "The End of Liberal International Order?" *International Affairs* 94 (1): 7–23.

Krastev, Ivan. 2017. *After Europe*. Philadelphia: University of Pennsylvania Press.

Lepore, Jill. 2018. *These Truths: A History of the United States*. New York: W. W. Norton.

Maoz, Ifat, and Clark McCauley. 2008. "Threat, Dehumanization, and Support for Retaliatory Aggressive Policies in Asymmetric Conflict." *Journal of Conflict Resolution* 51 (1): 93–116.

Mylonas, Harris. 2020. *Review Symposium on Alan Gamlen's "Human Geopolitics: States, Emigrants, and the Rise of Diaspora Institutions," Oxford University Press, 2019*. Global Citizenship Observatory, European University Institute. https://globalcit.eu/globalcit-review-symposium-alan-gamlen/6/.

Mylonas, Harris, and Maya Tudor. 2021. "Nationalism: What We Know and What We Still Need to Know." *Annual Review of Political Science* 24: 109–132.

Mylonas, Harris, and Ned Whalley. 2022. "Pandemic Nationalism." *Nationalities Papers* 50 (1): 3–12.

Schildkraut, Deborah. 2011. *Americanism in the Twenty-First Century: Public Opinion in the Age of Immigration*. New York: Cambridge University Press.

Singh, Prerna. 2022. "How Exclusionary Nationalism Has Made the World Socially Sicker from COVID-19." *Nationalities Papers* 50 (1): 104–117.

Smith, Rogers M. 1997. *Civic Ideals: Conflicting Visions of Citizenship in US History*. New Haven, CT: Yale University Press.

Snowden, Frank M. 2019. *Epidemics and Society: From the Black Death to the Present*. New Haven, CT: Yale University Press.

Theiss-Morse, Elizabeth. 2009. *Who Counts as an American?: The Boundaries of National Identity*. Cambridge: Cambridge University Press.

Vacca, John R., ed. 2019. *Online Terrorist Propaganda, Recruitment, and Radicalization.* Boca Raton, FL: CRC Press.

Weindling, Paul. 2000. *Epidemics and Genocide in Eastern Europe, 1890–1945.* Oxford: Oxford University Press.

Wilkinson, Steven. 2006. *Votes and Violence: Electoral Competition and Ethnic Riots in India.* Cambridge: Cambridge University Press.

Wimmer, Andreas. 2013. *Ethnic Boundary Making: Institutions, Power, Networks.* Oxford: Oxford University Press.

# Index